& Imbersid

Egg Uses and Processing Technologies

New Developments

Egg Uses and Processing Technologies
New Developments

Edited by

J.S. Sim
Department of Animal Science
University of Alberta
Canada

and

S. Nakai
Department of Food Science
University of British Columbia
Canada

CAB INTERNATIONAL

CAB INTERNATIONAL
Wallingford
Oxon OX10 8DE
UK

Tel: Wallingford (0491) 832111
Telex: 847964 (COMAGG G)
Telecom Gold / Dialcom: 84: CAU001
Fax: (0491) 833508

A catalogue record for this book is available from the British Library.

ISBN   0 85198 866 0

Printed and bound in the UK by Biddles Ltd, Guildford

Contents

Part III: Non-Food Egg Uses

Part IV: Functional Modification of Egg Products

Part V: Nutritional Modification of Egg Products

Preface

The rapidly declining per capita consumption of eggs over the past three decades is a major challenge faced by the egg industry. To meet this challenge, researchers around the world have begun to look at the egg beyond its traditional food value, to explore new product concepts and innovative processing technologies. However, transfer of this new information to the egg processing industry is lagging behind.

This monograph has been developed from The First International Symposium on "Non-Conventional Egg Uses and Newly Emerging Processing Technologies" which was held in Banff, Canada on April 22-25, 1992. The aim of this symposium was to allow the exchange of information on egg research among all sectors involved, including representatives from universities, industry, and public institutions. Some 35 research articles and reviews were presented by leading authorities from 15 countries. A wide range of topics was discussed, including current and future world trends, separation technologies for egg chemicals, non-food uses, and functional and nutritional modification or designing of egg products. These papers have now been updated, reviewed, and edited to form this volume. Since the goal of the symposium was to expand the utilisation of this valuable resource, it is anticipated that this book will serve as a key reference not only for egg researchers and the egg industry, but also more broadly for researchers in related disciplines including agriculture, nutrition, medicine, pharmacology, and biotechnology.

The editors feel that the initial goals of the symposium were successfully met and that the information presented in this monograph represents not only a state-of-the-art review of each topic, but also a starting point and catalyst for future research in these important areas.

The editors wish to express their sincere thanks to Professor R.J. Hudson, Dr Jerome Martin, Dr Peter Hunton, and members of the organising committee for their commitment and for making the Symposium a most memorable experience. Thanks are also due to many referees for their enormous effort and expert advice; without them, peer-review would not have been possible. Last, but not least, we would like to thank each of the authors for their contribution and for their cooperation and patience in revising and updating their papers. This monograph is the result of a special team effort, of which all involved should be proud.

Jeong S. Sim
Shuryo Nakai
Editors

Acknowledgements

We gratefully acknowledge the following agencies and industry patrons who generously donated funds to support the symposium:

Canadian Egg Marketing Agency (CEMA)
Natural Sciences and Engineering Research Council of Canada (NSERC)
Industry, Science and Technology Canada
Alberta Agricultural Research Institute (AARI)
University of Alberta

Burnbrae Farms Ltd
Canadian Lysozyme Inc.
Export Packers Co. Ltd
Highland Produce Co. Ltd
Hypaz International
Ovo-Biotechnica Inc.
Shaver Poultry Breeding Farm Ltd
Star Egg Co. Ltd
Sunnyglen Eggs Ltd
Taiyo Chemicals (Japan)
Villetard Eggs

Chapter One

The Egg Industry in the Year 2000

G. Fairhurst

Shaver S.A., Elevage de Selection Avicole, St Loup d'Ordeon, 89330 St Julien du Sault, France

Abstract Historical analysis gives a basis for prediction. Sophisticated breeding methods, together with significant changes in husbandry techniques, have resulted in the modern intensive egg production industry. There has been a stagnation in the demand for shell eggs in the developed countries and a corresponding decrease in profitability. Egg products are reaching a significant level and beginning to replace shell eggs in these countries. Any expansion in egg production during the next decade will probably be restricted to the developing countries. Climatic advantages, together with the availability of the main ingredients of poultry feed, could result in a relocation of the main centres of egg production. The ability of the products industry to convert raw egg into bacteria-free egg products with an enhanced shelf-life has facilitated global distribution. Investment at all levels of the industry will depend on the potential economic industrial application. New technologies will only be incorporated in animal breeding if they are cost-effective. A dramatic change in poultry feed technology would be required to arrest the relocation of egg production.
(Key words: hybrid, controlled environment, cages, pharmaceuticals, relocation, cost-effective products)

Introduction

Guessing the future is, of course, a dangerous exercise mitigated only by the fact that time alone can prove one wrong. It is through the examination of the history of our industry that we are able to identify trends, thereby adding a degree of credibility to any projections about the future. Although the commercial use of layers has been in existence for almost a century, the modern poultry industry has evolved as a result of many changes, the most dramatic of which have had their effect in the past 50 years. The foremost influential ones which have dictated the shape of today's global egg production are:

1. The development of the hybrid layer.
2. Controlled environment housing.
3. Cage systems.
4. The use of pharmaceutical and biological products.

The Development of the Hybrid Layer

The development of the hybrid layer brought with it a move from the 'art of breeding' to the 'science of genetics', resulting in today's sophisticated analytical basis for any change in bird genetic potential. In the early days of the industry, there were literally speaking hundreds of breeders, each one marketing its product in very limited geographical areas. As producers persisted with their demand for improved stock performance and with the increased dissemination of knowledge regarding the profitability of specific breeds, the number of breeders declined and the competition for market share between the survivors increased. This is still the case today when the number of internationally recognised breeding companies can be counted on one hand with fingers to spare! The competition for market share intensifies, as it is only through increased sales that income can be generated to meet the ever-growing financial demands of research and development.

Investment of this hard-earned income by the breeding companies is scrutinised at all stages to ensure that it is used to best effect, with the sole purpose of maximising the competitive status of company products - this is the case today and will undoubtedly be the case in the year 2000. The breeding companies will continue to develop the hybrid layer and they will continue to consider all new technologies which may be utilized to improve the rate of genetic improvement, but the primary question will be the one of cost-effective investment.

Controlled Environment Housing

The development of controlled environment housing was a natural result of attempting to maintain poultry populations in those areas of the world where the climate was not ideally suited for good poultry husbandry. It was also developed in an atmosphere of intense interest in meeting the physiological requirements of the bird in an attempt to reach that elusive characteristic known as genetic potential. Most of the early development took place during a period of expansion in the industry as a result of increasing demand for eggs and good profitability. Consequently, it carried with it a distinct penalty in the form of the high investment levels associated with such housing. The levels of sophistication available today for the control of the micro-environment to which we subject our stock were unthinkable 20 years ago.

One question remains, however. During a period of low level expansion and consequently reduced profits, is it cost-effective investment? I will return to this question later.

Cage Systems

In order to justify fully the high investment in controlled environment housing, it was necessary to maximise stocking density; this was achieved by the use of the cage system. Here again, continued development has resulted in novel and sophisticated methods for the distribution of feed and water, for the removal of manure, and for the collection and internal transport of eggs, all giving the added advantage of reduced labour requirements. The systems were and still are, unfortunately however, often subject to abuse and probably instigated the animal welfare interest in intensive systems.

The matter of animal welfare and the associated concept of animal rights will undoubtedly have an effect on the shape of our industry in the year 2000 and beyond.

The Use of Pharmaceutical and Biological Products

The increase in livestock intensity brought with it certain negative features, such as the increased exposure to disease challenge. By placing birds in cages, many of the diseases which had previously been a feature of birds housed on the floor were no longer of significant economic importance. Coccidiosis and worm infestations rarely occur in modern cage units. The viral and bacterial diseases, however, remain with us and have become increasingly important as the exposure levels increase. Such diseases as Marek's, gumboro, infectious bronchitis, and Newcastle disease can decimate flocks and production levels if not properly controlled. This challenge has been met with a high degree of success over the years by the application of the pharmaceutical and biological companies to the development of the appropriate medications and vaccines.

As production units become larger and more complex, the risk of increased exposure levels become a real danger, especially where multi-age sites are involved. Under such conditions, the standard of management and effectiveness of sanitation programmes may decide whether many of these facilities will still be economically viable in the year 2000. We can, hopefully, look to the molecular biologists for help in our struggle to produce birds which can repel these pathogens, or at least to minimise the adverse effects. To eliminate the pathogens completely will be considerably more difficult, although genetic engineering techniques may be able to alter the pathogenicity.

When reviewing the literature and statistics pertaining to global changes in consumption and production of eggs, one gets the distinct impression that the glory days are over, at least in the developed countries, and that any expansion of production will be mainly confined to the developing countries. This prognosis is, of course, based on historically known facts, but it also assumes certain future conditions. We know from recent experience how difficult and dangerous it is to forecast changes on a global basis. I am sure that we are all aware of dramatic changes which can completely nullify any well conceived prognosis for specific areas of the world. We have examples in the Gulf War, the civil war in the former Yugoslavia, and the on-going reorientation of the Commonwealth of Independent States. The picture is further complicated by a multitude of political manoeuvring which may have considerable effects on any specific market. I refer to such niceties as imports, exports, import taxes, export taxes, producer subsidies, and supply management. Wouldn't an agreement on tariffs and trade be a good idea?

The stagnation of production expansion in the developed countries can, of course, be explained by an equivalent trend in demand in those countries, and it has not been economically attractive enough to set up shell egg production centres entirely dedicated to export to any great degree. A contributing factor is the characteristic of the egg and its shell.

Whilst the egg shell has proved over the years to be an excellent container in which to incubate an embryo, it does have definite limitations as a packing material for human food, especially if that food has a limited shelf-life and very specific conditions of storage if it is to maintain an acceptable quality.

We can reasonably assume that the increasing demand for shell eggs in the developing countries will not be met by increased production in the developed countries. What are the limiting factors therefore, to expansion in the areas of increased demand?

The first and overriding factor is the financial status of the developing countries; this can be alleviated to some extent by aid from the developed countries, but there must be a realistic financial basis on which to build, together with an adequate buying power in the human population.

The second limitation is the availability of raw materials to produce feed for the stock. Here again, help can come from those countries fortunate enough to be able to produce those raw materials. The help can be in the form of complete feed, or in the form of grains and premixes. This concept sounds fine, but in the long run it is dependent on the country having the hard currency to pay and a commitment by that country to apportion funds to this end. I have personal experience of the difficulties some countries have in sticking with that initial commitment. Work in the field of animal nutrition is helping with its investigations into the use of alternative raw materials which may be available at the local level.

The third limitation is usually the lack of infrastructure to support both the production and the distribution of the final product.

Finally, we have the all-important factor of know-how. This can also be made available and must be backed up by extension work.

I have already stated that I attribute the lack of expansion in developed countries to an equivalent lack of demand. It would be remiss of me not to investigate further the reasons for this reduced demand - and there are many.

Three of the Most Important Reasons for Reduced Demand

With the increase in mechanisation, there is a corresponding change in the working habits of the population which, in turn, results in a change in eating habits. There is an awareness of what is popularly referred to as 'healthy eating' and the whole cholesterol story has had a long-term and damaging effect on our industry.

A more recent scare has been the *Salmonella* situation and whilst as an industry, we must never become complacent about the safety of our product, we appear always to draw the short straw when a problem occurs, which is in a specialist field outside the normal knowledge of our industry. Cholesterol and *Salmonella enteritidis* are two examples.

Another area of concern, which will demand our increased consideration well past the year 2000, is the environmental issue. For some years now, the awareness of our industry's role in environmental pollution has only been seriously considered in areas of high human population density and, in many cases, the issue has been approached with an attitude of confrontation rather than cooperation. In view of the real concern about our future environment and the undeniable fact that all stages of our industry, from hatchery through production and handling of final product, have the potential to pollute the environment in one form or another, we should examine practices which would allow for expansion, accepting that location may not be the prime site originally envisaged.

With this brief historical view of pertinent factors, let us now take stock of the situation and look to the future.

Assessment of Structure and Potential of the Industry in the Year 2000

I have suggested that the hybrid layer is here to stay and that the rate of genetic improvement may increase if the inputs from new technologies are cost-effective. There will be a few large breeding companies who may be forced by financial constraints to reduce the number of product lines available to the producer.

Expensive controlled environment housing may be restricted to those regions where the climate is too inhospitable for the bird to perform adequately, but even where the value of the final product can justify the high investment, chasing genetic potential may not be cost-effective. Where cages are the chosen method, the level of automation will be determined by the availability and cost of labour and also by the flow rate requirements of in-line operations. Treatment and handling of manure may be a prerequisite dictated by environmental considerations. Stocking densities and, indeed, the use of cages, will almost certainly be dictated by animal welfare regulations.

There will be an increased use of pharmaceutical products (in this group I include vaccines), tempered by any inroads that biotechnology can make on bird susceptibility, or the pathogenicity of the invading organisms. Officials in charge of food and safety regulations at regulatory agencies will be keeping a watchful eye on both bacteria levels and pharmaceutical residues in the final product.

The projection that any expansion of production will take place mainly in the developing countries is based on increases in population, together with improved economies and, therefore, improved purchasing power. With what we have seen in recent years, we cannot disregard social and political instability and we must continually reassess the percent of the increasing human population in the market economy.

With so many constraints and unknown variables, to paint a picture of the industry today and the future would only involve two colours - black and grey! But there must be other more positive changes taking place to add some colour to the picture and, of course, there are, but they may imply some radical changes in direction.

Earlier in this article, I referred to the glory days. These were characterised by unreasonable profit levels and the continued survival of inefficient operations: here I include all segments of our industry from breeders to hatcheries to producers to processors to allied industries, as far as including extension work and research.

We are all aware that, when pruning fruit trees, we have to cut into some of the good wood, as well as removing the dead branches. The position of each cut can be difficult and mistakes are made, but the result is usually blossom and fruit. This is also the case in the rejuvenation of both companies and industries.

There are clear indications that this process has already started. Here I must point out that some of the governmental cuts made in the financing of research and extension services internationally may have been unjustifiably hard pruning, but we must, at the same time, concede that there was an excess of dead wood and that some of the branches were growing in the wrong direction.

Recently I have had the opportunity to discuss the problems of our industry in countries as far apart economically, socially, and climatically as South America and northern Scandinavia and, whilst they each have their own particular problems, there is a keen awareness that as expansion in the industry slows down, rationalisation must take place to counteract the effects of oversupply and decreased income.

The first area to come under the microscope is cost of production. In Sweden there has been a state of political neutrality for many years and this political stand brought with it a requirement for a high level of self-sufficiency in many of the staple foods - eggs included.

Whilst this approach stimulated some interesting research into the use of home-grown feedstuffs and even into the use of home-grown hybrids, it created an environment of high levels of investment, backed in many cases by state support. This appears not to have been the case in many areas of South America.

To produce eggs in northern Sweden, with long winter nights and -35°C, controlled environment houses and added energy in the form of heat are an absolute necessity; this appeared not to be the case in many areas of South America where an open-sided structure with an insulated roof appeared to be entirely adequate.

Although a future entry by Norway, Sweden, and Finland into the European economic community will give access to world market prices for grain, they will have considerable difficulty competing with the feed prices available in those countries producing maize and soya. There is a strong animal welfare lobby in Sweden (the proposal that cages are banned before the year 2000 is still on the cards). This attitude was not immediately evident in South America. Sweden imposes rigorous environmental constraints, together with regulations concerning bio-security, food hygiene, and quality control which, whilst laudable objectives, were not evident to the same extent in Brazil.

Whilst there has been a consistent decline in egg production in Europe and Central and North America during the last five years, there has been considerable increase in production of eggs from South America during the same period. The indications are clear that this is not entirely due to a corresponding increase in domestic demand for shell eggs. What we are seeing is the start of a relocation of livestock to climates which meet the physiological requirement of the bird with very little modification and, consequently, a relatively low investment requirement per bird. At the same time and, in most cases, in the same place, relocation is occurring to areas where the main ingredients of poultry feed can be easily and economically produced. There are also areas of reduced environmental constraints since the pressures of high density human populations, although existing in the cities, do not cover the vast areas of open countryside to be found in Brazil.

Although I have used Scandinavia and South America as examples support-ing a geographical relocation of production centres, similar comparisons can be made in many other parts of the world. We may, therefore, be forced to reassess or indeed, reword the prognosis as stated earlier that 'any expansion of egg production during the next decade will mainly be confined to developing countries', perhaps it should read 'the main expansion of egg production during the next decade will be confined to those areas where the climate and availability of the main ingredients of poultry feed are conducive to cost-effective egg production and will probably be in areas within a band 30° above and below the equator'.

Perhaps the two statements are quite compatible. We must, however, consider what will happen in ten years' time in the less hospitable climatic areas. Many of the expensive facilities will require replacement - will the economic basis exist for high replacement costs, or will there be a further decline in production?

Of course, this potential relocation will not take place overnight and many will seek their future in a different direction and for different reasons. We have a good example in the proposed expansion by the Pohlmann Company of 5.6 million layers in north-eastern France. This is in a location which will give ready access to areas of high population densities, and which concurs with my previous suggestion that there are limitations to the successful transport of shell eggs in distance and time.

Many people will question the advisability of chasing least cost production if a high proportion of the potential consumers are located outside these areas. This is also true if we limit our thoughts to the shell egg market. I suggest, however, that we should all take notice of another indication of change which will undoubtedly have far-reaching effects on the shape of our industry in the year 2000 and well into the twenty-first century. I refer to the egg breaking and processing industry, backed up by innovative thinking with regard to the range and marketing of products.

I believe our industry owes a big 'thank you' to the engineers and technicians who have developed the modern breaking and processing equipment. We should always remember that the breaking industry was initially developed to remove down-graded products from the shell egg market and the equipment was, therefore, designed to handle such materials, especially with regard to poor shell quality. Today we are looking at an industry which no longer depends on poor quality raw materials and certainly no longer produces poor quality products. The processors have been extremely successful in promoting and marketing their products in those areas where the decrease in shell egg consumption was not caused by a decrease in the buying power of the human population. The *Salmonella* scare has certainly had a negative effect on consumer demand for shell eggs, but with the processors' ability to produce pathogen-free pasteurised product, they were quick to capitalise on this new opportunity. Growth in the share of total egg consumption attributable to the product industry has been noticeable in such diverse countries as, for example, Canada, Denmark, Italy, Switzerland, and in the USA.

It may be significant to point out that none of these countries feature in the 30° latitude band which I referred to earlier. In other words, in many of the areas where I consider cost-effective egg production to be difficult, there is already a growing acceptance of eggs in various product forms. We can, therefore, add the colours of white and yellow to brighten up that sombre grey picture which I painted earlier. So, what is missing from the formula?

Let us consider the negative aspects of geographical relocation. One of the problems encountered in the areas where cost-effective production could be considered is the fact that whilst the climate is well suited to bird performance, it is also conducive to the survival of many of the pathogens associated with both avian and human diseases. This is further aggravated by a very elementary approach to bio-security and general hygiene, when compared with normal praxis in the cooler climes. The negative effects can, in time, be reduced by improvements in management, social, and educational standards and by the development of pharmaceutical products specifically designed for these areas. Quality control in the food processing part of the chain will be more easily attained since the design of facilities and equipment will be based on experience gained from discerning and sophisticated markets. Investment will, of course, be required to establish modern breaking and processing operations, as is also the case for improvement in transport and communications. However, the advantages of cost-effective production and opportunities for expanded exports would, hopefully, justify this investment.

Another area of concern which is not specific to warmer climates, but of the utmost interest to the breeding companies, is the characteristic of the raw materials required by the breaking and processing industry, namely the egg.

To the non-discerning layman, the egg has few characteristics, the most obvious being size and shell colour. To the chef or housewife, a further number of characteristics are considered, such as shell quality, albumen quality, yolk colour, and presence or absence of inclusion bodies, i.e. blood spots and meat spots. To these can be added taints and odours, which although less common, are highly significant in the general context of egg quality. For the breaking and processing industry we must add further criteria, such as total fresh egg yield, total dry egg yield, yolk to albumen ratio, percent dry matter and protein of the albumen, percent dry matter and protein and fat of the yolk and, no doubt, many others!

Since we know that there are statistical differences between breeds for many of these traits, I send a strong plea to the industry to quantify the characteristics required by the industry as described above, so that selection criteria can be incorporated in breeding programmes to meet these requirements. Whilst many of the so-called quality traits are already subject to selection pressures, new techniques of measurement and increased levels of analytical sophistication will have to be devised to take account of new traits as they become economically important. We then have the time factor to consider between pure line selection and commercial production and it is only six years to the year 2000!

In this volume, we await fascinating presentations from the departments of food science, biotechnology, and many research institutes, each of which will have their own criteria regarding the ideal egg and, as breeders, we will have to draw our own conclusions in assessing both the feasibility and economic significance of any change in breeding direction - a fascinating prospect!

It has been suggested that biotechnology will, in all likelihood, give us the tools to create whatever genetic diversity is required by the industry. As our knowledge of the workings of control mechanisms increases, it should be possible to turn genes on and off by a variety of external factors, thus giving us the ability to change drastically the composition of eggs for production purposes, thereby coming closer to the requirements of the added value sections of our industry. We must, however, have the ability to reverse the process at pure line level in order to produce the next generation of both chickens and breeders. Again, as I have mentioned so many times before, the application of these technologies must be cost-effective.

The funding of pure research will, I suspect, be as difficult in the year 2000 as it is today. It is most unlikely that government funding will ever revert to the glory days of the research scientists. Private funding will be available, but here again, more account will be taken of the economic viability of an industrial application than to the funding of academic ingenuity.

Since I have now added the breaking and processing industry to my list of dramatic changes affecting the shape of our industry, I wonder if it is realistic to look for a sixth change by the year 2000. I am sure it has been noticed that the poultry feed industry does not feature amongst the first five discussed above.

From time to time, the section of agriculture associated with animal production is accused of inherent inefficiencies in the amount of arable land required to produce a unit of protein for human consumption and the world's human population would be better served if the same land was devoted to production of plant proteins for direct human consumption.

This topic has been the subject for discussion in many previous symposia and I will not elaborate on the subject at this juncture. It is well known in stock management that, for quick and effective treatment of birds, products can be administered by way of the drinking water. This applies to vaccines, antibodies, vitamins, and other pharmaceutical products. Can we look forward to the day when, on ordering feed, we get a delivery of low cost roughage in the feed silo and a tankful of liquid containing the essential nutrients to be administered in the water system? Is the concept totally unattainable? Or, has the investment in milling facilities already been too high for the feed industry to contemplate an alternative feeding system? Just think what that could do to the geographical distribution of egg production by the year 2000.

We should remember, however, that at this point in time we still have at the base of our industry an animal, *Gallus domesticus*, which produces a macro-lethical egg for the reproduction of its species and a human population which has various levels of sophistication in their demands - the less fortunate amongst us just want food.

In this chapter, I have tried to assess the structure and potential of our industry in the year 2000. Some of the radical changes I have suggested may never reach fruition.

Chapter Two

Innovative Egg Products and Future Trends

R.C. Baker

Rice Hall, Cornell University, Ithaca, NY 14853, USA

Abstract Developing new food products is a complex process involving uncontrollable factors as well as controllable elements. The product developers must understand the process to enhance the success of the intended product. Pioneering studies in 1960 by Cornell University researchers of a stagnant egg market indicated the need for new, convenient egg products. Consumers cited preparation time, children's dislike of eggs, no longer serving a hearty breakfast, and the messiness of egg dishes as reasons for lowering their consumption. Concern regarding cholesterol was minimal at that time. Researchers determined the following necessary stages to guide product developers through the development task: idea submission, development, taste-panelling, consumer sampling, shelf-life evaluation, packaging, production, and market-testing. Case studies of both successful and unsuccessful new products are discussed. The prediction of marketing experts regarding future marketing trends are based on a list of expected consumer demands.
(*Key words:* egg products, future trends, product development, taste panel, market testing)

Introduction

Developing successful new food products has never been easy, and the competitive environment of the future will make it even more difficult. The following are some of the external, thus uncontrollable challenges that face product developers: (a) inflation which causes the cost of development to skyrocket, thereby squeezing profit margins, (b) government regulations with unpredictable impacts, (c) markets that are becoming more narrowly segmented, (d) new technologies that threaten the position of established products (but also create possible opportunities), (e) demographic shifts, (f) increasing size and sophistication of domestic and foreign markets, and (g) the market uncertainties created by the changes in many foreign government structures.

A knowledge of simple basics is essential in today's complex and high-pressure product development climate. The uncontrollable elements mentioned earlier can certainly contribute to the success or failure of a new food product. Fortunately, there are some controllable basic internal elements that also determine the success of a company's product development effort. Some of these bases include generation of ideas, screening, preliminary evaluation, development and testing, and commercialisation. These key stages in product development must be supported by top management, make use of multidisciplinary teams, and have well-defined objectives, operational controls, and continuous market input to ensure that new products are potential winners.

At Cornell University we started developing new egg products in 1960. At that time egg producers were losing money primarily because egg consumption was going down quite rapidly. Why? Egg producers felt that the decrease was due to adverse publicity regarding the high amount of cholesterol in eggs. This was true but we also felt that another major reason was the lack of convenient egg products on the market.

To obtain information on diminished consumption we conducted a survey of 1000 families in Syracuse, New York. We sought an answer to the question of why consumers were not eating as many eggs as they did ten years before. The answers given (in order of frequency) were:

1. Lack of time to prepare eggs. Many women said they were working and did not have time to prepare eggs for breakfast and they considered eggs a breakfast food.
2. Children did not care for eggs. We were surprised to learn that consumption of eggs was very low for young children and teenagers.
3. Because of frequent coffee breaks, it was felt that a hearty breakfast including eggs was not entirely necessary.
4. Eggs were thought to be a messy food. Many said that egg yolk stuck to forks, dishes, frying pans, etc. and created extra work.

We concluded from this survey that a major cause for the decline in egg consumption was lack of convenience. Cholesterol content of eggs was mentioned by a few people but the preparation time element was of much more concern. Many said that they felt eggs were very nutritious and, if possible, they would like to serve then more often. As a result of this survey, we decided that we could help the egg industry by developing new, convenient egg products.

Before getting into product development we had to decide on the steps or stages involved in developing a new food product. We found that it was difficult to obtain agreement on just what these steps or stages should be. For this paper I will list and discuss the stages we use at Cornell University in developing new food products; we find that these same stages are generally used in the industry. They are (i) idea stage, (ii) development stage, (iii) taste-testing stage, (iv) consumer sampling stage, (v) shelf-life studies stage, (vi) packaging stage, (vii) production stage, and (viii) market-testing stage. These stages may not always fall in the same order. For example, it is entirely possible that a firm should not gear up for production until results of the market test are known. Some of the stages may be combined with others and some firms may not include all of them.

Idea Stage

This stage must come first, and it is extremely important. It involves 'cloud nine' dreaming and making every effort to second-guess what product or products the consumer will purchase on a trial basis and then continue to purchase. This stage should involve the research and development people, top management, and by all means, the marketing group. Every effort should be made to ensure the success of a new product when it enters the marketplace. It may take months or even years to make the final decision. The following questions need to be answered:

1. Will the product return a profit?
2. Does the product satisfy a consumer need?
3. Will the product be acceptable to the consumer, the wholesalers, and the retailers?
4. Is it unique and does it have distinctive characteristics that offer a totally new service to the consumer?
5. Does it fit? A new product must fit the capabilities of the company from the standpoint of production and marketing skills as well as complement the existing product line.
6. Is the new product anticipated to be novel or is it a 'me-too' product?
7. Does the company in question have the technology to develop the product?
8. How much capital is needed to develop the product, and is the capital available?

Taking time to make good judgements on the potential of a new food product does not cost a great deal of money. What does cost money are the stages that follow the idea stage. A product failure can cost thousands upon thousands of dollars. One should not hurry the idea stage but be as sure as possible that the new food product will be successful.

Development Stage

The development stage is very important in the creation of new products. Some think that since this stage involves cookery, if one is a good cook, then good products will surely result. This could be true by chance, but developing a new food product for the general public is much more complicated than producing a tasty dish for the family dinner. Experience is needed for the development stage because, attributes such as colour, flavour, texture, tenderness, juiciness, and shelf stability are often more difficult to maintain during commercial processing and distribution. Preparing a food product for a family meal is not the same as preparing one that will satisfy the multitudes and also have a long shelf-life. We have learned in developing new foods that many detrimental chemical reactions can take place during the shelf-life of products. This means that product developers must have a good background in chemistry and microbiology.

The development stage must coincide with the taste-panelling or sensory evaluation stage. As changes are made during the development stage, the product must be checked with experienced tasters. In the early stage of developing a new food product, an experienced taste panel will likely find many faults which the developers must solve. The taste panel members may indicate that the product does not have visual appeal. The developer must then come up with a product that is more appealing. Texture or mouth feel may be inferior, indicating that the ingredients must be changed to improve this parameter. Tenderness or toughness is very important in our enjoyment of a food. If the food is too tough, rubbery, or chewy, steps must be taken to make it more tender. Sometimes it is too tender and described as being mushy, and this, too, must be corrected.

Juiciness is a very important attribute of food and a product can be too juicy or too dry so here again knowledge is needed to produce a 'just right' product. Good flavour is vital and one must strive to obtain the ideal flavour without sacrificing the other attributes. Flavour is often difficult to distinguish from other attributes such as texture. If a food product has an inferior mouth feel, a taster may say it has a poor flavour. If a food does not taste good, the problem could be flavour, but it also may be inferior texture, tenderness, or juiciness.

Satisfying the taste panel accomplishes one hurdle, but proper shelf-life must also be achieved. If a product is to be sold as a refrigerated item, all parameters must be evaluated as storage progresses. If the product deteriorates rapidly, something must be done to improve shelf-life. This requires a knowledge of additives. If the food is to be canned, one must determine the safe processing time and temperature and their effect on the quality of the product. Ingredients may have to be changed. If the product is to be frozen, shelf-life must be checked at monthly intervals.

There are numerous possible complications. Here are a few common ones: (a) Is 'browning reaction' a problem; if so, can it be solved? (b) If the product is frozen, what is the effect of crystal size? (c) Is light a factor in deterioration, and if so what can be done? (d) Is rancidity a problem? (e) Will bacteria, moulds, yeasts, or pathogens be a problem?

The length of time needed to develop food products varies tremendously.

Some products may take only a few weeks, and others may take months. It is extremely important to develop a product that has appeal not only when it is first produced but throughout the channels of trade and during storage in the home.

Taste-Panelling Stage

Taste-panelling should run concurrently with formula development. The product being developed should be checked in its various stages to determine acceptability. Ideally, a food company developing new products should have two taste panels. One is a semi-trained or experienced panel for checking the product in various stages of development. This panel must have enough experience or training to distinguish good from undesirable flavour, proper texture, degree of tenderness, and degree of juiciness. Some people may do an excellent job with flavour, for example, but are unable to distinguish the ideal texture. The second panel should be a small group of consumers who can help developers produce products that will be popular with the consuming public. This taste panel can be made up of anyone who has a flexible work schedule and is interested.

Consumer Sampling Stage

Consumer sampling is not practised by all companies developing new food products, but it is an important part of food product development. Valuable information can be obtained at a relatively low cost by checking with a small population to see what they think of the product. Such sampling gives an indication of the product's potential success. Sampling may also indicate that the product could be more successful if changes are made. At Cornell, when it is felt a good product has been achieved, a sampling of one hundred families in Ithaca, New York is chosen. The selection is designed to include one-third with low income, one-third with medium income, and one-third with high income, because it is important to appeal to consumers in all economic groups. A sample of the product is given to each family and they are told that an interviewer will be back in a few days to find out what they think of the new product. Most people are very cooperative, enjoying the opportunity to try a new food at no cost.

Shelf-Life Stage

Shelf-life is an extremely important consideration of new food products whether they are sold refrigerated, frozen, or canned. It is imperative to have an understanding of how long a product will keep under a variety of temperatures and other environmental conditions.

To determine and improve shelf-life, an understanding of microbiology is essential. Many food companies have their own microbiologists, but smaller firms contract out their microbiology work. To determine shelf-life, it is important to determine the total counts of microbes that cause spoilage. In addition to determining the number of microbes, one must know whether the problem is with bacteria, moulds, or yeasts because this will vary with different food products. It is also essential to understand the potential of pathogens in the food product. Familiar names of pathogens of concern with most food products are *Salmonella*, *Staphylococcus aureus*, *Listeria*, *Campylobacter*, and *Clostridium*.

The relationship between rancidity and the shelf-life of a product is another important consideration. Many foods, especially meats, contain fat which can become rancid. There are two different kinds of rancidity, oxidative and hydrolytic. With most foods, oxidative rancidity is more of a problem than hydrolytic.

Packaging Stage

Good packaging is an extremely important consideration in product development and is becoming more important each year. Before the days of convenience foods, packaging was not of great concern, but today the food industry sells processed foods rather than commodities. Now and even more in the future, food packaging and processing decisions are related and an understanding of these relationships is vital. Consumers often buy the package, and the food is rather incidental. With some food products, the package costs more than the food and often much more. In the future, consumers will demand attractive, convenient packaging of food and will be willing to pay the price. Companies that can produce a package that the consumer wants will stay in business and make a profit.

The colour used in the packaging materials of food products is of extreme importance, and it varies with different foods. We have found that in packaging chicken products most consumers in the United States prefer yellow because they prefer chicken with yellow skin. This is not true in countries where the preference is for white skin. Using red in food packaging is smart because red attracts the eye. Many consumers associate clear film with freshness. Consumers in the United States prefer fresh chicken, not frozen, so if a milky film is used, many consumers feel the product is, or has been, frozen.

In addition to attractiveness and convenience, many other attributes of packaging are of importance, including protecting the food, not imparting flavour, resistance to tearing, ease of application, lightness of weight, not reacting chemically with the food, and economy. Of increasing importance is the consideration of the impact of packaging material upon the environment.

Production Stage

At some point in new food product development, a production line must be set up. If the new product is to be market-tested, it is not wise to arrange for full-scale production until after a successful market test. Some companies use a small-scale pilot plant for the production of the food product for market-testing.

Establishing a production line is not an easy task because the equipment needed varies with each product. The production line should be set up by a trained engineer because mistakes can be very costly. Some of the factors to keep in mind when establishing a production line are cost of equipment, cost of energy to operate, yield, safety, saving labour, sanitation, ease of cleaning, and adherence to government regulations. The ideal is to be able to produce the best product possible at the lowest cost. This must be done to meet the competition.

Market-Testing Stage

Market-testing is not done on all new food products nor by all food companies. It is normally done by the larger food companies because they want to reduce the risk of having an expensive failure with a national introduction. Many small food companies that sell locally do not go to the expense of a market test. They gamble, and their introduction of a new food product into a limited area is a form of market-testing. If the product fails, something else has to be tried. Market-testing is by-passed by some companies if similar products have been successful or if the companies have good track records.

Market-testing is done to obtain more accurate information on the product's sales potential. If the product succeeds in initial sales and repeat purchases, a larger, even a national introduction, can be tried. If the product fails, it can be either dropped or examined to determine the reasons for failure. If the marketing group can correct the failure, the product can be placed in another test market to confirm its ability to succeed.

There are many different ways to do a market test, and the strategy varies from company to company. The selection of the market-test site is important; better results are obtained in some cities than in others, because they effectively represent the intended market, e.g. particular ethnic group(s) or specific income group or lifestyle. It is important to select a site in which the population is made up of many different ethnic groups with a broad spectrum of income. If, for example, the population was made up only of Germans and Greeks, one would not know how other ethnic groups felt about the product. If the population was of high income, one would not know how the product would sell with other income groups. It is important to obtain all the information possible on both the initial and the repeat sales of any new product.

Over the years we have developed 58 new poultry meat and egg items at Cornell University of which 28 have been market-tested. Due to the cost not all new food products are tested. In addition 28 new products from underutilized species of fish have been developed of which 15 have been market-tested. Following the market test of 12 weeks, all of the details are written up in a brochure which is available, free of charge, to anyone interested.

Cornell Convenient Egg Products

A number of new convenience egg products designed to meet the needs of the home and the institutional market have been developed at Cornell and many of them have been enthusiastically received in the market test. Although institutions account for approximately a quarter of the dollar value of civilian food consumption, little has been done to promote the use of eggs in this market, nor has much effort been made to increase the use of eggs in the home. If eggs are to be consumed in greater volume, they must be offered to consumers and institutions in new and enticing convenient forms.

A convenience egg product for household and institution markets should possess most of the following characteristics:

1. More convenient to use than shell eggs.
2. Reduce preparation time.
3. Available at a competitive price.
4. Mass produced at a reasonable cost.
5. Easy to handle in distribution channels.

Some of the egg products developed at Cornell University include Easter eggs, Kid's Pak, shell-less eggs, modified atmosphere eggs, Tren (apple juice - egg drink), frozen egg omelettes, instant French toast, liquid egg, chiffon pies, 500 calorie pies, Hi-Pro cookies, hard-cooked egg roll, and egg crust pizza.

Frozen Omelettes

Often omelettes are not served in the home because of the time required for preparation or the lack of enticing additional ingredients. A wide variety of restaurants include omelettes on their menus, but preparation poses a problem especially during rush hours. Also, each chef prepares omelettes in a slightly different way, so standardisation may be achieved by using a frozen product. A frozen omelette manufactured at a central plant would minimise this variation and also could be packaged in standard sized portions. Since frozen products are common, marketing frozen omelettes is not a major problem. Large-scale purchasing and manufacturing can make the product available at a competitive price.

Meats other than traditional ham can be used in the manufacture of omelettes. Smoked turkey, smoked broilers, or turkey ham used in omelettes are difficult to distinguish from the usual pork ham. This discovery was considered important because many people who do not eat pork for religious or health considerations can eat poultry meat; therefore we made omelettes from turkey hams as well as pork ham. The omelettes produced from poultry meat were called Catskill (a local mountain range); those with pork ham, Western (Fig. 2.1).

Fig. 2.1. Egg omelettes, Western and Catskill.

The ingredients used included frozen whole egg, cooked ham (pork or turkey), fresh onions, fresh green peppers, salt, and monosodium glutamate. The market test for both omelettes was successful in both household and institutional market settings.

Apple Egg Drink

Drinks using various fruit juices combined with eggs were developed at Cornell. Since New York State is a large apple-producing area, a drink made of apple juice and eggs was considered appropriate for market-testing. The name chosen for the product was Tren - a combination of tree and hen, suggesting the trend to drink rather than eat breakfast (Fig. 2.2).

It was known as a 'breakfast in a glass' since the egg contains all nutrients except vitamin C and this vitamin was easily added to the apple juice. The ingredients included apple juice, fresh egg (one egg for 8 ounces of juice), sugar, citric acid, and apple essence.

The market-testing was done at the retail level and in vending machines on the Cornell campus. Some people had a psychological problem when trying Tren for the first time.

Fig. 2.2. Tren, an apple juice - egg drink.

Hi-Pro Cookies

Our Syracuse study pointed out that the consumption of eggs by young children and teenagers was very low. We all know that children like cookies so our aim was to develop a tasty cookie high in egg content. We were able to produce a product that contained one-half egg per cookie (Fig. 2.3).

Fig. 2.3. Hi-Pro cookies, molasses, peanut butter, and chocolate.

With the addition of a small amount of soy flour the High-Pro cookies contained 12% protein. To cover any hint of high egg content three varieties were produced: chocolate, peanut butter, and molasses. The Hi-Pro cookies were market-tested in Rochester, New York with great success.

Hard-Cooked Egg Roll

Years ago hard-cooked egg slices were part of salads in most restaurants and hotels. With high labour costs for cooking and peeling (a slow task, as very fresh eggs do not peel well) hard-cooked egg slices became scarce. Hard-cooked egg rolls may be the answer to having egg slices in salads again.

Several different casing materials were tested and rated for appearance, ease of use, prevention of greenish-black yolks, and peeling quality. The amount of moisture and flavour loss of eggs cooked in the various materials was considered, as was the cost of the material. Polypropylene performed the most satisfactorily of the materials tested. Experimental results showed that a desirable packaging material would be a 1 mm thick coloured film, non-permeable to carbon dioxide and water vapour, and available in tubular form. Coloured film minimizes browning reaction and visibility of air bubbles. A built-in zip tape to speed removal of the casing is desirable.

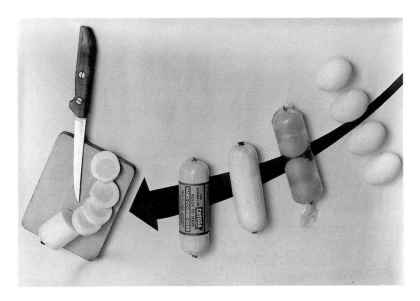

Fig. 2.4. Hard-cooked egg roll.

The hard-cooked egg roll (Fig. 2.4) was market-tested both at retail and institutional levels with great success. Once commercialized, this product at the commercial level has found its greatest success in the institutional market. It is popular in Europe, where it is known as the 'long egg'.

Egg-Crust Pizza

The popularity and success of frozen pizza as a meal or snack are undisputed. In 1981, the pizza industry experienced its first billion-dollar year, with unit sales increasing by 10% annually. If the egg industry could capture only 5% of the pizza market, annual consumption of eggs in the United States would increase at the equivalent of 40 eggs per capita.

In preliminary work done in our laboratory, an egg-crust pizza was developed primarily with whole eggs. Sensory evaluation with high school students showed that they were receptive to the product. The New York State Department of Agriculture and Markets became interested and served the pizza at the New York State Fair for two years with great success. Reports indicated that students liked the egg-crust pizza and no psychological barriers appeared to be associated with consumption. The school systems indicated that they would purchase the product if it were available ready to re-heat.

Despite the success of the whole-egg crust pizza, it was felt that it would have trouble competing economically with conventional flour-crust pizza. As a result, developmental emphasis changed to using egg albumen in the pizza crust. Our final product in which the pizza crust contained 85% egg albumen and 15% flour proved to be the most popular (Fig. 2.5).

Fig. 2.5. Egg-crust pizza.

Discussion

When we started developing and market-testing new poultry meat and egg products in 1960, the market was practically devoid of these items. The turkey industry was the first to start developing and marketing new products and this has continued. Turkey consumption in the US has more than doubled in just a few years. The broiler industry soon followed by developing new convenience items with tremendous success. The egg industry in North America has been slow to join the turkey and broiler industries in developing and marketing new food products but the spark is being ignited and the fire should spread in years to come.

Marketing experts in the US predict that in the future consumers will demand more snack and finger foods, more foods engineered through modification by processing, more food especially for the aged, more attention to nutrition, more attention to avoiding additives, more prepared foods, more individualized packing, better packaging to prevent additives, more aseptic packaging, more modified atmosphere packaging, and more retort pouches.

References

Baker, R.C., Hahn, P.W., and Robbins, K.R. (1988) *Fundamentals of New Food Product Development*. Elsevier Science Publishing Co., New York, pp. 25-36 and pp. 223-234.

Brochures on the formulation and manufacture of the following products can be obtained from Dr. R.C. Baker, Room 112, Rice Hall, Cornell University, Ithaca, New York 14853: *Frozen Omelets, Apple Egg Drink, Hi-Pro Cookies, Hard-Cooked Egg Roll,* and *Egg Crust Pizza.*

Chapter Three

Innovative Egg Products and Future Trends in Europe

R.W.A.W. Mulder, C.A. Kan, and A.F.P. van Leeuwen[1]

Spelderholt Center for Poultry Research and Information Services (COVP-DLO), PO Box 15, 7360 AA Beekbergen and [1]Product Board for Poultry and Eggs, PO Box 502, 3700 AM, The Netherlands

Abstract Approximately 13% of the eggs produced are processed for the egg industry. As European production is more than self-supporting, the export market is important for several countries. It is estimated that 95% of the egg products are used for traditional applications in food products and 5% finds its way into non-food items in a number of industries. Products aimed at the catering market, one and two litre long shelf-life packs with liquid egg products, are becoming better known and more popular. The production of lysozyme from egg white is growing; the production of avidin on the other hand has just started. The effect of EC directives on animal welfare, hygiene, and health aspects of marketing eggs on the structure of the egg products industry is discussed. Programmes for a total integrated quality (management) control system in the whole production chain are introduced by the egg industry. Consumer demands will become more important in the future with respect to new products and ways of presentation. (*Key words:* egg production, egg consumption, influence of EC directives, hygiene, animal welfare, quality control, consumption, research needs)

Introduction

Consumers appreciate table eggs for their nutritional value, their taste and their numerous applications in preparing foods. However, egg products are relatively unknown; although liquid egg products have the same product properties, the individual consumer does not recognise egg products as valuable items in the shops. This is due to the fact that egg products are only known by their traditional applications for specific purposes.

It is surprising that in different countries of the developed and developing world, where the demand for animal protein is high, the production of egg products is so different. North America, Europe, and Japan produce approximately 40% of all eggs in the world and some 14% are converted to egg products. Although 60% of the world's egg production is located in other areas, only 3% of the world's egg products are produced in those countries. The 300 egg processing plants in the world are roughly located as follows: 100 in North America, 100 in Europe, and 100 in other countries.

Trends in Europe

Production Data

The main egg-producing countries in Europe are former USSR, Germany, France, Spain, Italy, The Netherlands, UK, and Poland. Table 3.1 presents the production data for the year 1989. The data for 1990 for these countries showed an increase in production. In 1991 however this increase disappeared in the overall picture of egg production in Europe. The main cause was the low production in the new German states. In most EC countries and especially in the UK and in France, egg production increased.

Table 3.1. European egg production (1000 tonnes) in 1989 (per major producing country).

Country	Egg production
Former USSR	4680
Germany	1059
France	891
Spain	773
Italy	699
Netherlands	643
UK	612
Poland	448

EC countries tend to produce more and more brown eggs. In Portugal, UK, France, and Italy about 95% of the eggs produced have brown shells. Denmark is the only real exception with 90% white eggs. Spain is the only other EC country now where more white eggs than brown eggs are produced.

Consumption Data

Total table egg consumption decreased in 1991 in EC countries. In Germany the number of eggs consumed per head decreased from 252 in 1990 to 240 in 1991. In The Netherlands there was a decrease of 6 eggs per head, resulting in a per capita consumption of 170 in 1991. The lowest table egg consumption is recorded in Portugal, 164. However, the consumption per head in that country is still increasing.

Egg Products

The development of the egg products industry started in the early 1950s and shows similarities to the development of the food industry in general. The food industry is the major user of egg products and therefore this development is not surprising. Aspects of convenience, consistency of composition, and product safety play an important role in the use of egg products in the production of other foods. On the other hand the food industry uses the specific product characteristics of the eggs for several applications, thus making the egg multi-applicable.

EC countries produce far more egg products than they sell or use in their home markets. Therefore the main producing countries have an enormous export volume, The Netherlands and Belgium being market leaders.

Figure 3.1 shows the export volume of EC countries of liquid and frozen egg products. Figure 3.2 shows the data with respect to exports of dried egg products from EC countries. Here The Netherlands and Germany are the main exporters.

The Netherlands is the main exporting country of egg products in the world. Therefore some more detailed data on the purchase and sales of egg products are given (Fig. 3.3). Figure 3.3 shows the number of eggs purchased by the Dutch industry in the years 1988-1991. In the last year 92,300 tonnes of eggs were converted to egg products, which made 1991 a record year. Figures 3.4 and 3.5 summarize the total production of whole egg, egg yolk, and egg white and the tremendous increase of production and exports during the last 25 years.

Exports of egg products from the top exporters, The Netherlands and Belgium, are mainly to other EC countries.

EC Directives

The structure of the European egg industry will be influenced by the EC directive (EEC, 1989) on hygienic and health aspects of the marketing of egg products, which became effective on January 1, 1992. This regulation is in force now and the main points are: a certification procedure for processors, to be allowed to produce egg products; a set of standards for the microbiological status of products; and the ban on the use of 'loss' eggs.

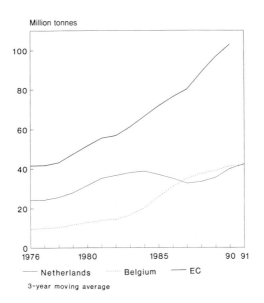

Fig. 3.1. Export of egg products (liquid/frozen) from the EC.

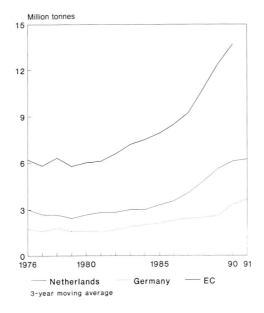

Fig. 3.2. Export of dried egg products from the EC.

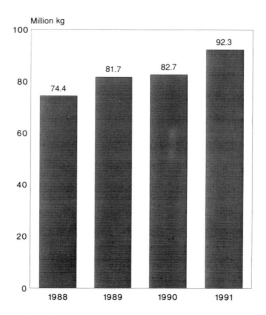

Fig. 3.3. Purchase of eggs by Dutch egg products industry.

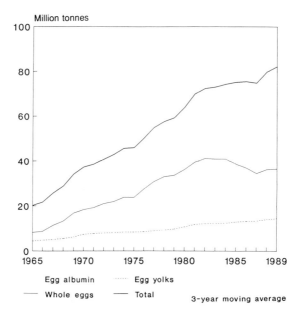

Fig. 3.4. Production of egg products in The Netherlands (dried products converted into liquid equivalents).

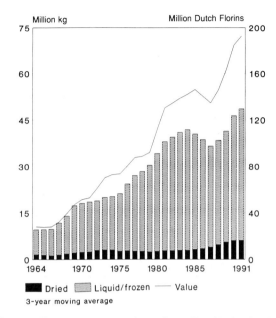

Fig. 3.5. Export of egg products from The Netherlands.

The standards for microbiological contamination do not distinguish between *Salmonella enteritidis* and other *Salmonella* spp. The other remarkable aspect is the fact that standards for microbiology, based on traditional cultural techniques, have been partially replaced by analyses for the contents of succinic acid and lactic acid, products which might be formed during storage, handling, and processing of the egg products and which are a good indicator for the hygienic level of the initial egg product processed. The content of 3-hydroxy-butyric acid, as an indicator for the use of incubator reject eggs, is also regulated.

'Loss' eggs are not allowed to be processed and included in products for human consumption any more. For all of Europe, it is estimated that 100 million kg 'loss' egg product is affected by this prohibition. The egg producing industry therefore has to increase its percentage output of first grade eggs, to be realised by investments, guidance of farmers, and the distribution chain in a total integrated quality (management) control programme. The larger poultry integrations in The Netherlands and other European countries are already implementing such a quality control programme. In such programmes all links in the production chain have to cooperate in order to produce safe and good quality eggs.

Thus, in the immediate future not enough first grade eggs can be purchased for processing in the traditional producing countries. As foreseen in Europe 92 as being well known in the Schengen Treaty (also in the current Maastricht Treaty), the borders between countries would disappear completely, as well as for egg production and processing. Therefor, it is possible that the location of egg production and egg processing will undergo drastic changes.

EC Directive on Animal Welfare

Another directive, which may heavily influence the structure of European egg production, is the directive with respect to animal welfare aspects of housing systems for layers. The possibility that the use of battery cages for layers might be replaced by an alternative and more animal-friendly housing system is being discussed. A final decision is not expected within the next two years.

Products

In Europe a whole product range of liquid, frozen, and dried egg products, with or without addition of sucrose, salt, or other components, is produced. Furthermore, cake mixes, omelette mixes, cooked and peeled eggs have been developed and these products are finding their way into the consumer markets.

Egg products are still mainly used in the food industry. Traditional applications are mayonnaises and salad dressings, biscuits, pastry and bakery products, pasta, ice-cream, chocolate, baby foods, and meat products.

A special product of The Netherlands is advokaat or 'egg nog'. The product is made of egg yolk or whole egg, mixed with sucrose and ethanol, which results in a final alcohol percentage between 14 and 18%. The traditional product is gellified and the drink has to be 'eaten' - together with whipped cream - with a small spoon. Nowadays, there are a lot of advocaat-type products on the market with a much lower alcohol percentage, sometimes even below 6%. It is known from research that the alcohol percentage in the advocaat strongly influences the survival of pathogenic bacteria. At alcohol percentages below 8% *Salmonella* bacteria - if present - may survive and therefore this development, although more kilograms of egg products can be marketed, is unsatisfactory from a public health point of view. This applies even more when the initial egg products used are not produced under good hygienic circumstances.

95% of egg products are used for food applications, while 5% find their way into non-food items in the pharmaceutical, biotechnology, and leather industries. The egg products industry is continuously searching for new ways, in the development as well as in the presentation of products, to market their products. After the *Salmonella* crisis of 1988, new markets were found in institutions (kitchens, restaurants, canteens, hospitals), air-line catering, and bakeries, where - because of product safety aspects - the traditionally used table egg has been replaced by pasteurised products. Several companies in Belgium, Denmark, Germany, and The Netherlands now sell 'long shelf-life' packaging containing one or two litres of pasteurised liquid whole egg, egg albumen, egg yolk, egg yolk with 20% sucrose, and even a low-cholesterol type egg product. The distribution of these products has - until now - not been through the retail channel, so the individual consumer is not yet aware of the existence of these types of products. If these products are introduced for the household market, smaller packages (0.25 or 0.125 litre) should be produced. In the UK, dried products of albumen and whole egg are sold (to our knowledge) in consumer packages in supermarkets. Also a large variety of omelette mixes is available in consumer units.

In several European countries research has started to investigate the consumers' purchasing attitude towards these type of products. By investigating consumer attitudes and adjusting to their wishes and demands, by offering new products and new ways of presentation, the egg industry is trying to stop the decrease in total egg consumption. One example is the development of a series of trendy 'sportdrinks', which have a healthy image, and are based on milk, soy, and egg proteins already available in the UK. More products are being developed in which egg proteins are used together with vegetable proteins.

Egg Components

Lysozyme is one of the components produced on a large scale in Europe. The first production processes were set up according to the Italian patent being hold by the SPA company; nowadays several other processes have been developed. The important aspects are the choice of the starting material, the purity of the isolated lysozyme, and the utilisation of the remaining egg albumen. Besides the application of lysozyme in the food industry either as a preservative or clarifying agent in wine-making, lysozyme is used in many pharmaceutical preparations. Avidin isolation and production on a commercial scale has started recently, but the application in the pharmaceutical/biochemical industry has not yet led to a high market share. One application is a screening kit for mould toxins. Isolation of ovalbumin and conalbumin or of other proteins and compounds from the egg interior is as far as we know not commercially applied in Europe.

In Europe the extraction of cholesterol by supercritical carbon dioxide has not led to an enormous commercial production. So far the process has been studied mainly with dried egg products as the starting material; new research is aiming at extraction from liquid products. To market the remaining cholesterol-free product, solutions are being sought for the rancidity problems occurring in these products after the removal of cholesterol and other fats.

Other Applications

Several other egg products are used in the cosmetics industry, ranging from application in egg shampoo to lipsticks and moisture regulating creams.

In the biotechnology industry, egg albumen is used as carrier material for immobilised bacteria and yeasts in the production of new drugs etc.

Other applications are in the flavour and the leather industries. The variety of applications is wide and it is expected that in the near future the market share of egg products in the non-food sector will increase considerably but the volume will still be low.

Research

European producers and processors who enter the 21st century will be those who accept and use the highest grade of agricultural technology, applicable through all production and processing phases. Therefore, research is needed to solve short-term problems and to be prepared for new products or applications.

In the short term these problems are:

1. The hygienic conditions of the production of eggs and egg products.
2. The solution of the environmental problems caused by egg shells.
3. The validation of the total volume of 'loss' eggs.

With regard to the hygienic conditions there are several systems and solutions to be implemented. They need, however, high investments and also a change in the mentality of the management by farmers and processors. Integrated quality control programmes are a tool in this process. The problem of the occasional contamination of eggs with *Salmonella enteritidis* cannot be solved easily. A policy of killing poultry flock found to be contaminated with *Salmonella*, as followed by several European governments, can only be successful if at the same time research is done to clarify the fundamental mechanism of infection of hens and eggs by this organism.

The environmental problems caused by dumping egg shells should be solved immediately. Also in this case, there is a solution which needs investment. The installation of egg shell dryers and additional sterilisation units can solve the hygienic problem. By using such processes egg shells can be made *Salmonella*-free and used again as a calcium source for layers or for fertilising soil.

The problem of validation of the 'loss' eggs volume is also difficult to solve, if the profit is to be similar to that of products fit for human consumption. The short-term solution is to stabilise the products and to use them in liquid or dried form in aquaculture or even in home aquaria to feed fish. The latter is done by some companies, also using waste from hatcheries. To use these products economically, the valuable components could be isolated and the remaining product used in animal feed or in pet food. Research in this direction is ongoing in several European institutes and industries.

The development of new products and new ways of presentation of products should be performed after investigating the attitudes and demands of consumers. The development of new products is the responsibility of the industry; research institutes and universities can only provide knowledge and equipment to define the areas and their scientific feasibility.

The main topics of research in this respect, which are under way in Europe, concern the following areas:

1. Improvement of extraction techniques to separate albumins from globulins in egg white.
2. The development of hypo-allergenic egg products.
3. The estimation of the influence of technology on egg albumen protein denaturation and the effects on functional quality after application in different food systems.

Reference

437/EEC (1989) Council directive on hygiene and health problems affecting the production and the placing on the market of egg products. *Official Journal of European Communities* 32 (L212):87.

Chapter Four

Innovative Egg Products and Future Trends in Japan

R. Nakamura

Department of Food Science and Technology, Nagoya University, Nagoya, Japan

Abstract Three examples of newly developed, value-added egg products in Japan are discussed. The first product is sialic acid separated from chalaza and yolk membrane. The utilisation of sialic acid as a reagent for biochemical research and as a starting material for the manufacture of derivatives of sialic acid for pharmaceutical uses has been developed. The second product is ovomucoid separated from egg white for use as a ligand in HPLC columns used in chiral separation of a variety of pharmaceutical compounds. The third product is yolk low density lipoprotein for use as a medium for serum-free culture of animal cells.

(*Key words:* chalaza, yolk membrane, ovomucoid, low density lipoprotein, LDL, sialic acid)

Introduction

Since Japan is one of the largest egg-consuming countries in the world, a variety of egg products are now in commercial-scale production. However, the main interest of the egg-processing industry in Japan is in the production of 'value-added' products, especially for medical and pharmaceutical uses. This trend may not be changed for a long time to come. Biologically important components of eggs are being separated for specific purposes. For example, lysozyme, avidin, and egg lecithin have been already isolated in large industrial scale. In this presentation, three value-added products recently developed in Japan will be discussed.

Separation of Sialic Acid from Chalaza and Yolk Membrane

Interest in sialic acid has increased rapidly in recent years in Japan, especially after recognition of its roles in the regulation of a variety of important biological phenomena. The generic name sialic acid comprises all *N*- and *O*-acyl derivatives of neuraminic acid (Neu) isolated from natural sources. Many different derivatives of sialic acid are present in vertebrates and certain invertebrates as components of glycoproteins, glycolipids, oligosaccharides, and polysaccharides, as well as in free form. The most common form of sialic acid is *N*-acetyl-neuraminic acid (Neu5Ac or NANA) followed by *N*-glycolylneuraminic acid (Neu5Gc) as shown in Fig. 4.1.

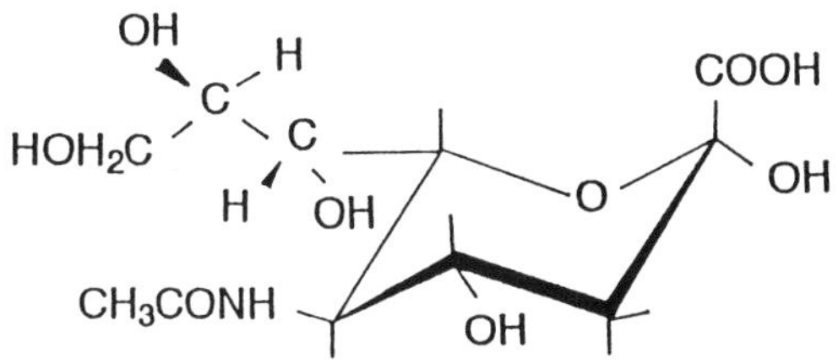

N-Acetylneuraminic acid
(Neu5AC, NANA)

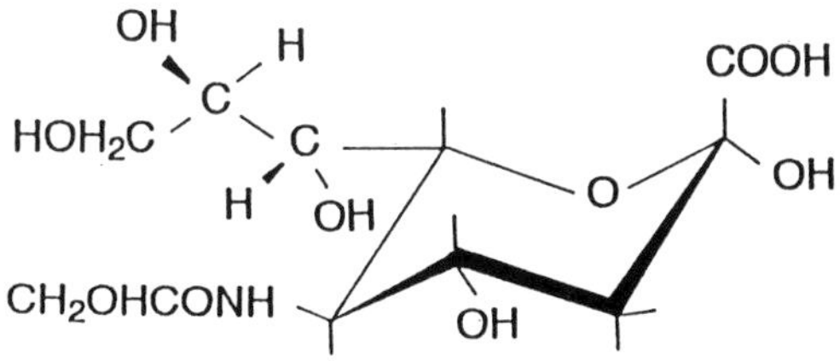

N-Glycolylneuraminic acid
(Neu5GC)

Fig. 4.1. Structure of natural sialic acids.

The wide distribution of sialic acids on the surface of macromolecules and cells, their strong electronegative charge (pK around 2), and the existence of multiple, tissue-specific forms suggest their involvement in cellular function. In fact, enzymic removal of sialic acid results in marked changes in the biological behaviour of cells and molecules. Five main functions of these saccharides are: functions based on negative charge, influence on macromolecular structure, antigens, components of receptors, and prevention of receptor recognition. Of these functions of sialic acid, anti-recognition is considered the most interesting and probably the most important. Although sialic acid is an essential component of receptors, it has been found to prevent the recognition of receptors by the corresponding ligands, or vice versa, as well as the recognition of antigenic sites by components of the immune system.

The utilisation of sialic acids as reagents for biological research and as starting material for the synthesis of derivatives for pharmaceutical use is increasing. Therefore, the economical large-scale supply of sialic acid is desirable. As the chemical synthesis of sialic acids is complicated and expensive, their isolation from natural sources is preferable. Extraction from only a few natural sources, e.g. swallow's nest, milk proteins, and microbial fermentation, has been explored for the industrial production of sialic acids. Recently, Professor Itoh of Tohoku University and his colleagues discovered the presence of a sialic acid-rich component in the chalaza of chicken eggs. They also found that egg yolk membrane contained a large amount of sialic acid (Table 4.1). Both chalaza and yolk membrane are separated from eggs by filtration at egg processing plants prior to the processing for egg products and usually discarded without utilisation. To utilise these washed by-products of egg processing, sialic acid is separated from chalaza and yolk membrane mixture.

Table 4.1. Sialic acid contents in hens' egg fractions.

Fractions	Sialic acid, % dry matter
Shell	0.27
Shell membrane	0.43
Egg yolk	0.41
Egg white	0.31
Egg yolk membrane	1.75
Chalaza	2.77

As shown in Fig. 4.2, a crude mixture of egg chalaza and yolk membrane (800 kg) is washed with water and the residue (125 kg) is homogenised with three volumes of water. The suspension is acidified to pH 1.4 with 3 M H_2SO_4 and heated for 1 h at 80°C. After cooling, saturated $Ba(OH)_2$ solution is added until pH 5.0 is attained, and the mixture is filtered. The filtrate is applied to a column of Dowex HCR-W2 (20-50 mesh), followed by a column of Dowex 1-X8 (200-400 mesh). The latter column is washed with water and then eluted with a linear gradient of HCOOH from 0 to 2 M. The Neu5Ac is eluted over a narrow range at approximately 0.8 M HCOOH. The eluates containing Neu5Ac are evaporated to dryness at 40°C under a reduced pressure. The residue obtained decolorised with activated charcoal powder and then lyophilised.

^{13}C-NMR spectra of sialic acid prepared from a mixture of chalaza and yolk membrane (Fig. 4.3) agree well with that of the standard Neu5Ac. The yield of Neu5Ac by this procedure is about 300 g. One of the large egg-processing companies in Japan is now planning to produce sialic acid by this procedure.

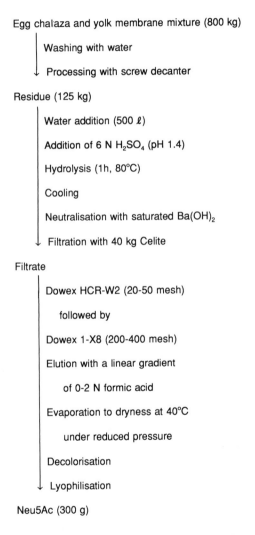

Egg chalaza and yolk membrane mixture (800 kg)

    Washing with water

    Processing with screw decanter

Residue (125 kg)

    Water addition (500 ℓ)

    Addition of 6 N H_2SO_4 (pH 1.4)

    Hydrolysis (1h, 80°C)

    Cooling

    Neutralisation with saturated $Ba(OH)_2$

    Filtration with 40 kg Celite

Filtrate

    Dowex HCR-W2 (20-50 mesh)

      followed by

    Dowex 1-X8 (200-400 mesh)

    Elution with a linear gradient

      of 0-2 N formic acid

    Evaporation to dryness at 40°C

      under reduced pressure

    Decolorisation

    Lyophilisation

Neu5Ac (300 g)

Fig. 4.2. Flow-chart of large-scale preparation of sialic acid.

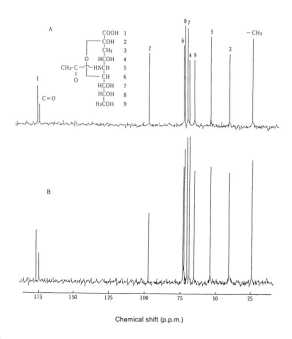

Fig. 4.3. ^{13}C-NMR spectra of sialic acid. A: Neu5Ac standard, B: sialic acid from chalaza prepared on a large scale.

Use of Ovomucoid as a Column Ligand in HPLC

Recent studies on liquid chromatographic chiral resolution showed that the excellent resolution of chromatography using protein-bound stationary phases makes it a useful analytical method. Bovine serum albumin (BSA)-conjugated columns and α_1-acid glycoprotein columns were developed based on the idea that those serum proteins which take part in enantio-selective drug transport *in vivo* must also have a chiral recognition capacity in their immobilised state. These columns have been employed for chiral resolution of a large number of compounds, almost all of which are of pharmaceutical importance. For example, chlorprenaline is a chemical compound widely used as a bronchodilator, but its activity is only present in the (-)-form. The chiral resolution mechanism of proteins is still not clear, although some hypothetical mechanisms involving racemic solutes and low-molecular-weight ligand-conjugated columns have been proposed. Furthermore, it may be true that almost all proteins could be enantio-separators, so that more efforts should be made to find proteins with properties of chiral recognition.

Since ovomucoid is one of the most stable proteins present in egg white, it might be useful for the preparation of a chiral recognition column. Based on this idea, an ovomucoid column was prepared and its application to the resolution of racemic compounds was investigated using three different racemic compounds provided by Dr Miwa of Esai Company (Fig. 4.4). Resolution of the acidic compound dibenzoyllysine was achieved with a 15 cm column eluted with 20 mM potassium phosphate buffer, pH 6.0, as shown in Fig. 4.5.

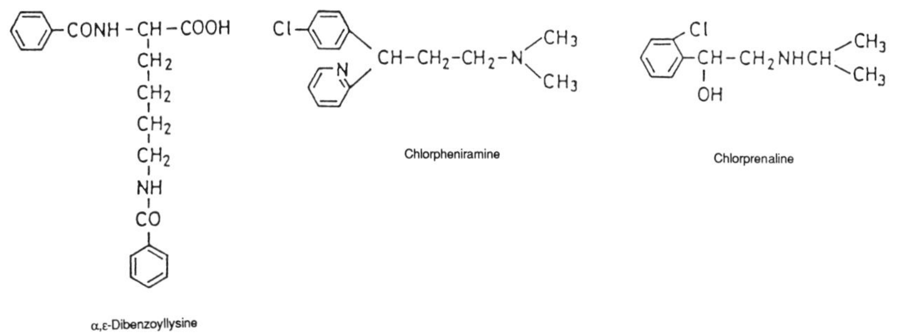

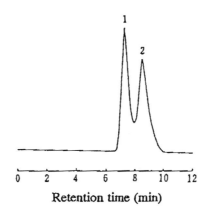

Fig. 4.4. Structures of compounds resolved by HPLC on an ovomucoid-conjugated column.

Fig. 4.5. Separation of enantiomers of α,ε-dibenzoyllysine. Mobile phase, 20 mM potassium phosphate buffer, pH 6.0; detection, 220 nm; flow rate, 1.0 ml/min. The earlier eluted peak is that of *d*-α,ε-dibenzoyllysine.

Amines (chlorpheniramine and chlorprenaline) were resolved in the same column and with the same buffer at different pH and different 2-propanol concentrations (Figs 4.6 and 4.7). In these experiments, the ovomucoid-conjugated column was stable over a wide range of pH and to organic solvents. Furthermore, this column was also stable at room temperature for more than three months.

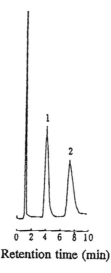

Retention time (min)

Fig. 4.6. Separation of *d,l*-chlorpheniramine on an ovomucoid-conjugated column. Mobile phase, 20 mM potassium phosphate buffer, pH 5.5, containing 6% 2-propanol; detection, 220 nm; flow rate, 1.2 ml/min; sample amount, 2.5 µg of the mixture.

To study the correlation between the chiral recognition properties and structure of the ovomucoid molecule, effects of neuraminidase treatment and chemical deglycosylation were studied. The neuraminidase-treated ovomucoid column had the same capacity and ability to chirally separate amines as the untreated column, although it had a lower retention of acidic solutes than the untreated column. However, the deglycosylated ovomucoid column did not show chiral recognition of chlorpheniramine and ketoprofen. Since chemical deglycosylation of ovomucoid did not alter its trypsin inhibitory activity, the effect of treatment on the protein conformation seems to be small. These results seem to show that the sugar chain of ovomucoid is essential for its enantiospecific interaction with solutes.

Miwa and his colleagues further developed an ovomucoid-conjugated polymer column instead of an ovomucoid-conjugated silica column and were able to determine the concentration of enantiospecific chlorprenaline in plasma (Table 4.2). These ovomucoid-conjugated columns are now produced by a company in Japan and are being used in various fields. Some thousand grams of ovomucoid are now being used for the production of these specific columns in Japan.

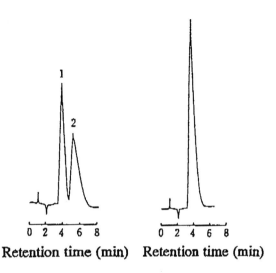

Retention time (min)   Retention time (min)

Fig. 4.7. Separation of *d,l*-chlorprenaline and chromatogram of *l*-chlorprenaline on an ovomucoid-conjugated column. Mobile phase, 20 mM potassium phosphate buffer, pH 6.1; detector, 210 nm; flow rate, 1.2 ml/min; sample amount, 1 µg of the mixture.

Table 4.2. Recovery of chlorprenaline enantiomers from plasma.

Concentration	Recovery % (mean ± SD, n = 3)	
(ng/ml)	(-)-Chlorprenaline	(+)-Chlorprenaline
50	93 ± 2	90 ± 7
100	92 ± 4	90 ± 6
500	96 ± 4	95 ± 1
1000	99 ± 6	98 ± 4

Egg Yolk Low Density Lipoprotein as an Additive for the Serum-Free Culture of Animal Cells

Development of serum-free media for the growth of mammalian cells is important not only for research purposes to identify macromolecules secreted by mammalian cells *in vitro*, but also for the production of medically important macromolecules. Conventional medium with serum contains various types of serum protein in such large amounts that separation and purification of the cellular products have been seriously hampered. Furthermore, serum is a complex mixture of components which may vary according to the age, sex, and species of its source. Dr Sato and his colleagues have systematically investigated the hormone and growth factor requirements of a large number of cell lines. They described completely defined serum-free media in which cells maintain their differentiated characteristic properties.

Professor Murakami and his colleagues at Kyushu University found that a mixture of insulin, transferrin, ethanolamine, and selenite (ITES) stimulated serial growth of many plasmacytomas and hybridomas of both murine and human origin in a serum-free medium. However, a mouse myeloma cell line and some other cells failed to multiply in this serum-free medium. They tried to find growth-stimulating substances effective for the mouse myeloma cell and found that the low density lipoprotein (LDL) of egg yolk had growth-promoting activity on a variety of cell lines in a serum-free medium supplemented with ITES.

They further fractionated LDL and obtained the most active yolk lipoprotein (YLP pI 7.5) fraction by chromato-focusing. As shown in Fig. 4.8, the minimum requirement of YLP pI 7.5 was 4 µg/ml for the mouse myeloma X63-653 cell line. Higher concentrations of YLP pI 7.5 did not inhibit cell growth.

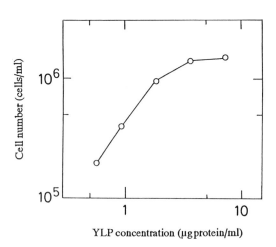

YLP concentration (µg protein/ml)

Fig. 4.8. The serum-free growth response of X-63-653 cells to YLP pI 7.5. The growth-promoting activity of YLP pI 7.5 was tested in serum-free eRDF supplemented with ITES.

The growth-promoting activity of YLP pI 7.5 in serum-free medium was further confirmed on various cell lines including plasmacytomas, epithelial and normal fibroblast cells (Table 4.3). Wakamatsu et al. (1987) also assessed the effect of unfractionated LDL. Although the amount required was much larger than that of YLP pI 7.5, unfractionated LDL promoted the growth of various cell lines when used in combination with ITES (Fig. 4.9).

The growth-promoting mechanism of LDL is not clear. Although cholesterol and phospholipids are known to stimulate growth of some cell lines, the lipid fraction of LDL did not promote cell growth. Fatty acids stimulate the growth of fibroblast and myeloma cells if used in an albumin complex to reduce toxicity caused by peroxides released from fatty acids.

Table 4.3. Growth-promoting activity of YLP pl 7.5 on various cell lines.

Cell lines	Cell number (x 10^4/ml) in eRDF supplemented with				Growth ratio (%)
	ITES	YLP pl 7.5	ITES-YLP pl 7.5	Serum	ITES-YLP pl 7.5/se
Human B cell lines					
Bri-7	22	28	73	100	73
HMy-2	35	20	80	100	80
RPMI-6666	16	14	20	30	67
Oda	9	9	16	71	22
NL-M	8	7	9	70	13
K103	6	10	30	53	57
WIL-2	26	85	123	150	82
K562	16	25	30	50	60
Human T cell lines					
HSB-2	10	10	43	50	86
CEM	14	43	76	82	92
Molt-4	12	10	42	65	65
Human B cell hybridoma					
HB4C5	91	18	149	140	106
Human lung adenocarcinoma					
PC-8	5	3	13	16	83
Human lung squamous cell carcinoma					
QG-56	17	5	17	22	80
Human melanoma					
Bowes	38	10	48	28	171
Human breast adenocarcinoma					
ZR75-1	16	6	34	20	170
Human stomach adenocarcinoma					
MKN-28	16	28	44	49	90
Human normal					
Flow-11000	2	2	3	8	35
Mouse myeloma					
P3U1	13	20	98	126	78
X-63-653	10	23	100	131	76

Cell were inculated at a density of 5 x 10^4/ml in eRDF medium. The cell number was counted after 5 days except for ZR75-Flow-11000 cells (10 days).

However, a complex of fatty acids and BSA did not promote the serum-free growth of X63-653 cells. Neither the lipid nor the protein fraction of LDL showed growth-promoting activity, indicating that an associated structure of lipids and proteins is still necessary for the lipoprotein to exhibit the activity.

The LDL sample is now produced by a pharmaceutical firm in Japan as an additive for the serum-free culture of animal cells.

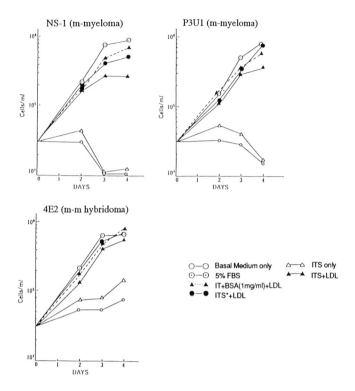

Fig. 4.9. Effect of LDL on growth of different suspended cells. Basal medium: Ham's F-12 plus Iscove's modified Dulbecco's medium (1:1), LDL: 200 µg/ml, I: 5 µg/ml insulin (bovine), T: 5 µg/ml transferrin (human), S: 5 ng/ml selenium, ITS*: CR-ITS⁺ Collaborative Res., Inc. (6.3 µg/ml insulin, 6.3 µg/ml transferrin, 6.3 ng/ml selenium, 1.3 mg/ml BSA, 5.4 µg/ml linolenic acid).

References

Allernmark, S. (1988) In: Krtsulovic, A.M. (ed.), *Chiral Separation of HPLC.* Wiley, New York, pp. 285-315.

Allernmark, A., Bomgren, B., and Boren, H. (1983) Direct liquid chromatographic separation of enantiomers on immobilized protein stationary phases. *J. Chromatogr.* 264:63-68.

Barnes, D. and Sato, G. (1980) Methods for growth of cultured cells in serumfree medium. *Anal. Biochem.* 102:255-279.

Fujii, D.K. and Gospodarowicz, D. (1983) Chicken egg yolk supplemented medium and the serum-free growth of normal mammalian cells. *In Vitro* 19: 811-817.

Hammansson, J. (1983) Direct liquid chromatographic resolution of racemic drugs using α_1-acid glycoprotein as the chiral stationary phase. *J. Chromatogr.* 269:71-80.

Itoh, T., Munakata, K., Adachi, S., Hatta, H., Nakamura, T., Kato, T., and Kim, K. (1990) Chalaza and egg yolk membrane as excellent sources of sialic acid (*N*-acetylneuraminic acid) for an industrial-scale preparation. *Jap. J. Zootechnol.* 61:277-282.

Juneja, L.R., Koketsu, M., Nishimoto, K., Kim, M., Yamamoto, T., and Itoh, T. (1991) Large-scale preparation of sialic acid from chalaza and egg-yolk membrane. *Carbohydr. Res.* 214:179-186.

Kawamoto, T., Sato, J.D., Le, A., and McClure, D.B. (1983) Development of a serum-free medium for growth of NS-1 hybridoma. *Anal. Biochem.* 130: 445-453.

McClure, D.B. (1983) Anchorage-independent colony formation of SV40 transformed BALB/c-3T3 cells in serum-free medium: role of cell- and serum-derived factors. *Cell* 32:999-1006.

Miwa, T., Ichikawa, M., Tsuno, M., Hattori, T., Miyakawa, T., Kayano, M., and Miyake, Y. (1987) Direct liquid chromatographic resolution of racemic compounds. Use of ovomucoid as column ligand of ovomucoid-conjugated columns in the direct liquid chromatographic resolution of racemic compounds. *Chem. Pharm. Bull.* 35:682-686.

Miwa, T., Kuroda, H., and Sakashita, S. (1990) Characteristics of ovomucoid-conjugated columns in the direct liquid chromatographic resolution of racemic compounds *J. Chromatogr.* 511:89-95.

Miwa, T., Sakashita, S., Ozawa, H., Haginaka, J., Asakawa, N., and Miyake, Y. (1991) Application of an ovomucoid-conjugated column for the enantio-specific determination of chlorprenaline concentrations in plasma. *J. Chromatogr.* 566:163-171.

Murakami, H., Masui, H., Sato, G.H., Sucoka, N., Chow, T.P., and Kano-Sueoka, T. (1982) Growth of hybridoma cells in serum-free medium: ethanolamine is an essential component. *Proc. Natl Acad. Sci. USA* 79:1158-1162.

Murakami, H., Okazaki, Y., Yamada, K., and Ohmura, H. (1988) Egg yolk lipoprotein, a new supplement for the growth of mammalian cells in serum-free medium. *Cytotechnology* 1:159-169.

Pirkle, W.H., Pochaspsky, T.C., Mahler, G.S., and Field, P.E. (1985) Chromatographic separation of the enantiomers of 2-carboalkoxyindolines and *N*-aryl-α-amino esters on chiral stationary phase derived from *N*-(3,5-dinitrobenzoyl)-α-amino acids. *J. Chromatogr.* 348:89-96.

Schauer, R. (1985) Sialic acids and their role as biological masks. *Trends Biochem. Sci.* 10:357-360.

Shipley, G.O. and Ham, R.G. (1983) Multiplication of Swiss 3T3 cells in a serum-free medium. *Exp. Cell Res.* 146:249-260.

Wakamatsu, T., Sato, Y., and Oshida, K. (1987) Effects of chicken egg yolk low density lipoprotein on the growth of various cell lines. *Soshikibaiyo Kenkyu* 6:70-73.

Chapter Five

Innovative Egg Products and Future Trends in China

S.C. Yang

Department of Food and Nutrition, Providence University, Shalu 43301, Taiwan, Republic of China

Abstract Several methods of preserving eggs have been developed in China for hundreds of years. These have resulted in a variety of traditional egg products such as alkaline-gelled eggs (pidan), salted eggs, and sour eggs. These types of egg products are processed separately by addition of sodium hydroxide, sodium chloride, and acetic acid. For better quality of the egg products, addition of flavouring substances and ageing are important. The Chinese people have a wide preference for traditional egg products, and the consumption of these egg products remains constant. For health concerns, several researchers have improved the manufacturing process of alkaline-gelled eggs to eliminate the contamination of heavy metals which cause food safety problems. In addition to providing the food industry with refrigerated, frozen, and dried eggs, the development of non-traditional, further processed egg products is an important factor in increasing egg utilisation. Several egg products which have been developed in Taiwan include egg tofu, egg drink, egg jelly, 'egg snack', puff dried eggs, texturised egg product, and transparent pidan.
(*Key words:* alkaline-gelled eggs, salted eggs, sour eggs, processed egg products)

Introduction

Since man's first recording of history, eggs have been regarded as an important foodstuff and they play a special role in man's daily livelihood. Taiwan's livestock was brought over by the Han people in the early years from mainland China, which explains the close resemblance in the preferences of the people on both sides for egg products. Since the early days, Chinese people have had various methods of storing eggs that resulted in a variety of popular traditional foods derived from the egg itself. These include pidan (ducks' eggs preserved in lime), salted eggs, sour eggs (eggs cured in vinegar and wine distillate), flavoured eggs, steamed eggs, and egg tofu, etc. Apart from these, poached eggs, scrambled eggs, boiled eggs, egg rolls, salad dressings, and mayonnaise, etc. have become very general household dishes and may be purchased at any supermarket.

In China, according to statistics, the yearly egg consumption per person was approximately 119 eggs in 1990. In Taiwan, yearly egg consumption increased from 40 eggs per person in 1963 to 220 eggs per person in 1990, with steady ducks' egg consumption all along. In the early days hens' eggs were mostly used in fresh foodstuffs with ducks' eggs used to make traditional egg products. Today, besides being used in fresh foodstuffs, hens' eggs are used to produce liquid eggs and egg powder, or to replace ducks' eggs in the making of traditional pidan and salted eggs. Furthermore, there are even more new food products under development which use eggs as the major ingredient.

The major objective of this report is to describe in brief the variety of traditional Chinese egg products, their processing and properties, and to discuss the prospect of new developments in egg products.

Quality and Processing of Various Traditional Egg Products

Pidan (Alkaline-Gelled Eggs)

Pidan is the oldest egg product of China. It is a food derived from alkali-treated duck eggs and has a special flavour. Although there are many ways and recipes for producing pidan, the basic production principle is the same. This is to make use of strong alkali to gel the albumen. Cunningham and Cotterill (1962) have given a report about the phenomenon of alkaline-gelled albumen. Lin and Liao (1973) who studied the alkaline gelation of duck egg albumen observed that when the pH of albumen exceeds 11.8, the gel will be translucent, and the time required for gelation will become less as the pH value rises. However, the gelled albumen will become a viscous liquid as self-liquefaction occurs when the pH is above 13.6. If at room temperature, the pH value lies between 12.0 and 12.8, the gelation will last for one month without being liquefied. Maintenance of gelation is of prime importance for the quality of pidan. The relationship between the change of viscosity and pH value for albumen of duck eggs is specified in Fig. 5.1. The viscosity of albumen increased rapidly as the pH value exceeded 11.8. Chang (1979) reported that ovalbumin and conalbumin can be gelled individually upon treatment with alkali, while other protein components do not have such characteristics. For the gelation of ovalbumin and conalbumin, protein molecules will first be unfolded. When the concentration of protein has reached a certain level, the unfolded protein molecule will be gelled as a fibrous network with water absorbed within it, and subsequently forms a transparent, elastic gel.

According to Su and Lin (1991), the albumen components of fresh duck eggs interact upon alkaline treatment, thereby forming high molecular weight compounds. However, these high molecular weight compounds were clearly dissociated at a pH of > 13.0. Based on the result of SDS-PAGE, conalbumin began to disappear when the pH exceeded 12.2. As the pH value reached 12.8, the ovalbumin also began to disappear.

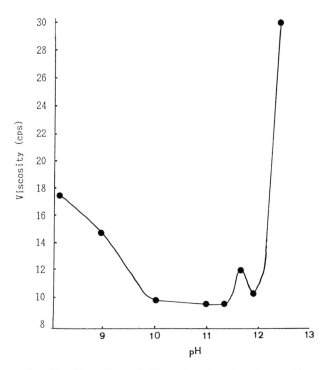

Fig. 5.1. The effect of pH on the viscosity of egg white.
From Lin and Liao (1973).

There are three different methods for manufacturing pidan. They are:

1. Coating method: lime and the required constituents are mixed to form a mud, which is used to coat the egg's surface before tightly sealing the eggs into a container. The eggs age naturally upon storage for a certain period of time.
2. Immersion method: the eggs are immersed in a soaking solution containing alkali and flavourings for a certain period of time before being taken out for packaging and storage.
3. Combination method: a combination of the above methods. First, the eggs are immersed in soaking solution for some time before being taken out and coated with yellow mud formulated with the soaking liquid and other coating materials.

The eggs are kept in a clean, dry container which is then sealed and stored in a cool place.

Of the three methods, the immersion method is the most convenient and quickest means of making pidan. Figure 5.2 illustrates the schematic process for making pidan with the immersion method.

Traditionally, duck eggs are preferable to chicken eggs in the making of pidan. The reason for this is that the chicken egg shell is thicker, which slows down the penetration of alkaline material, taking longer for soaking, and lowering productivity.

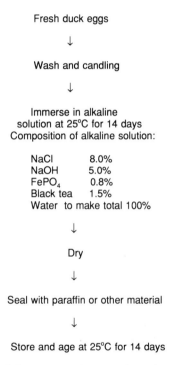

Fresh duck eggs

↓

Wash and candling

↓

Immerse in alkaline
solution at 25°C for 14 days
Composition of alkaline solution:

NaCl   8.0%
NaOH   5.0%
$FePO_4$ 0.8%
Black tea   1.5%
Water to make total 100%

↓

Dry

↓

Seal with paraffin or other material

↓

Store and age at 25°C for 14 days

Fig. 5.2. Flow scheme of the process for manufacturing pidan by the immersion method. Candling is the process used to detect and remove cracked or abnormal eggs, such as an egg with a blood spot.

Due to the alkaline reaction, the albumen of pidan turns dark grey with the egg yolk turning a dark greenish colour. This property has nothing to do with the addition of black tea, which only adds to the flavour and taste. Pidan itself has a very strong smell of hydrogen sulphide and ammonia, which is due to the content of sulphur-containing amino acids, such as cysteine, which is hydrolysed by desulphydrase and deaminase, hence producing the characteristic smell.

$$HS\text{-}CH_2\text{-}CH(NH_2)\text{-}COOH \quad \xrightarrow{\text{desulphydrase}} \quad CH_2\text{=}C(NH_2)\text{-}COOH$$

cysteine   amino unsaturated propionic acid
                                              + H_2S (hydrogen sulphide)

$$CH_2\text{=}C(NH_2)\text{-}COOH + H_2O \quad \xrightarrow{\text{deaminase}} \quad CH_3\text{-}CO\text{-}COOH + NH_3$$

amino unsaturated   pyruvic acid + ammonia
propionic acid

The formation of hydrogen sulphide is also related to the egg yolk turning a dark greenish colour. Ho *et al.* (1975) studied the colour changes in the egg yolk and proved that the dark greenish colour was due to the combination of hydrogen sulphide and ferrous ions generated in the egg yolk, to become ferric sulphide.

The addition and penetration of metallic salts into the pidan can, apart from giving the pidan its typical colour, also prevent contamination by micro-organisms. Without the addition of metallic salt, the pidan will become either light green or light yellow, a result of the alkaline reaction. During the processing of pidan when metallic salts are not added, the colour of pidan can be prevented from lightening by sealing the pidan that has undergone soaking and is at the stage of ageing, in an absolutely tight packaging to prevent the loss of hydrogen sulphide gas. Lead oxide used to be added in order to prevent the liquefaction of pidan, but it later turned out that lead oxide reacted with the hydrogen sulphide to become lead sulphide which appears as dark specks on the egg shell. Such practice has been prohibited in Taiwan for reasons of food safety, due to the toxicity of heavy metals. However, the process is still in use in China but is regulated: lead residues must not exceed 3 p.p.m.

After the process of soaking, the pH of egg albumen is between 12.4 and 12.6. The pH is lowered to around 11 as the alkali, together with water, gradually transfers into the egg yolk. After about one month of ageing, Sunghua (the Chinese name for a pine tree) crystals are formed in the albumen, and hence the name of Sunghua pidan, a product of popular preference and of higher price. Sunghua crystals form between days 4 and 7 of soaking with the addition of ferric phosphate to the alkaline solution (Liao and Lin, 1974). The pattern of formation involves the crystals spreading out in branches from a centre like pine-needles, with the crystals themselves appearing in units of yellowish brown long hexagonal cylinders (Fig. 5.3).

Upon completing the process of soaking, the outer layer of egg yolk turns light green and gradually solidifies while the inner part of the egg yolk maintains its yellowish semi-solid state. During ageing, the alkali and salt in the albumen gradually permeate into the egg yolk along with water. Hence, the egg yolk begins to solidify from the outer layer to the inner core and, upon full ageing, the pH value reaches 9-10.

Fig. 5.3. (a) Sunghua of alkaline egg white gel induced by ferric phosphate; (b) Sunghua of pidan (100×). The pattern of formation being the crystals spreading out in branches from a centre like pine-needles; (c) the unit crystal of Sunghua from pidan (320×).

In a scanning electron microscopic study of duck egg yolk, Yang and Hsu (1989) observed the difference between the microstructure of the inner and outer layers of the yolk of pidan (Fig. 5.4).

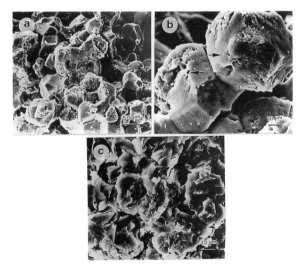

Fig. 5.4. (a) Surface structure of the outer part of yolk of pidan showing numbers of yolk grains adhering together; (b) higher magnification of panel a showing destroyed grains with resulting holes; (c) surface structure of the inner part of yolk of pidan showing random layers of material.

In the harder outer layer, numbers of yolk grains were found adhering together with many holes, indicating the gradual destruction of its many constituents. In the softer inner core, all the yolk grains had disintegrated into layers of material. This clearly explains the influence of yolk grains on the hardness of solidified pidan egg yolk.

In terms of nutrient content, there is no critical difference between pidan and fresh duck eggs. However, due to immersion in alkaline solution, pidan has higher contents of sodium, iron, calcium, and ash (Table 5.1). Since an excess amount of sodium in the diet may cause hypertension, reduction of sodium in pidan is of interest to health-conscious consumers.

Table 5.1. Chemical composition of duck eggs and duck egg pidan (per 100 g edible part).

Composition	Duck egg	Duck egg pidan
Energy (kcal)	192.0	179.0
Moisture (g)	70.2	67.2
Protein (g)	13.0	14.5
Lipid (g)	14.8	12.0
Carbohydrate (g)	0.5	2.0
Ash (g)	1.5	4.1
Inorganic substances (mg)		
Calcium	65.0	84.0
Phosphorus	232.0	198.0
Iron	2.6	3.1
Sodium	120.0	850.0
Potassium	130.0	65.0
Vitamins		
A (IU)	740.0	750.0
B (mg)	0.17	0.14
B (mg)	0.36	0.09
Niacin (mg)	0.10	0.10
C (mg)	0	0

From Tung *et al.* (1961).

Reduced-sodium pidan is made with soaking solution containing 50% sodium chloride and 50% potassium chloride rather than 100% sodium chloride. Reduced-sodium pidan has processing yield and texture characteristics which are comparable to those of traditional pidan, although the colour is darker (Su and Lin, 1992). According to the report of Ho *et al.* (1975), the content of higher molecular weight proteins decreases and the amount of lower molecular weight proteins increases during the process of converting duck eggs into pidan as a result of NaOH. Analysis of the amino acid composition of pidan and duck egg is shown in Table 5.2. It can be seen, either in egg yolk or albumen, that the content of individual amino acids in pidan is lower than that in fresh duck eggs.

The nutritional value of pidan is identical to that of fresh duck eggs, and pidan has become a very popular appetiser in Chinese cuisine owing to its unique taste. Even though the lead content of pidan has been reduced, from health and food safety aspects, care should be also taken to control the content of other heavy metals such as iron, lead, and aluminum in the production of pidan.

Table 5.2. Amino acid composition of pidan and raw eggs.

Amino acid	Egg white		Egg yolk	
	Raw egg (mg/g N)	Pidan (mg/g N)	Raw egg (mg/g N)	Pidan (mg/g N)
Lys	385	304	345	314
His	138	118	128	118
Arg	278	205	290	254
Asp	582	545	406	425
Thr	355	322	256	190
Ser	375	346	331	265
Glu	725	735	480	490
Pro	223	199	199	223
Gly	177	172	114	120
Ala	241	216	185	185
Val	203	170	154	154
Met	124	114	41	41
Ile	218	191	182	182
Leu	364	336	300	282
Tyr	214	176	188	176
Phe	252	210	178	147
Trp	57	57	42	42

From Ho *et al.* (1975).

Salted Eggs

As the name implies, these are eggs to which edible salt has been added as a preservative. The raw materials consist of duck eggs, with chicken eggs less frequently used. Traditionally, salted eggs are produced by the coating method. The egg is usually coated with a 1-2 cm thick layer of edible salt, red soil, wood ash, wine, and tea-leaves distilled with water into mud. The coated egg is kept at 20°C for about 30 days before cleaning away the coated layer and then cooking for serving.

The immersion method is a quicker and more convenient way of making salted eggs. The process involves adding 30% of edible salt to distilled water. Note that if the salt concentration is too low, the penetration process is slow and would only add to the possibility of the growth of contaminating microorganisms. A salt concentration of 30% or more is regarded as essential. When the salt has dissolved, the water is boiled for 15 min and then cooled to room temperature. Two percent of Kaoliang wine is added before immersing the eggs in the salted water and then storing at 20-25°C. After 45 days, the eggs are removed and dried naturally before storage at 10°C. The salted eggs are boiled for 15 min before serving. The egg albumen and yolk are solidified, hence giving a special taste.

Heated, salted eggs can be further kept for a week at a temperature of 25°C, or 2 weeks at 5°C. Lin *et al.* (1984) have reported that if the heated, salted egg is further soaked in a solution of polyvinyl acetate in 20% alcohol, the shelf-life is extended, thus salted eggs can be preserved for 2 weeks at 25°C or 5 weeks at 5°C.

Using the immersion method of making salted eggs, after 45 days of soaking, water content of the albumen is 81.2%, with 6.7% sodium chloride and 0.01% volatile basic nitrogen. For the egg yolk, water content is 22.5%, with 1.6% sodium chloride and 0.01% volatile basic nitrogen. A salted egg with fat oozing out of the egg yolk is most popular among consumers. The reason for the phenomenon of fat-permeated egg yolk is unknown, but it is possible that it has something to do with the salt content. The viscosity increases as salt penetrates to the egg yolk and, coupled with the low salt concentration (1.6%), the egg yolk becomes slightly hard. Yang and Hsu (1989) have reported that the diameter of yolk grain in the yolk of a salted egg is larger than that in those that have not been salt-treated. It is hence thought that the size of the grain and whether the yolk grain has disintegrated or not would affect the texture of the salted egg yolk.

The whole salted egg is edible but usually the egg albumen is discarded or used as animal feed because of its salty taste. The egg yolk is frequently used to decorate or wrap cakes and snacks. Chen (1991) submitted a patent for the manufacturing of salted egg yolk. The method used involves separating the egg albumen from the egg yolk. The egg yolk is then immersed in a solution containing 15% sodium chloride for 36-48 h. The finished product contains 2.6% salt, and is of comparable quality to those manufactured by the traditional method. In addition, since the egg albumen is not salted, it can be of additional use and not wasted.

The colour of egg yolk determines the desirability of salted eggs. Egg yolk of golden yellowish colour is demanded by most consumers. Research indicates that the feed has a strong influence on the colour of egg yolk produced. Tai *et al.* (1985) have found that synthetic oxycarotenoids such as citranaxanthin and canthaxanthin can produce a golden yellow colour in the duck egg yolk. Cantha-xanthin, in particular, is presently used in Taiwan as an additive to duck feeds.

Flavoured Eggs

Spices can be added to the egg soaking solution to create so-called flavoured eggs (Chang and Lin, 1986). Composition might include 25% edible salt, 2% black tea, 5-10% spices (such as clove, five spices mix, anise, pepper or combinations) with immersion at 25°C for a period of 20 days. Eggs are then taken out and boiled for 5 min before serving, or further smoked to produce smoked, flavoured eggs. Tea-flavoured eggs are also very popular. The difference between flavoured eggs and tea-flavoured eggs is that the eggs (chicken or duck) for tea-flavouring are first boiled and cooked and the egg shell is broken slightly. The eggs are then immersed in a solution of salt, tea-leaves, and spices and cooked for a few hours to create the unique flavour. The method does not require a long period of soaking.

Since Chinese people are very sensitive about the flavour of foods, it is not surprising that there are different tastes in salted, flavoured eggs in the various regions of China.

Sour Eggs

Eggs are immersed in a solution of vinegar and edible salt until the egg shell is partially dissolved or softened. The egg is then immersed in wine distillate with edible salt and a small amount of vinegar. Ageing is completed after 3-5 months of soaking. The sour egg is further heated to 80°C for 5 min before serving. Sour egg is a product of acid coagulation and it contains the unique sour and wine tastes.

Egg Tofu

A traditional Eastern food, egg tofu is made from whole egg mixed with water, salt, monosodium glutamate, sucrose, soy sauce, and fish protein extract, and then heated to form an egg gel. Its texture is identical to that of bean-curd (tofu), hence its name. Production of egg tofu is based on the principle of heat coagulation of egg albumen and egg yolk proteins. Factors affecting coagulation of egg tofu include the concentration of egg proteins, time and temperature of heating, and pH value.

According to Yang (1994), the hardness of egg tofu increases with increasing concentration of egg proteins, heating temperature, and heating time. Egg protein of 4.5% concentration which is heated at 90°C for 30 min is regarded as most acceptable with respect to hardness of the egg tofu. The microstructure of egg tofu observed under scanning electron microscope shows a characteristic three- dimensional network. As the protein concentration increases from 3 to 6%, the network construction is more intense with fewer empty spaces. Apart from that, heating temperature also affects the texture and appearance. Higher temperature tends to increase the hardness of egg tofu and it is hence important that temperature is controlled between 85 and 90°C in order to maintain its unique taste and colour.

New Developments in Egg Processing

In Taiwan, eggs are sold as raw eggs to the public, or to the food industry in the form of refrigerated, frozen, or dried eggs. Some of the eggs are made into traditional egg products such as pidan and salted eggs. However, the problem of excessive production of raw eggs still exists and creates an agricultural economic crisis for many farmers. Many researchers, who highly recommended and recognised the nutritional value of eggs, have studied every possible avenue for further processing eggs in an effort not only to solve the problem of over-production of raw eggs, but also to increase agricultural profit.

There are many new developments in egg processing; while some have been successfully developed, many are still in the research stages.

Transparent Pidan

In traditional pidan, the albumen is a translucent gel of a brownish colour. In order to satisfy the multiple taste and curiosity of consumers, transparent pidan was developed. According to Su and Lin (1993), the making of transparent pidan begins by immersing duck eggs in a solution that contains 4.2% NaOH and 5% NaCl, and is kept immersed at a temperature of 25°C for 8 days. During immersion, the pH value of egg albumen is approximately 12.0-12.2. Subsequently, the eggs are boiled in water at 70°C for 10 min with temperature increasing at a rate of 1.1 ± 0.2°C per minute. With heating conditions properly controlled, a clear transparent pidan can be obtained. Heating conditions play an important role in determining the degree of transparency of the pidan albumen. If the rate of protein denaturation exceeds that of protein coagulation, for example if heated at 95°C for 10 min, the gel becomes non-transparent and its texture becomes fairly tough. Transparent pidan is an innovation derived from traditional pidan. It offers and contributes to the various consumers a multiple choice of taste and appearance.

Egg White Drink

Egg beverages such as egg milk drink, egg drink powder, egg sour drink, etc. have been developed in some countries and a few have been patented. However, commercialisation of egg beverages has some drawbacks in that upon heating, the egg tends to coagulate, precipitate, or change in flavour. These problems can be solved by the reaction of additives and enzymatic hydrolysis. Chang (1986) described an egg white drink manufacturing procedure (Fig. 5.5).

Using this processing system, an egg white drink has been produced which is acceptable in terms of its viscosity, appearance, taste, flavour, and physical stability. In addition, storage at 4°C for one and a half months has been found to be possible without any sign of chemical or microbiological changes.

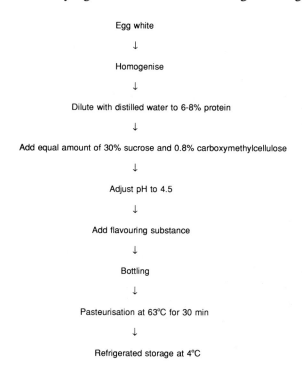

Egg white

↓

Homogenise

↓

Dilute with distilled water to 6-8% protein

↓

Add equal amount of 30% sucrose and 0.8% carboxymethylcellulose

↓

Adjust pH to 4.5

↓

Add flavouring substance

↓

Bottling

↓

Pasteurisation at 63°C for 30 min

↓

Refrigerated storage at 4°C

Fig. 5.5. Flow diagram of the process for manufacturing egg white drink.

Egg White Jelly

Heating temperature, heating time, concentration of egg proteins, and pH value are factors that influence the characteristics of heated egg white gels. Kitabatake *et al.* (1988a,b) have been successful in obtaining a transparent egg white gel that increases the multiple purposes of egg white processing, using a method involving two-stage heating with careful control of pH value and low salt concentration. Yang and Lin (1990) suggested adding 10% sucrose plus a stabilising agent such as agar to the dialysed egg white solution and applying heat for an hour at a temperature of 80°C. Upon cooling, a translucent egg white jelly is formed. Fruit extract or colourings may be added to produce varying taste and appearance. To produce a smoother flavour, the pH value of the egg white might be increased to 3.5 while at the same time increasing the sucrose content to 15-20%. Varying or improved texture might be obtained by the addition of agar-agar, gelatin, carrageenan, etc.; these are more suitable than guar gum and xanthan gum. For example, egg white jelly treated with 1.0% agar-agar or 3.5% gelatin (based on egg albumen weight) has a texture identical to that of fruit jelly. Since egg white jelly has a higher protein content than fruit jelly, it is a product worthy of further research and development.

'Egg White Snack'

The main ingredients of 'egg white snack' are egg white powder, potato starch, granulated sugar, and leavening agents mixed in water to make an egg white batter. The shaped batter is heated in an electric or microwave oven to make an 'egg white snack'. Spices and colourings may be added to enhance the appearance and taste (Huang and Yang, 1993).

The quality of the egg white snack is affected by the water content of the batter. The water content should be between 25 and 40%. The batter becomes too sticky and difficult to mix if the water content is below 25% and the snack becomes crumbly if the water content is above 40%. If 3% baking powder plus 0.5% gluconodeltalactone is added to the batter, it improves the quality and produces proper swelling and a higher yield of the snack. The quality of egg white snack is also affected by the relative proportions of egg white powder and potato starch. Figure 5.6 indicates that regardless of microwave vacuum or electric oven heating, the volume of egg white snack expands with the increase in the amount of egg white powder added. Swelling ratio is at its maximum when the percentage of egg white powder is 30% and that of potato starch is 20%, and at this ratio, breaking strength is optimised. Even though there is no significant difference in processing yield, the textural acceptability in sensory evaluation is at its highest.

Apart from the effect of composition, different heating methods also influence the quality of the egg white snack. Table 5.3 shows the differences in quality of batches of egg white snack that are of the same composition but heated with an electric oven, microwave vacuum oven, or a mix of both heating methods. Results showed that when the batter is heated with an electric oven, it swells least but has too high a breaking strength and is too hard. Microwave vacuum heating created the maximum volume but the texture became crumbly. The mixture that is first heated in an electric oven at 75°C for 32.5 min, until water content is 10%, then re-heated in the microwave vacuum oven at 0.5 kW, 160 Torr for 10 min, produces the best texture and appropriate volume. The final snack has a water content of 3.5-5.8%.

The appearance of these snacks treated with different heating methods is illustrated in Fig. 5.7. Besides starch, proper amounts of grain flour, soybean powder, or wheat flour may be added to produce varieties of patterns and tastes.

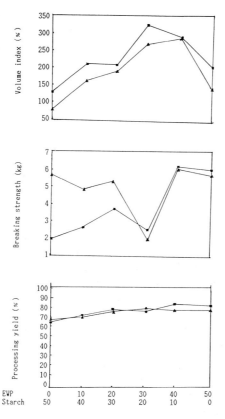

Fig. 5.6. Effect of the egg white powder (EWP): potato starch ratio on processing yield, volume index, and breaking strength of the egg white snack. ▲ - ▲ , electric oven 75°C for 15 min, then microwave 0.5 kW × 8 min, 160 Torr; ■ - ■ , electric oven 75°C for 50 min.

Texturised Egg Products

Extrusion technology has also been applied as a new development of egg processing. Hsu (1993) used the technique when after uniform mixing of different ratios of egg white powder, soybean powder, and isolated soy protein, a twin-screw extruder was applied to manufacture the texturised egg products. Results indicated that the moisture content, water holding capacity, hardness, cohesiveness, elasticity, brittleness, and brightness (Hunter's L value) of the texturised egg products were significantly affected by soybean meal/egg white ratio, feed rate, screw speed, and/or their interactions. Chewiness was slightly affected only by screw speed. Other quality attributes, including hardness, gumminess, breaking energy, tensile strength in tearing, and colour (Hunter's a and b values), were not significantly affected by the three factors (ingredients, feed rate, and screw speed) within the ranges tested.

Table 5.3. The processing yield, volume index, breaking strength, and textural acceptability scores of egg white snack[1] under various heating conditions.[2]

Heating	Processing yield (%)	Volume index (%)	Breaking strength (g)	Textural accessibility score
Oven 75°C for 70 min	75.5[b]	147[b]	3608[b]	3.4
Microwave 0.5 kW for 12 min	71.2[b]	582[a]	698[d]	3.3
Oven 75°C for 10 min then microwave 0.5 kW for 21 min	71.8[b]	120[b]	4810[a]	4.9
Oven 75°C for 32.5 min then microwave 0.5 kW for 10 min	80.8[a]	218[b]	2210[c]	5.5
Microwave 0.5 kW for 2.5 min then oven 75°C for 55 min	74.1[b]	183[b]	3505[b]	3.4
Microwave 0.5 kW for 7 min then oven 75°C for 32.5 min	74.6[b]	603[a]	323[d]	4.8

[1] Basic ingredients: 30% egg white powder and 20% potato starch (based on the weight of the batter).
[2] Microwave heating at 160 Torr of absolute pressure.
Each mean represents the average of ten observations.
The data with the different superscripts within the same column differ significantly ($P < 0.05$).

Fig. 5.7. The appearance of egg white snack under various heating conditions.

Conclusion

Through the years, there is no doubt that traditional egg products such as pidan, salted eggs, and flavoured eggs have maintained their popularity. However, apart from the traditional egg products and liquid eggs, new product development in egg processing is necessary to explore further the potential value of eggs. This report has briefly described the production and quality of Chinese traditional egg products, in addition to explaining the latest research and development carried out in the processing of egg products. Readers may find this report useful in understanding the developments of egg processing that are currently being performed in Taiwan.

References

Chang, H.S. (1979) Studies on the alkaline gelation of ovalbumin and conalbumin. *J. Chin. Soc. Anim. Sci.* 8:35-50.

Chang, H.S. (ed.) (1986) Eggs and egg products. In: *Egg Processing*. Hwa Shang Yan Publishing Company Inc., Taipei, Taiwan, pp. 560-563.

Chang, H.S. and Lin, S.M. (1986) Studies on the manufacturing of flavored chicken eggs. *J. Chin. Soc. Anim. Sci.* 15:71-82.

Chen, M.T. (1991) A new procedure for processing salted yolk. Patent pending.

Cunningham, F.E. and Cotterill, O.J. (1962) Factors affecting alkaline coagulation of egg white. *Poultry Sci.* 41:1453-1461.

Ho, W.T., Chang, Y.M., and Huang, P.C. (1975) Studies on composition and nutritive value of pidan. 1. Vitamin contents, protein characteristics and the nature of the dark green compound of pidan yolk. *J. Chin. Agric. Chem. Soc.* 13: 58-65.

Hsu, S.Y. (1993) Processing effects on the qualities of texturized egg products made of fresh egg white. *Food Sci. (Taiwan)* 20:161-167.

Huang, Y.C. and Yang, S.C. (1993) Studies on the effect of processing conditions on the qualities of egg white snack. *Food Sci. (Taiwan)* 20:291-300.

Kitabatake, N., Shimizu, A., and Doi, E. (1988a) Preparation of transparent egg white gel with salt by two-step heating method. *J. Food Sci.* 53:735-738.

Kitabatake, N., Shimizu, A., and Doi, E. (1988b) Preparation of heat-induced transparent gels from egg white by the control of pH and ionic strength of the medium. *J. Food Sci.* 53:1091-1095.

Liao, K.T. and Lin, C.W. (1974) Formation of Sunghua crystals in pidan. *J. Chin. Soc. Anim. Sci.* 3:39-42.

Lin, C.W., Chu, B.C., and Wang, S.W. (1984) Effect of coating conditions on the quality of cooked salted egg during storage. *J. Chin. Soc. Anim. Sci.* 13:55-63.

Su, H.P. and Lin, C.W. (1991) Effect factors of duck egg white's alkaline gelation and liquefaction. *Food Sci. (Taiwan)* 18:389-394.

Su, H.P. and Lin, C.W. (1992) Effect of low sodium salt on the quality of duck pidan. *Food Sci. (Taiwan)* 19:302-309.

Su, H.P. and Lin, C.W. (1993) A new process for preparing transparent alkalised duck egg and its qualty. *J. Sci. Food Agric.* 61:117-120.

Tai, C., Lee, S.R., and Hertrampf, J. (1985) Effect of Citranaxanthin and Canthaxanthin on duck egg yolk color. *Taiwan Livestock Res.* 18:241-249.

Tung, T.C., Huang, P.C., Li, H.C., and Chen, H.L. (1961) Compositon of foods used in Taiwan. *J. Formosan Med. Assoc.* 60:973-1005.

Yang, S.C. (1994) Effect of protein concentration and heating conditions on texture and microstructure of egg tofu. *Food Sci. (Taiwan)* 21:1-13.

Yang, S.C. and Hsu, H.K. (1989) Scanning electron microstructure of the yolk of duck egg and duck egg products. *J. Agric. Chem. Soc.* 27:460-472.

Yang, S.C. and Lin, Y.F. (1990) Studies of egg white gelation and development of egg white jellies. *Food Sci. (Taiwan)* 17:123-137.

Chapter Six

Innovative Egg Products and Future Trends in Korea

I.-J. Yoo

Korea Food Research Institute San 46-1, Baekhyun-Dong, Bundang-Ku, Songnam-Si, Kyonggi-Do 463-420, Republic of Korea

Abstract Egg production in Korea was 393,000 tonnes in 1990. More than 10,000 tonnes of eggs were imported and the amount imported has increased every year. Despite the tendency to consume more processed food, creation of additional demand is not likely because domestic egg consumption mainly depends upon table eggs. Processed eggs for marketing in Korea can be classified into two kinds. One is primary processed eggs including liquid eggs and egg powders. The other is secondary processed eggs which are further processed such as egg flake and egg curd. In addition to the above egg products, specific nutrient-fortified eggs are produced by modifying the feed formula and breeding techniques. The technologies developed so far including Korean patents are introduced. Convenience foods using egg and nutrient-controlled eggs are likely to be popular in the near future. For example, low cholesterol eggs and polyunsaturated fatty acid fortified eggs will be produced to meet the consumer demand. However, problems such as introducing egg quality grading systems and extending the short shelf-life of fresh eggs should be solved as well.
(Key words: Korea, nutrient fortified egg, processed egg, egg products)

Introduction

Since 1981, egg production has increased consistently and egg production in 1990 was 393,000 tonnes in Korea. Per capita consumption was 168 eggs, or 9.2 kg (Table 6.1). Egg consumption shows an increasing tendency, so egg production in Korea has a great future. In 1990, more than 10,000 tonnes of eggs were imported from other countries and the amount imported has increased every year. There are three egg-processing industries which have a production capacity of 90 tonnes per day. They produced only about 30 tonnes of eggs per day until 1988, but from 1989 the efficiency of operation of the egg-processing industry was improved to over 60% as shown in Table 6.2.

Table 6.1. Production and per capita consumption of eggs by year.

Year	Production		Per capita consumption	
	tonnes	million eggs	gram	eggs
1981	253,000	4431	6270	114
1982	257,960	4505	6325	115
1983	283,565	4939	6795	124
1984	283,194	4936	6695	122
1985	309,302	5390	7200	131
1986	332,000	6029	7977	145
1987	361,539	6573	8590	156
1988	397,124	7220	9460	173
1989	380,543	6919	8919	163
1990	393,305	7151	9191	168

From National Livestock Cooperatives Federation (1991).

Table 6.2. Production and production capacity of processed egg.[a]

Year	Production capacity (tonnes)	Production (tonnes)
1985	1080	167
1986	3625	1838
1987	14,451	2698
1988	15,070	5259
1989	84,920	58,834
1990	86,277	55,092

[a] Liquid egg including egg powder.
From Ministry of Health and Social Welfare (1985-1990).

Monthly egg prices in Korea fluctuate severely during the year because the production of processed eggs is too small to create a buffer (Table 6.3).

Confectionery products have recently changed from the usual hard biscuits to soft cookies; as a result, the egg content of the products has increased from 3-5 to 7-10%. The meat and fishery industries are increasing the production of imitation crab meat every year. Baking industries consume more than 50 tonnes of egg products per day, and this demand is expected to increase by 20-30% compared with present consumption in the near future (Yoo, 1990). However, domestic egg consumption mainly depends upon table eggs. Considering the tendency to increased demand for processed food, creation of additional demand for eggs is not anticipated. Therefore, ways of promoting egg consumption could include production of eggs retaining biological function, or egg products from liquid egg which could be raw materials for convenience food or intermediate materials. In this article, I would like to review the current status of the Korean egg market, new egg products and technology, and Korean patents registered including those applied for in the last decade.

Table 6.3. Change of monthly egg prices during the year.

Price	Monthly prices in 1990 (won/10 eggs)												Avg.
	Jan	Feb	Mar	Apr	May	Jun	Jul	Aug	Sep	Oct	Nov	Dec	
A	564	553	596	634	660	567	464	461	657	645	559	561	577
B	610	590	628	675	708	636	513	528	707	703	610	602	626
C	650	654	672	698	737	703	674	647	709	763	717	715	695
A/C	0.87	0.85	0.89	0.91	0.90	0.81	0.69	0.71	0.93	0.85	0.78	0.78	0.80
B/C	0.94	0.90	0.93	0.97	0.96	0.90	0.76	0.82	1.0	0.92	0.85	0.84	0.90

A: Farmer's price
B: Wholesaler's price
C: Consumer's price

Egg Products and Processing Equipment

Table Eggs Fortified with Special Components

Recently, several new kinds of table eggs have been marketed in Korea. Eggs fortified with micro-components such as minerals and vitamins are popular. Ginseng, known as a healthy food, is used to improve the quality of eggs. Attempts to modify fatty acid composition have been successful. Several specially fortified eggs have been marketed, as discussed below.

Nutrition eggs (produced by Purina Korea Co.). These eggs contain at least 20% more vitamin D, E, and B_{12} than ordinary eggs. The guaranteed shelf-life is 3 days which is decided by the producer for the purpose of quality control.

Vita-iodine eggs (produced by Mannawon Farm). These eggs contain more than 7 mg% vitamin E and 1.3 mg% iodine. According to the producer, ordinary eggs contain no iodine and 1.1 mg% vitamin E.

Omega eggs (produced by Mannawon Farm). 'Omega eggs' are fortified with ω3 fatty acids. The ω3:ω6 fatty acid ratio is modified to conform with requirements for good health. Omega eggs are produced by hens consuming a specially formulated feed containing a high level of ω3 fatty acids. While eggs produced by hens on ordinary feed contain 0.33% ω3 fatty acids, eggs produced by hens on specially formulated feed contain 2.30% ω3 fatty acids which is around 7 times higher than the former. The ω6:ω3 ratio of the omega egg was 6.5:1 which is close to the ideal ratio for health, while that of ordinary eggs was 39.8:1.

Iodine eggs (produced by Dongwha Livestock Co.). More than 0.3 mg of iodine is contained in each 'iodine egg'. According to the producer, active iodine which is contained in the iodine egg is easily digested and effectively absorbed. The iodine content is at least 20 times higher than that of ordinary eggs.

Ginseng eggs (produced by Mannawon Farm). Korean ginseng by-products from ginseng extract processing plants are used to produce 'ginseng eggs'. Consumers believe, but without reliable supporting data, that these ginseng eggs contain saponin, which is the effective component of Korean ginseng.

Others. 'Ginseng-herb egg' is produced by feeding ginseng leaves and medicinal herbs.

Egg Products from Liquid Eggs

Liquid egg including egg powder and frozen egg could be used widely by the food industry. The confectionery and bakery industries are the main consumers of liquid eggs. But aside from the confectionery and bakery industries, other uses exist for egg products. Egg flake, dried egg sheet, egg curd, and rolled egg sheet were developed several years ago. However, only egg flake and dried egg sheet are marketed today. The main uses of egg flake and egg sheet are as follows.

Egg flake. Egg flake is mainly used in instant noodles. The main component is egg yolk. It is also used in frozen fish paste (Kim and Yoo, 1990). Egg flake can be produced by two drying methods, either by hot air blast drying at 55-90°C or by extrusion technology.

Dried egg sheet. Dried and strip type egg sheet is often used in cooked noodles and other dishes. It is similar to egg flake (Kim and Yoo, 1990), but is produced in a slightly different way. Extrusion technology can be used to produce the egg sheet or a stuffer with a specially designed nozzle can be used.

Others. Egg curd and rolled egg sheet used to be marketed but they are no longer produced because of their short shelf-life and incomplete refrigeration systems. 'Jerky type' egg products similar to American jerky style beef, containing egg yolk, starch, and fish paste, has been developed by Lee *et al.* (1988), but are not yet available commercially.

Processed Shell Eggs

Injection of pickles into the egg through the egg shell pores under pressure was attempted in order to produce a seasoned shell egg (Lee and Park, 1989; Lee, 1989). Conventionally, seasoned shell eggs are produced by soaking the eggs in pickle solution for a relatively long period. But using high pressure, the processing time required to produce seasoned shell eggs can be reduced. Alkaline-gelled and fermented eggs used to be manufactured using ducks' eggs and are now produced using hens' eggs.

Seasoned shell eggs. The main composition of the curing solution is 30% salt, 20% soy sauce, and additives including spices. The process takes place at 50-120°C, 3 kg/cm^2. Seasoned shell eggs can be kept for 4 weeks by coating the shell with oil after the cooking process (Fig. 6.1).

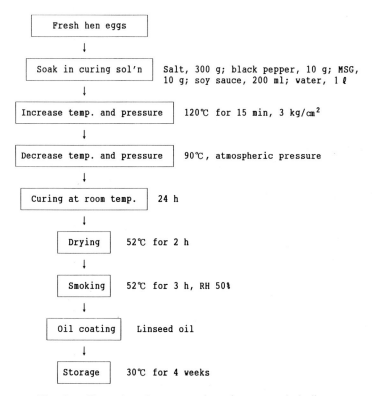

Fig. 6.1. Flow chart for processing of seasoned shell eggs.

Alkaline-fermented eggs. Egg albumen tends to gel in alkaline conditions. Traditionally duck eggs were used to make alkaline-fermented eggs using lime, tea-leaves, and other coating materials at ambient temperature. But such a method requires a long time to complete the fermentation and gelation. Now several alkalis, such as calcium hydroxide and calcium oxide, are used to shorten the process, combined with high temperature. Hens' eggs are now commonly used to make alkaline-fermented eggs.

Other Processed Eggs

In Korea, people have accepted vinegar-pickled eggs for many years; these eggs are manufactured by beverage companies and sold in small bottles. Quail eggs are marketed in this way, but the peeled cooked quail eggs stored in brine are not produced by the industry because peeling the quail egg is tedious work.

Vinegar-pickled eggs. Five or six days after immersing shell eggs in vinegar, the egg shell disappears. The product tastes somewhat sour, so one of the beverage companies has improved the quality, and produced a commercial product.

Pickled quail eggs. Cooked and peeled quail eggs with brine were marketed after sterilisation, packed in a retort pouch.

Cooked Egg Equipment

Cooked egg vending machines and devices for frying eggs have been developed and introduced recently. In the former, a microwave oven is incorporated in the vending machine, which will cook the egg in only 20 seconds. When the cooked egg is served from the machine, it is presented along with seasoned salt and a disposable spoon. The latter type of device is designed to cook two fried eggs at the same time in a microwave oven. It is made of plastic and was developed by the Korean Poultry Association to promote egg consumption.

Quality Improvement Technology

The main functional properties of eggs are coagulation, foaming, and emulsification. These functions are always very important in processing egg products. Attempts to improve the functional properties of eggs during the last decade in Korea are described here.

Frozen Eggs

The best way of preserving eggs for long periods while retaining quality is freezing. The problem when whole egg and egg yolk are frozen is the gelation which detracts from the functional properties. Kim *et al.* (1989) suggested that adding 5% NaCl to egg yolk instead of 10% NaCl would prevent the gelation of egg yolk at -8°C. They suggested that storage at -8°C could save energy when users do not want to keep the products for more than 6 months. Trials to depress the freezing point by using cryoprotectants were attempted by Lee and Lee (1988). They stored egg yolk and whole egg at -15°C in an unfrozen state by adding 45.2% and 70.3% cryoprotectants (45% fructose and 55% glucose), respectively. They also found that treating liquid egg with 0.15% papain could inhibit gelation during storage to some extent.

Egg Powder

Kim *et al.* (1980) found that if the pH of egg was increased before drying, the functional properties such as whippability and emulsion capacity were improved.

Retort Pouch Packed Whole Eggs

Prevention of the green-grey discoloration of retort pouch packed liquid whole eggs was accomplished by the addition of about 0.015% Na_2EDTA. Palatability of retort pouch packed eggs was not affected by the addition of 0.02% Na_2EDTA (Song *et al.*, 1984).

Mayonnaise

The minimum amount of egg yolk needed to prepare mayonnaise was found to be 6.5% (Cha *et al.*, 1988). Kim *et al.* (1990) reported that contamination of egg yolk with egg white decreased the viscosity, emulsifying capacity, and emulsion stability of egg yolk.

Lysozyme Separation

Egg white lysozyme separation was attempted by several researchers (Yoo *et al.*, 1989c; Lee *et al.*, 1989; Park *et al.*, 1990) and it was found that the enzyme activity increased by 20% when lysozyme was dissolved in 0.066 M phosphate buffer at pH 6.3 and then incubated at 37°C for 2 h (Lee *et al.*, 1990).

Lysozyme as a Preservative

The effect of lysozyme and sodium hexametaphosphate on the lysis of *Lactobacillus plantarum* was investigated in MRS broth and 0.06 M phosphate buffer (pH 6.3) at 32°C by Lee *et al.* (1991b). Lysozyme and sodium polyphosphate have synergistic effects on the growth inhibition and lysis of *L. plantarum*. Lee *et al.* (1992) found that the growth of *L. plantarum* could be almost completely inhibited in 120 p.p.m. lysozyme with at least 0.8 mM EDTA. Kim *et al.* (1991)

reported that the addition of 100 p.p.m. lysozyme and 0.2% sodium poly-phosphate was no different from that of 1.5% potssium sorbate when used to extend the shelf-life of pressed ham. In the preservation of Vienna sausage, addition of lysozyme and polyphosphate was effective in extending shelf-life (Lee et al., 1991a).

Heat Coagulation of Egg Albumen

Yoo et al. (1990a) reported that heat-set albumen gel showed the maximum hardness at pH 4.5-5.0 and pH 9.0. Yoo et al. (1990b) found that hardness of albumen gels was decreased by the addition of over 2.5% sucrose. Yoo et al. (1990c) reported that the initial heat denaturation temperature of egg albumen was increased by 11°C by acetylation, by 12.5°C by maleylation, and by 14.5°C by succinylation.

Heat Sensitivity of Egg Albumen

Yoo (1988) studied the effect of heating time and temperature, pH, and NaCl concentration on heat sensitivity of egg albumen during heat treatment in terms of turbidity and whippability. In order to reduce the heat sensitivity of egg albumen, effects of metal ions were investigated on its functional properties (Yoo et al., 1989a). Based on these studies they reported that the maximum foaming capacity was obtained at 8.3% protein concentration; however, foaming capacity as well as foam stability decreased by diluting albumen before and after heat treatment at 60°C for 5 min (Yoo et al. 1989b).

Korean Egg Patents

Title & Patent No.: Manufacturing Method for Aseptic Shell Eggs, Korean Patent 89-4363 (Lee and Kim, 1989). Principle and purpose: for manufacturing of seasoned shell eggs by penetration, seasoning is penetrated into the shell egg through shell pores under pressure followed by smoking and coating processes. Procedure: there are two ways to manufacture aseptic seasoned shell egg. One is heating followed by seasoning; the other is seasoning followed by heating. The latter method is as follows: washing (20°C) → pickling (pH 4-6) → inversion → equilibrium (50-60°C, 15-20 min) → 1st seasoning (60°C, 15-60 min, 1.5-3.0 kg/cm^2) → 2nd seasoning (60°C, 12-36 h, 1 atm) → 1st heating (100°C, 5 min) → 2nd heating (60°C, 15-20 min) → dipping (15-20 min in water) → drying (30°C, 20 min, 40°C, 20 min, and 50-60°C, 2-3 h) → smoking (50-60°C, 2-4 h, RH 65-75%) → coating (spraying with linseed oil or mineral oil).

Title & Patent No.: Semi-cooking Method of Egg, Korean Patent 177 (Sim, 1968). Principle and purpose: manufacturing semi-cooked egg suitable for eating without egg shell damage after cooking procedure. Procedure: to prevent egg contents flowing out, the contents are partially coagulated by dipping the eggs in 50°C water for 5 min to harden the shell membrane. The eggs are then dipped in 70-72°C water for 10 min.

Title & Patent No.: Manufacturing Method for Fermented Egg, Korean Patent 79-1 (Ueda, 1979). Principle and purpose: dipping in alkali solution not only shortens the coagulating period but also accelerates the process of manufacturing the fermented and uniformly coloured eggs. Procedure: the white of shell eggs is coagulated by dipping in 24-45°C alkaline solution for about 15 days; the eggs are then removed from the solution and heated up to 58°C for 10 min to 2 h.

Title & Patent No.: Processing Method for Nutrient Fortified Beverage of Egg and Soymilk, Korean Patent 82-1701 (Seo, 1982). Principle and purpose: to eliminate the flavour of soybean milk or to restrain the action of lipoxygenase which causes a fish-like smell, soybeans are peeled, ground through 30-40 mesh, and heated in a rotary toaster; soybean protein is then extracted by dipping in water at 50-100°C. Procedure: mix coagulate-separated soybean milk from 0.05-0.5 wt% alginate and coagulated egg → stir the mixture for 20-30 min with peptiser (e.g. phosphate, carbonate, or polyphosphate) → homogenise at 80°C, 100-250 kg/cm^2 → UHT pasteurisation (121°C, for 30 min, 1 kg/cm^2).

Title & Patent No.: Processing Method of Egg, Korean Patent 72-161 (Kang, 1972). Principle and purpose: manufacturing jelly type fermented shell eggs with tea, pine needles, and other seasonings in 100°C water, which has characteristic flavour and long-term preservation ability. Procedure: boil water with tea and pine needles → mix other seasonings in 100°C water → ferment shell egg at 15°C for 50 days in the mixture in a sealed bottle → coat with 3-5 mm of clay to prevent evaporation of moisture.

Title & Patent No.: Seasoning Method of Egg, Korean Patent 85-375 (Yoo, 1985). Principle and purpose: since cooked eggs are inconvenient to handle, season, keep clean, and regulate salt content. Liquid seasoning is injected into the air cell of the egg to achieve a desirable salt content. Procedure: inject liquid seasoning through the egg shell with fine needled injector → stand at 15°C for over 8 h → cook at 70-75°C for 30 min for semi-cooking method or at 75-80°C for 40 min to 1 h for full-cooking method.

Title & Patent No.: Manufacturing Method of Processed Egg, Korean Patent 91-6921 (Hayashi, 1991). Principle and purpose: manufacturing good tasting and 'scrambled' processed egg product which is restored to the original state by adding hot water. Procedure: cook protein, starch, grain, and their processed material → mill the edible part under 4 mesh → separate 4-100 mesh granules which will include >60% of gross granule weight → mix granules with non-denatured egg → extrude at a rate of 5-60%.

Title & Patent No.: Egg Yolk Centering Method, Korean Patent 89-1560 (Yoo, 1989). Principle and purpose: the specific gravity of the avian shell egg is about 1.08-1.09 and the shell egg floats in 12% saline solution; as the specific gravity decreases day by day, so the air cell faces upwards at this salt content. The egg is heated at 80°C for 3-5 min to centre the egg yolk because ovomucoid is coagulated at 80°C. Procedure: after the first cooking of avian shell eggs in 80°C, 12% saline solution, for 3-5 min, the second cooking in a 95°C salt solution for 5-20 min will centre the egg yolk.

Title & Patent No.: Condensation Method of Egg, Korean Patent 89-3251 (Namkoong, 1989). Principle and purpose: to improve the quality of semi- dry powdered egg yolk, eggs are boiled at 1 atm, then shell-free hard-boiled eggs are dipped in wood vinegar and other seasonings. The wood vinegar causes the moisture in the white to move to the yolk, so the white has a sticky 'mouth feel' and the yolk is moist. Procedure: shell-free boiled eggs are dipped in 90% wood vinegar and 5% spice and seasoning for 10-20 h. After the solution has penetrated to the inside of the egg, the egg is boiled in a pressure jar at 0.70-1.4 kg/cm^2, for 40-70 min.

Title & Patent No.: Egg Shell Removing Method for Semi-Cooked Egg, Korean Patent 90-4273 (Namkoong, 1990). Principle and purpose: to remove egg shell from semi-cooked eggs, the best cooking method is to boil at 69-71°C. The concentration of acetic acid is generally 4-5%, but when dipping in 20-40% acetic acid, the calcareous egg shell dissolves completely without damage of egg white. Procedure: semi-cook the egg in 70°C water for 1-3 min → cool → dip in 20-40% acetic acid and 3% saline solution for 30-60 min → remove shell.

Title & Patent No.: Processing Method for Duck Egg, Korean Patent 91-2480 (Hyun, 1991). Principle and purpose: process for solidifying duck eggs not by boiling but by dipping in alkaline solution to prevent the loss of nutritive value and to gain good flavour. Procedure: dip duck egg in 1300-1500 ml water with 30-50 g $MgCO_3$, 45-60 g Na_2CO_3, 25-35 g NaCl → leave at 23-28°C for 18-21 days → dry at 40°C for 2-3 min.

Title & Patent No.: Isolating Method for Specific Antibody Containing Materials from Egg, Korean Patent 91-5409 (Dogor, 1991). Principle and purpose: method of manufacturing materials containing specific antibody obtained from eggs from a hen which has immunity against selective antigens and method of isolating materials containing activated specific antibody which can be obtained easily and inexpensively during the year. To immunise a hen, antigens are injected or given by oral administration. Procedure: make emulsified antibody from homogenised or vigorously stirred egg yolk (or white, or whole egg) → make powder by spray drying or freeze drying → manufacture powdered material containing antibody used for oral administration.

Title & Patent No.: Manufacturing Method of Fish Meat Egg, Korean Patent applied 84-8757 (Kim, 1984). Principle and purpose: method of manufacturing two-coloured extruded fish meat of similar shape to sliced cooked eggs. The centre of this product is yellow like egg yolk and the outer part is white like egg white. Procedure: a mixture of 52% fish meat, 41% yolk, 2.3% white, 2.6% starch, and other additives is heated, sliced, and extruded using a sausage extruder.

Title & Patent No.: Manufacturing Method for Egg Yoghurt and Egg Beverage, Korean Patent applied 85-5758 (Kim, 1985). Principle and purpose: method for manufacturing egg yoghurt and egg drink made from eggs and skim milk by the process of mixing, stirring, homogenisation, pasteurisation, cooling, and fermenting. Procedure: separate whole egg, egg white, or egg yolk → pasteurise at 60-64.5°C, 2-3 min → mix and homogenise egg white or egg yolk → pasteurise at 90-93°C, 10 min → add starter culture (0.1-1% *Lactobacillus bulgaricus*) → ferment at 32-44°C, 4-18 h → stir curd → homogenise at 150-200 kg/cm^2 → mix with syrup and pasteurised water (egg soft drink) or mix with fruit (egg fruit drink).

Title & Patent No.: Manufacturing Method for Egg Yolk Oil, Korean Patent applied 87-6843 (Kim, 1987). Principle and purpose: method of manufacturing non-denatured egg yolk oil preservative of non-offensive smell and good quality by extraction of pasteurised eggs with a steriliser. Procedure: add 200 p.p.m. sodium chlorite → separate yolk from white → homogenise and filter yolk under vacuum → pasteurise (60-70°C plate heat exchanger) → dry (150°C in spray drier with N$_2$) → yolk powder (2-5% moisture content) → stand for several hours (soak in *n*-hexane, anaerobic state) → centrifuge (removes non-fat yolk and organic solvent) → vacuum evaporation → product (transparent yellow egg yolk oil).

Title & Patent No.: Seasoning Method for Cooked Egg, Korean patent 88-2454 (Kim, 1988). Principle and purpose: method of seasoning cooked egg by using osmosis of brine through the egg shell and shell membrane to the egg yolk and white, and then cooking eggs in the brine. Procedure: dip fresh eggs for 10-15 min in 1000 ml of Baumé 10-24 brine → rotate eggs from time to time at 20°C for 7-10 days → product.

Title & Patent No.: Dried Egg Flake for Instant Noodles, Korean patent 85-1808 (Lee, 1985). Principle and purpose: manufacturing dried egg flake for instant noodles suitable for Korean taste, the structure of which can be restored simply by adding hot water after freezing at -14°C for 12 h and drying for 30 min with hot air. Procedure: boil the shell-free eggs and other seasonings → freeze at -14°C for 12 h → dry for 30 min with hot blast.

References

Cha, G.S., Kim, J.W., and Choi, C.U. (1988) A comparison of emulsion stability as affected by egg yolk ratio in mayonnaise preparation. *Korean J. Food Sci. Technol.* 20(2):225-230.

Dogor, H. (1991) Isolating method of specific antibody containing materials from egg. Korean Patent 91-5409.

Hayashi, G. (1991) Manufacturing method of processed egg. Korean Patent 91-6921.

Hyun, D.W. (1991) Processing method for duck egg. Korean Patent 91-2480.

Kang, K.W. (1972) Processing method of egg. Korean Patent 72-161.

Kim, S.S. (1984) Manufacturing method of fish meat egg. Korean Patent applied 84-8757.

Kim, H.K. (1985) Manufacturing method for egg yoghurt and egg beverage. Korean Patent applied 85-5758.

Kim, J.Y. (1987) Manufacturing method for egg yolk oil. Korean Patent applied 87-6843.

Kim, S.S. (1988) Seasoning method of cooked egg. Korean Patent 88-2454.

Kim, K.S. and Yoo, I.J. (1990) Study on development of egg flake product. I 1021-0104, *Research Report of Korea Food Research Institute.*

Kim, K.S., Song, I.S., Kang, T.S., and Song, K.W. (1980) Studies on the diversified utilization of egg. *Annual Report of Food Research Institute of AFDC Republic of Korea.*

Kim, K.S., Yoo, I.J., and Kang, T.S. (1989) Studies on the egg storage technology. *Korean J. Poultry Sci.* 16(4):233-238.

Kim, J.W., Hong, K.J., Cha, G.S., and Choi. C.U. (1990) Changes in physical properties of salted egg yolks as affected by refractive index during frozen storage and their effects on functionalities in mayonnaise preparation. *Korean J. Food Sci. Technol.* 22(2):162-267.

Kim, Y.B., Lee, S.K., Kim, K.H., and Yoo, I.J. (1991) Effect of the addition of lysozyme and sodium ultraphosphate on the shelf-life of press ham. *Korean J. Anim. Sci.* 33(2):176-184.

Lee, Y.S. (1985) Dried egg flake for instant noodles. Korean Patent 85-1808.

Lee, S.K. (1989) Studies on the shell crack of chicken eggs during cooking. *Korean J. Anim. Sci.* 31(7):449-452.

Lee, S.K. and Kim, Y.M. (1989) Manufacturing method for aseptic seasoned shell egg. Korean Patent 89-4363.

Lee, Y.C. and Lee, K.H. (1988) Freezing preservation of liquid egg by freezing point depression. *Korean J. Food Sci. Technol.* 20(4):594-599.

Lee, S.K. and Park, J.H. (1989) A study on the development and storage of pickled shell-eggs. *Korean J. Anim. Sci.* 31(8):519-526.

Lee, S.K., Yoo, I.J., and Kim, Y.M. (1988) Studies on the processing of seasoned product containing egg yolk. *Korean J. Poultry Sci.* 15(1):45-51.

Lee, S.K., Yoo, I.J., and Min, B.Y. (1989) Studies on the lysozyme isolation by ion-exchange chromatography. *Korean J. Anim. Sci.* 31(12):780-787.

Lee, S.K., Yoo, I.J., Kim, K.S., and Kim, Y.B. (1990) Studies on the activity and stability of egg white lysozyme. *Korean J. Poultry Sci.* 17(2):109-114.

Lee, S.K., Kim, K.H., Kim, K.S., Kim, Y.B., and Yoo, I.J. (1991a) Effect of lysozyme and sodium ultraphosphate on the shelf-life of vienna sausage. *Korean J. Anim. Sci.* 33(1):78-84.

Lee, S.K., Hong, S.P., Kim, Y.B., Choi, S.Y., and Yoo, I.J. (1991b) Effect of lysozyme and sodium ultraphosphate on the lysis of *L. plantarum*. *Korean J. Anim. Sci.* 33(1):73-77.

Lee, S.K., Kim, I.H., and Yoo, I.J. (1992) Effect of lysozyme, glycine, lysine and EDTA on the growth of *Lactobacillus plantarum*. *Korean J. Food Sci. Technol.* 24(1):11-13.

Ministry of Health and Social Welfare (1985-1990) *Production Amount of Food and Additives*. Seoul, Korea.

Namkoong, R. (1989) Condensation method of egg. Korean Patent 89-3251.

Namkoong, R. (1990) Removing egg shell method of half-cooked egg. Korean Patent 90-4273.

National Livestock Cooperatives Federation (1991) *Materials on Price, Demand and Supply of Livestock Products*. Seoul, Korea.

Park, S.J., Kim, H.S., Kim, H.W., and Ahn, T.H. (1990) Continuous separation of lysozyme from egg white by ion exchange column chromatography. *Korean J. Food Sci. Technol.* 22(6):711-715.

Seo, J.S. (1982) Processing method for nutrient fortified beverage of egg and soymilk. Korean Patent 82-1701.

Sim, Y.S. (1968) Semi-cooking method of egg. Korean Patent 177.

Song, I.S., Yoo, I.J., Kang, T.S., and Min, B.Y. (1984) Prevention of the green-grey discoloration in retorted liquid whole eggs. *Korean J. Poultry Sci.* 11(1):27-32.

Ueda, S. (1979) Manufacturing method for fermented egg. Korean Patent 79-1.

Yoo, B.E. (1985) Seasoning method of egg. Korean Patent 85-375.

Yoo, I.J. (1988) Studies on heat sensitivity of egg albumen. I. Effects of heating time and temperature, pH and NaCl concentration on heat sensitivity of egg albumen. *Korean J. Poultry Sci.* 15(1):39-44.

Yoo, I.J. (1989) Egg yolk centering method. Korean Patent 89-1560.

Yoo, I.J. (1990) Egg processing industry. *Bulletin of Food Technology*, Korea Food Research Institute. 3(2):5-11.

Yoo, I.J., Lee, S.K., and Kim, Y.B. (1989a) Studies on heat sensitivity of egg albumen. II. Effect of pH and/or the addition of metal ions on heat sensitivity of egg albumen. *Korean J. Poultry Sci.* 16(1):17-22.

Yoo, I.J., Kim, K.S., and Song, K.W. (1989b) Studies on heat sensitivity of egg albumen. III. Effect of egg albumen concentration and addition of sugars on heat sensitivity of egg albumen. *Korean J. Poultry Sci.* 16(1):23-28.

Yoo, I.J., Lee, S.K., Kim, K.H., and Min, B.Y. (1989c) Effects of egg white and ion exchange resin pretreatment on separation of egg white lysozyme. *Korean J. Poultry Sci.* 16(3):157-167.

Yoo, I.J., Kim, C.H., Han, S.H., and Song, K.W. (1990a) Studies on heat stability of egg albumen gel. I. Effects of heating time and temperature, pH and NaCl concentration on heat stability of egg albumen gel. *Korean J. Poultry Sci.* 17(2):127-133.

Yoo, I.J., Kim, C.H., Han, S.H., and Song, K.W. (1990b) Studies on heat stability of egg albumen gel. II. Effect of egg albumen concentration and addition of sugars on heat stability of egg albumen gel. *Korean J. Poultry Sci.* 17(3):211-216.

Yoo, I.J., Kim, C.H., Han, S.H., and Song, K.W. (1990c) Studies on heat stability of egg albumen gel. III. Changes of heat stability of egg albumen gel by chemical modification. *Korean J. Poultry Sci.* 17(3):217-223.

Chapter Seven

Separation, Purification, and Thermal Stability of Lysozyme and Avidin from Chicken Egg White

T.D. Durance

Department of Food Science, University of British Columbia, 6650 NW Marine Drive, Vancouver, British Columbia, Canada V6T 1Z4

Abstract Lysozyme and avidin isolates from chicken egg white are commercially important products. Preferred techniques allow their extraction without impairment of food quality or functionality of the residual egg white. Isolation and processing of avidin and lysozyme are aided by their unusual resistance to thermal inactivation and ease of analysis. Efficient affinity methods have been developed for both proteins but have not received widespread industrial use because of high costs, regulatory restrictions, and concerns about resin life and reliability of resin supply. Recent studies have renewed interest in membrane separation techniques but at present the methods are limited by poor yield and purity of the isolate. The classic lysozyme precipitation method is still employed in some operations but has often been supplanted by ion exchange methods. Both bulk and column cation exchange methodologies are described. Early avidin isolation methods were dedicated to the single product and destroyed the food quality of the residual egg white. Modern methods are often integrated with lysozyme extraction to reduce costs. A simultaneous lysozyme-avidin extraction method is described. Secondary purification of avidin may be accomplished by a variety of techniques but, again, cation exchange has proven to be efficient due to high capacity, good resolving power, and reasonable price.
(*Key words:* lysozyme, avidin, chicken, egg, white, separation, purification, extraction, heat, stability)

Introduction

Lysozyme and avidin are two proteins found in egg white. They are commercially available as isolates, they have similar charge properties, and they can be isolated together. Lysozyme is defined as a 1,4 ß-N-acetyl muramidase, an enzyme which cleaves the glycosidic bond between C-1 of N-acetylmuramic acid and the C-4 of N-acetylglucosamine in bacterial peptidoglycan (Jollès and Jollès, 1984). Lysozymes are common defensive proteins in nature which are produced by a wide variety of animals, insects, plants, and fungi. However, almost all commercial lysozyme is chicken lysozyme from egg white. Lysozymes from other sources, such as turkey egg white (La Rue and Speck, 1970), duck egg white (Kondo et al., 1982), bacteriophage T4 (Szewczyk et al., 1982), human milk (Jollès and Jollès, 1971) and bovine milk (Vakil et al., 1969) are available in research quantities but are much too expensive to compete with egg white lysozyme in the foreseeable future.

Avidin is also widely distributed in nature but the commercial scene is again dominated by the avidin from chicken egg white. Avidins are glycoproteins with a strong and specific affinity for the vitamin biotin. The avidin-biotin interaction is characterized by a dissociation constant of 10^{-15} M, making it one of the strongest non-covalent ligand-protein interactions known (Green, 1963). Avidin was first detected as a factor in raw hen egg albumen which caused an unusual dermatitis in rats when fed as their sole protein source (Boas, 1927). Subsequent studies established that the symptoms were due to biotin deficiency (Gyorgy et al., 1940). Although its biological function has never been unequivocally established, avidin is believed to be a defensive agent which inhibits microbial growth by preventing uptake of biotin (Korpela et al., 1981; Green, 1975). Avidin is known to remain active even when associated with the surface of a bacterium by charge interaction, a fact which may enhance its antimicrobial activity. Recent studies have led to suggestions of avidin involvement in biotin nutrition of the chick embryo (Bush and White, 1989).

Structurally distinct, avidin-like proteins have also been isolated from hen egg yolk (Murthy and Adiga, 1984) and from cultures of several *Streptomyces* species (Chaiet et al., 1963; Bayer et al., 1986). Although neither streptavidin nor the biotin binding protein of yolk has as high an affinity for biotin, they may be preferred in some applications because of their lower charge in neutral buffers. Avidin's exceptionally high isoelectric point can lead to non-specific binding to anionic groups in some applications.

Physical and Chemical Properties

Egg white is essentially an aqueous solution of proteins, of which lysozyme and avidin are relatively minor examples (Table 7.1). It is a relatively inexpensive source of specialized food and pharmaceutical proteins because little, aside from protein, is present to interfere with extraction. Also, eggs are sanitary, familiar food products, cheaply available in large quantities. The antimicrobial activity of lysozyme has been exploited somewhat by the food industry and appears to have potential for much wider use. In Europe, lysozyme has been extensively studied and used to prevent a defect known as late blowing or late gas in hard cheeses. Numerous references to the use of lysozyme as a food preservative have appeared in the scientific and patent literature. These have been reviewed recently (Cunningham *et al.*, 1991; Banks *et al.*, 1986).

Table 7.1. Composition of chicken egg white.

Component	Concentration (g/l)
Water	883
Proteins	
ovalbumin	56.7
ovotransferrin	12.6
ovomucoid	11.6
ovomucin	3.7
lysozyme	3.6
G_2 globulin	~4
G_3 globulin	~4
ovoinhibitor	1.6
ovoglycoprotein	1.0
ovoflavoprotein	0.8
ovomacroglobulin	0.5
cystatin	0.05
avidin	0.05
Carbohydrates	
glucose (free)	5
glycoside groups (bound to proteins)	4
Minerals	6
Lipids	0.3

In addition to the formal definition stated above, lysozyme has a number of other distinctive characteristics which are important in identification and isolation of the protein (Table 7.2). Both lysozyme and avidin are basic proteins, a trait which sets them apart from other egg white proteins, all of which have acidic isoelectric points. The low molecular weight of lysozyme can cause problems during dialysis since many ultrafiltration and dialysis membranes with nominal molecular weight cut-offs below 14,000 daltons will allow passage of some lysozyme.

The thermal stability of lysozyme is remarkably high, especially at acid pH. Thermal destruction of activity was evaluated in a stirred reaction cell at temperatures from 71°C to 94°C. Initial lysozyme concentration was 4.3 mg/ml (3350 U/ml) in 0.066 M potassium phosphate buffer, pH 6.24. Residual activities after various times at these temperatures are summarised in Fig. 7.1.

Table 7.2. Important physical characteristics of avidin and lysozyme.

	Avidin	Lysozyme
Molecular weight	68,000	14,300
No. of subunits	4	1
% Carbohydrate	8.50	
$E^{1\%}$ at 280 nm	15.4	26.4
pI	10	10.7
Thermal D^a at 93°C	118[b]	110[c]

[a] D is the time required at a particular temperature to destroy 90% of activity.
[b] Phosphate buffer, 0.1 M, pH 7.3 (Durance and Wong, 1992).
[c] Phosphate buffer, 0.02 M, pH 6.24.

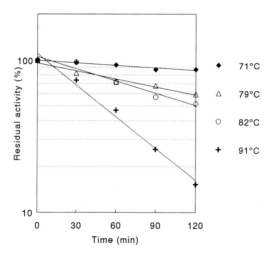

Fig. 7.1. Thermal inactivation of lysozyme in a stirred reactor, pH 6.24, 0.066 M potassium phosphate buffer, from 71°C to 91°C.

Higher temperatures, from 94°C to 124°C, were evaluated by heating lysozyme in sealed glass 100 µl capillary tubes. Inactivation appeared to approximate first order reaction kinetics. The relationship between D values, the time associated with a decimal reduction in activity, and incubation temperature, is illustrated in Fig. 7.2. In the stirred reactor, the decimal reduction temperature or z value was estimated to be 19.9 ± 0.81 C° (standard error, Celcius degree).

In the capillary tubes, the z value appeared somewhat higher at 22 ± 1.6 C°. When the two sets of data were combined z was estimated at 20.8 ± 0.7 C°. Lysozyme can be expected to survive pasteurising or even some sterilising heat treatments, at least below neutral pH.

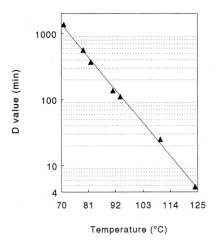

Fig. 7.2. Decimal destruction times of lysozyme from 71°C to 124°C, in potassium phosphate buffer, pH 6.24.

Avidin is also very heat stable. Although once thought to be inactivated rapidly above 85°C, it has now been shown to be even more heat stable than lysozyme (Durance and Wong, 1992). D values of activity loss at pH 7.3 are illustrated in Fig. 7.3. The z value was estimated to be 33 C°. Thus avidin solutions, like lysozyme, may be heat pasteurised without substantial loss of activity. However, avidin is relatively susceptible to oxidation.

Tryptophan residues are essential components of the active site and their oxidation results in complete inactivation. Exposure to strong light and even long frozen storage of the dry product may lead to activity loss.

Assays

The most popular assay of lysozyme activity is the turbidity assay of *Micrococcus lysodeikticus* lysis (Jollès *et al.*, 1965). Much more sensitive radioimmunoassays (Yuzuriha *et al.*, 1978; Thomas *et al.*, 1981) and a fluorometric method with a fluorescent peptidoglycan (Maeda, 1980) have also been described. Muzzarelli *et al.* (1978) described a colorimetric lysozyme assay which employed Remazol Brilliant Blue - glycochitin as a substrate.

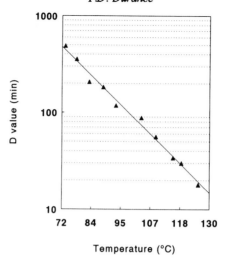

Temperature (°C)

Fig. 7.3. Decimal destruction times of avidin in 0.02 M sodium phosphate buffer, pH 7.3, from 73 to 126°C.

The earliest methods of avidin determination were microbiological, based on the binding of biotin in bacterial nutrient medium and the subsequent inhibition of growth of *Lactobacillus arabinosus*, a species with a strict requirement for biotin (Hertz *et al.*, 1943). This remains the most sensitive method, measuring avidin concentrations as low as 2 ng/ml, but it is extremely time consuming. An enzyme linked assay described by Bayer and co-workers was somewhat less sensitive but much more convenient and precise (Bayer *et al.*, 1985). Radiolabelled [14C]biotin methods are possible but expensive in terms of reagents and time (Green, 1963).

Green (1965) later described a spectrophotometric assay based on the binding of an anionic dye, 2-(4'-hydroxyazobenzene) benzoic acid (HABA). Binding of HABA to avidin resulted in a new absorbance band at 500 nm and dye binding was reversed by binding of biotin to the avidin. Back-titration of the avidin-HABA complex with biotin allowed assay of avidin down to 20 µg/ml. This method has since been modified to increase sensitivity down to 3 µg/ml and to allow its application to cooked egg products (Durance, 1991). Precision and accuracy were shown to be excellent.

A fluorescent band of avidin can also be used for assay purposes. The 290 nm excitation, 350 nm emission band due to tryptophan was largely quenched by the binding of biotin and titration with biotin to an endpoint of minimum fluorescence gave a measure of avidin concentration (Lin and Kirsh, 1977). The method was capable of estimating biotin concentrations as low as 1.5 µg/ml. However, the presence of other tryptophan-containing proteins interfered with resolution. A fluorescent dye binding method of assay was also described (Mock *et al.*, 1985). 2-Anilino naphthalene-6-sulphonate exhibited a large increase in fluorescence in the presence of avidin while addition of biotin resulted in a proportional reduction. These methods can provide useful additional methods of avidin determination, for those who have access to a spectro-fluorometer. However, for routine isolation work, some version of the HABA assay is recommended while the enzyme linked assay mentioned above is best for very low avidin concentrations.

Purity of avidin and lysozyme has routinely been evaluated by sodium dodecyl sulphate polyacrylamide gel electrophoresis (SDS-PAGE). Lysozyme migrates slightly faster than avidin. Occasionally a minor band appears slightly below avidin, even when purity otherwise appears high. This may represent avidin which has lost some or all glycosidic groups, resulting in lower molecular weight but unimpaired biotin-binding activity.

Lysozyme Isolation Methodologies

Direct Precipitation

The classic industrial isolation of lysozyme was based on direct crystallisation from egg white (Alderton and Fevold, 1964). Five percent sodium chloride was added to albumen at about pH 9.5, close to the isoelectric point of lysozyme. Lysozyme crystallises, precipitates, and may be recovered by filtration or centrifugation. The precipitate may be washed with salt solution, redissolved, and ultrafiltered to reduce salt content. Lysozyme recoveries typically range from 60 to 80%. Several re-solubilisation and re-crystallisation cycles may be performed to achieve higher purity. The final protein product may be freeze-dried in buffer at neutral pH or acidified to pH 3 with hydrochloric acid and spray-dried as lysozyme chloride.

Patents have been granted for processes and additives to increase the rate and extent of lysozyme crystallisation (Matsuoka *et al.*, 1968; Hasagawa, 1977). Another paper explored the possibility of precipitating lysozyme from egg white with polyacrylic acid (Sternberg and Hershberger, 1974). However, the major drawback to the direct method is not the yield or time factor but the fact that the residual egg white is too salty for most food applications.

Membrane Separations

Ultrafiltration (UF) has been investigated as an aid in lysozyme isolation but with limited practical success. Twofold concentration of egg white proteins did not improve lysozyme crystallisation yields or purity (Chang *et al.*, 1986). One might expect UF separation of lysozyme from most other egg white proteins because of its relatively low molecular weight. However, Chang *et al.* (1986) reported that only 2-4% of lysozyme passed through a 30,000 nominal mol. wt cut-off UF membrane when egg white was concentrated twofold. Apparently, lysozyme's well known capacity to form non-covalent complexes with other proteins greatly reduced the rate of diffusion through the membrane. Only by repeated dilutions of the retentate with 1% sodium chloride solution was it possible to recover a majority of the lysozyme in the permeate. A recent patent claimed that a microfiltration membrane with pores of 0.02 to 10 μm was successful for lysozyme separation (Lepienne *et al.*, 1986). However, while lysozyme recovery by UF may be feasible, yields will probably remain low.

Affinity Methods

Cherkasov and Kravchenko (1969) showed that lysozyme would bind to chitin by means of a substrate recognition mechanism and developed affinity methods based on this activity. Because lysozyme hydrolyses chitin incompletely and very slowly, chitin was a feasible support. Glucochitin (deaminated chitin) produced from crab shell was also a suitable material. The hydrolytic activity was also eliminated by the use of a deacylated chitin, chitosan (Muzzarelli *et al.*, 1978). A 1987 patent described a process in which chitosan was immobilised in a gel matrix and used for batch recovery of lysozyme (Reid, 1987). Weaver *et al.* (1977) recommended the use of deaminated squid chitin as an affinity adsorbent for lysozyme recovery. Deamination reduced ionic charge on the chitin and thereby reduced non-specific binding. Elution was achieved by an acid pH shift, a less than ideal method for large-scale applications since re-equilibration and rinsing of the resin can be a lengthy process. McKay *et al.* (1982) used an immunoadsorbent affinity method to isolate human lysozyme. Szewczyk *et al.* (1982) purified bacteriophage T4 lysozyme by an affinity method, utilising a competitive inhibitor of lysozyme as the ligand.

Ion Exchange/Precipitation

Some popular large-scale methods employ cation exchange and isoelectric precipitation in tandem. A cation exchange resin such as carboxymethyl cellulose (CMC) may be stirred with egg white for a period of minutes to hours, until the bulk of the lysozyme has been adsorbed. Residual egg white may then be decanted or filtered off and used for a variety of food applications. CMC-lysozyme may be washed with water before lysozyme is eluted by suspension in a 5% sodium chloride. CMC is recovered and reused. The lysozyme solution is then adjusted to a suitable pH for crystallisation, between 9.5 and 10.8. The salt solution may also be reused.

Ahvenainen *et al.* (1980) reported a similar procedure except that CMC was replaced by the cation exchanger Duolite C-464 and lysozyme was released from the resin by the addition of ammonium rather than sodium salt. Duolite C-464 is formed from a co-polymer of methacrylic acid and divinylbenzene and has a rigid macroporous structure and excellent flow properties. These workers reported better yields in recovered lysozyme when eluted with ammonium than sodium or potassium salts. However, special safety precautions may be required for handling ammonium hydroxide and a final ion exchange step may be required to remove traces of ammonia from lysozyme.

Wilkinson and Dorrington (1975) reported a procedure for removing lysozyme from egg white centrifuged out of waste egg shells. Heat- and acid-precipitated albumen proteins were removed by filtration and lysozyme was recovered from the watery filtrate by cation exchange. Although unsuitable for treatment of food grade albumen, this method may be attractive for extraction of non-food lysozyme from waste streams.

Column Ion Exchange Methods

Column ion exchange is inherently superior to bulk ion exchange both in terms of yield and purity of isolates. This arises from the fact that both egg white feed and isolation eluents are exposed to a large number of theoretical plates in a column, as compared with a single equilibrium ion exchange environment in a stirred tank. Until relatively recently, resins were not available with the macro-porous structure and mechanical rigidity necessary for column chromatography of egg white. However, this situation has changed.

A very effective column chromatography method using cation exchange was described by Li-Chan *et al.* (1986). They compared seven different resins for lysozyme capacity, recovery of lysozyme, egg white flow rates, and potential for clogging. Duolite C-464 was best suited for this application.

Viscosity of untreated egg white may impede flow through a column. Many studies (Muzzarelli *et al.*, 1978; Heney and Orr, 1981; Piskarev *et al.*, 1990) have diluted egg white to reduce viscosity but this practice is undesirable because the added water will probably have to be removed before the residual egg white can be utilised. Li-Chan *et al.* (1986) examined various means of homogenising albumen to reduce viscosity prior to column application including vacuum filtration, high speed blending, and sonication. The most successful and practical method for large-scale applications was passage through a single stage milk homogeniser at 1000 p.s.i. This treatment reduced egg white viscosity by about 60%.

Affinity methods have not generally been adopted for large-scale isolation of lysozyme from egg white because of the high costs of the affinity supports or ligands or because of poor flow rates of undiluted egg white through affinity resins. Non-specific adsorption of other albumen proteins, expensive or time consuming elution procedures, and column clogging have also been sources of difficulty with some systems. Producers have generally opted for ion exchange resins which are commercially available, moderately priced, of consistent quality, and have long useful life. However, the glucochitin supports derived from seafood by-products may deserve a second look.

Avidin Isolation Methods

Classical Method

The earliest reported large-scale extraction of avidin was that of Dhyse (1954). His method began with acetone precipitation, followed by filtration, washing of the precipitate with acetone, and extraction of the precipitate with sodium chloride solutions to solubilise avidin, which was then re-precipitated with ethanol. Product purity was about 50% active avidin and yield was 38%. This method successfully exploited the remarkable solubility and stability of avidin to separate it from the other egg white proteins but entailed destruction of the residual egg white, was very lengthy, and utilised solvents which require special handling in large operations.

Rhodes *et al.* (1958) pioneered the use of CMC for separation of egg albumen proteins. They recovered 39% of avidin activity in a minor elution peak but did not determine purity. Melamed and Green (1963) used a batch adsorption on CMC followed by three repeated chromatographic separations on CMC columns and a final ion exchange on Amberlite CG-50 to achieve avidin purity of approximately 93% and a yield of 27%. Gatti *et al.* (1984) used the method of Melamed and Green together with a gel filtration step to obtain an apparently homogeneous avidin sample for X-ray crystallography studies.

Two successful affinity methods for avidin have been described. Cuatrecasas and Wilchek (1968) bound avidin to a Biocytin-Sepharose column and eluted with 6 M guanidine-HCl at pH 1.5. Avidin was completely denatured but could be renatured by dilution, with a 90% yield. Iminobiotin, which binds avidin at pH 11 but not at pH 4, was also employed as an affinity ligand. Yields of 95% and 98% have been reported (Heney and Orr, 1981). Both methods use expensive affinity supports with limited useful lifetimes. Furthermore, only ion exchange processes have been approved by the US FDA for treatment of food grade albumen.

Cation exchange extraction of lysozyme from egg white often results in co-extraction of avidin (Fig. 7.4) due to the charge similarities of the two proteins (Table 7.2). If lysozyme is subsequently eluted in high salt buffer and purified by crystallisation, avidin will remain in the spent salt solution because avidin is much more soluble than lysozyme at alkaline pH. Secondary purification of the avidin may be achieved by one of the procedures described below. Thus, avidin can be a by-product of lysozyme extraction.

Avidin may be isolated simultaneously with lysozyme from undiluted egg white, by column chromatography (Durance and Nakai, 1988a). Li-Chan *et al.* (1986) found that avidin co-eluted with lysozyme when the two were adsorbed on Duolite C-464 (Fig. 7.4). Simplex optimisation was employed to refine elution conditions such that avidin eluted at a higher salt concentration than lysozyme (Fig. 7.5).

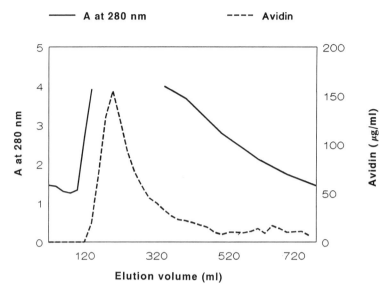

Fig. 7.4. Co-elution of lysozyme and avidin from a cation exchange resin, Duolite C-464.

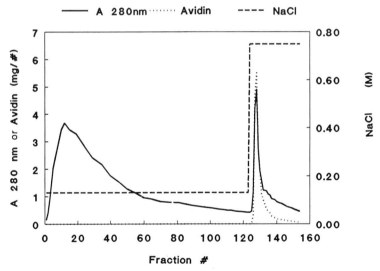

Fig. 7.5. Optimised elution of lysozyme and avidin from a cation exchange resin, Duolite C-464.

Although this procedure gave acceptable purity and yield of lysozyme, avidin yield was less than 70% and purity was typically less than 10%. The simultaneous isolation scheme is illustrated as a flow chart in Fig. 7.6. Egg white was homogenised, then applied to a column of Duolite C-464 in the sodium

T.D. Durance

form, at pH 7.9. Following washing of the column with water, bound lysozyme was eluted with 0.15 M salt, after which fresh egg white was applied to the column and another cycle of lysozyme extraction was completed. Avidin was allowed to accumulate on the ion exchange resin through many cycles of egg white application and lysozyme elution (Fig. 7.7).

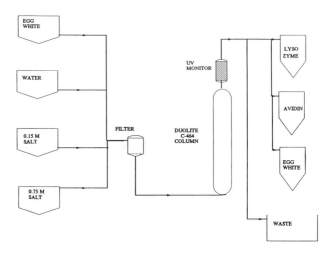

Fig. 7.6. Flow diagram of procedure for simultaneous column extraction of lysozyme and avidin from egg white.

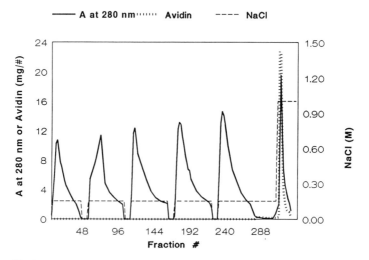

Fig. 7.7. Elution profile achieved by an ion exchange procedure for simultaneous extraction of lysozyme and avidin from egg white.

When finally eluted, avidin yield was 74% or better and purity was as high as 41%. The highest previously reported purity of a primary avidin fraction was 24% from CMC (Green and Toms, 1970) and 10% purity is typical. Furthermore, even after 16 cycles of egg white application and lysozyme elution, over a period of 23 days the purity of the lysozyme and avidin fractions remained high and the column did not clog.

Secondary Purification of Avidin

Only the affinity methods produce avidin of high purity in one step and these methods suffer from limitations of high costs, short useful life, and/or poor product activity. As biochemical applications of avidin frequently require 80-92% purity, additional purification steps have been employed to refine bulk avidin extracts. An HPLC method for purification of avidin has been described (Piskarev *et al.*, 1990). Primary isolation from egg white was achieved by ammonium sulphate and ethanol precipitation. Final purification was performed on an anion exchange HPLC column (Protein PAK DEAE-5PW, Millipore Waters). For large-scale production of avidin, HPLC is impractical. Other low pressure column techniques can be effectively employed however.

CMC is an accepted medium for purification of avidin, as demonstrated by its utilisation by Melamed and Green (1963), Green and Toms (1970), and others. In a recent study, avidin purity was increased substantially by this treatment, from 26% to 87% purity, with activity yields of >97% (Table 7.3). Avidin loading of 38 mg/ml CMC volume was achieved without loss of purity or yield (Durance and Nakai, 1988b). Tris-HCl buffer, pH 9.0, was selected as most suitable as it was sufficiently alkaline to resolve avidin from its most common contaminant, lysozyme, and was non-toxic and economical.

Table 7.3. A comparison chromatographic methods for secondary purification of avidin.

Method	Yield (%)	Final avidin purity (mg/mg protein)
Cation exchange (CM cellulose)	98.3	0.87
Gel filtration (Sephadex G75)	98	0.49
Hydrophobic interaction (C-10 aliphatic)	87	0.50
Hydrophobic interaction (phenyl Sepharose)	84	0.62
Metal chelate interaction (IDA-BDE)	92	0.75

Gel filtration, hydrophobic interaction chromatography (HIC) on aliphatic HIC or phenyl Sepharose columns, and metal chelate interaction chromatography (MCIC) each increased avidin purity substantially (Table 7.3). However, these methods have more limited commercial application because of their lower capacity.

Utilisation of Lysozyme and Avidin Reduced Egg White

Egg white depleted of lysozyme and/or avidin may often be used for the same food applications as native albumen. Li-Chan *et al.* (1986) showed that neither homogenisation nor removal of these proteins by ion exchange reduced gel strength or whippability of egg white. In fact, in most cases, functional properties were improved. The US FDA allows the use of lysozyme and avidin reduced egg white in manufacture of food grade dried egg white, provided an approved ion exchange resin is used in the process. Dried product must be identified as lysozyme and avidin reduced but, when it is used as an ingredient in fabricated foods, the statement 'lysozyme and avidin reduced' may be omitted from the label (Anon. 1986). In Canada, the removal of lysozyme and avidin need not be indicated on the label if egg white constitutes less than 5% of the weight of the final product.

Conclusions

Isolation methods for both lysozyme and avidin have been widely investigated around the world and over many years with the result that many effective methods have been developed. Industrial acceptance of a method requires not only high efficiency and purity of the product but also low operating costs, long resin life, regulatory acceptance of products and depleted egg white, and a constant, reliable supply of extraction materials. At present these criteria appear best met by ion exchange resins. However, future developments could conceivably tip the balance in favour of either affinity methods or membrane technologies.

References

Ahvenainen, R., Heikonen, M., Kreula, M., and Linko, P. (1980) Separation of lysozyme from egg white. *Food Process Engr.* 2:301-310.

Alderton, G. and Fevold, H.L. (1964) Direct crystallization of lysozyme from egg white and some crystalline salts of lysozyme. *J. Biol. Chem.* 164:1-5.

Anonymous (1986) Lysozyme and avidin reduced dried egg whites: amendment of the standard of identity. *Federal Register* 51:11434.

Banks, J.G., Board, R.G., and Sparks, N.H.C. (1986) Natural antimicrobial systems and their potential in food preservation of the future. *Biotech. Appl. Biochem.* 8:103-147.

Bayer, E.A., Ben-Hur, H., and Wilchek, M. (1985) A sensitive enzyme assay for biotin, avidin and streptavidin. *Anal. Biochem.* 154:367-370.

Bayer, E.A., Ben-Hur, H., Gitlin, G., and Wilchek, M. (1986) An improved method for single-step purification of streptavidin. *J. Biochem. Biophys. Meth.* 13(2):103-112.

Boas, M. (1927) Effect of desiccation upon nutritive properties of egg white. *Biochem. J.* 21:712.

Bush, L. and White, H.B. (1989) Avidin traps biotin diffusing out of chicken egg yolk. *Comp. Biochem. Physiol. B* 93:543-547.

Chaiet, L., Miller, T.W., Tausig, F., and Wolf, F.G. (1963) Antibiotic MSD-235. II Separation and purification of synergistic components. *Antimicrob. Agents Chemother.* 1963:28-32.

Chang, C.T., Chen, L.H., Sung, H.Y., and Kao, M.D. (1986) Studies on the purification of lysozyme from egg white by ultrafiltration. *J. Chin. Agric. Chem. Soc.* 24(1):86-93.

Cherkasov, I.A. and Kravchenko, N.A. (1969) Improved method for isolating lysozyme by enzyme-substrate chromatography. *Biochimiya* 34:885-886.

Cuatrecasas, P. and Wilchek, M. (1968) Single step purification of avidin from egg white by affinity chromatography on biocytin-Sepharose. *Biochem. Biophys. Res. Commun.* 33:235-239.

Cunningham, F.E., Proctor, V.A., and Goetsch, S.J. (1991) Egg-white lysozyme as a food preservative: an overview. *World's Poultry Sci. J.* 47(2):141-163.

Dhyse, F.G. (1954) A practical laboratory preparation of avidin concentrates for biological investigation. *Proc. Soc. Exp. Biol. Med.* 85:515-517.

Durance, T.D. (1991) Residual avidin activity in cooked egg white assayed with improved sensitivity. *J. Food Sci.* 56:707-709.

Durance, T.D. and Nakai, S. (1988a) Simultaneous isolation of avidin and lysozyme from egg albumen. *J. Food Sci.* 53:1096-1102.

Durance, T.D. and Nakai, S. (1988b) Purification of avidin by cation exchange, gel filtration, metal chelate interaction and hydrophobic interaction chromatography. *Can. Inst. Food Sci. Technol. J.* 21:279-286.

Durance, T.D. and Wong, N.S. (1992) Kinetics of thermal inactivation of avidin. *Food Res. Internat.* 25:89-92.

Gatti, G., Bolognes, M., Coda, A., Chiolerio, F., Filippini, E., and Malcovati, M. (1984) Crystallization of hen egg-white avidin in a tetragonal form. *J. Mol. Biol.* 178:787-789.

Green, N.M. (1963) Avidin: the use of [^{14}C]biotin for kinetic studies and for assay. *Biochem. J.* 89:585-591.

Green, N.M. (1965) A spectrophotometric assay for avidin and biotin based on the binding of dyes by avidin. *Biochem. J.* 94:23c-24c.

Green, N.M. (1975) Avidin. *Adv. Protein Chem.* 29:85-133.

Green, N.M. and Toms, E.J. (1970) Purification and crystallization of avidin. *Biochem. J.* 118:67-70.

Gyorgy, P., Rose, C.S., Hofmann, K., Melville, D.B., and du Vigneaud, V. (1940) A further note on the identity of vitamin H with biotin. *Science* 92:609.

Hasagawa, M. (1977) Lysozyme. Jpn. Kokai Tokkyo Koho 77 82,784.

Heney, G. and Orr, G.A. (1981) The purification of avidin and its derivatives on 2-iminobiotin-6-aminohexyl-Sepharose 4B. *Anal. Biochem.* 114:92-96.

Hertz, R., Raps, R.M., and Sebrell, W.H. (1943) Induction of avidin formation in the avian oviduct by stilbestrol progesterone. *Proc. Soc. Exp. Biol. Med.* 52:15-17.

Jollès, J. and Jollès, P. (1971) Human milk lysozyme: unpublished data concerning the establishment of the complete primary structure; comparison of lysozymes of various origins. *Helv. Chim. Acta.* 54:2668-2675.

Jollès, P. and Jollès, J. (1984) What's new in lysozyme research? *Mol. Cell. Biochem.* 63:165-189.

Jollès, P., Charlemagne, D., Petit, J.F., Maire, A.C., and Jollès, J. (1965) Biochimie comparée des lysozymes. *Bull. Soc. Chim. Biol.* 47:2241-2245.

Kondo, K., Fujio, H., and Amano, T. (1982) Chemical and immunological properties and amino acid sequences of three lysozymes from Peking duck egg white. *J. Biochem.* (Tokyo) 91:571-587.

Korpela, J.K., Kulomaa, K.S., Elo, H.A., and Tuohima, P.J. (1981) Biotin binding proteins in eggs of oviparous vertebrates. *Experimentia* 37:1065-1066.

La Rue, J.N. and Speck, J.C. (1970) Turkey egg white lysozyme: preparation of the crystalline enzyme and investigation of the amino acid sequence. *J. Biol. Chem.* 245:1985-1993.

Lepienne, A., Maubois, J., Thireau, M., and Piot, M. (1986) Process for extracting lysozyme from egg whites by micro-filtration on suitable membrane. French Patent 2569722.

Li-Chan, E., Nakai, S., Sim, J., Bragg, D.B., and Lo. K.V. (1986) Lysozyme separation from egg white by cation exchange column chromatography. *J. Food Sci.* 51:1032-1036.

Lin, H.J. and Kirsh, J.F. (1977) A rapid, sensitive fluorometric assay for avidin and biotin. *Methods Enzymol.* 62:287-289.

Mackay, B.J., Iacono, V.J., Zuckerman, J.M., Osserman, E.F., and Pollock, J.J. (1982) Quantitative recovery, selective removal and one-step purification of human parotid and leukemic lysozymes by immunoadsorption. *Eur. J. Biochem.* 129:93-98.

Maeda, H. (1980) A new lysozyme assay based on fluorescence polarization or fluorescence intensity utilizing a fluorescent peptidoglycan substrate. *J. Biochem.* (Japan) 88:1185-1191.

Matsuoka, Y., Hidaka, Y., and Takahashi, G. (1968) *Accelerated crystallization* of lysozyme in egg white. 68, 24,445 (Cl.36BO), Eisai Co., Ltd. Japan. Patent appl. 1966.

Melamed, M.D. and Green, N.M. (1963) Avidin 2: purification and composition. *Biochem. J.* 89:591-599.

Mock, D.M., Langford, G., Dubois, D., Criscimagne, N., and Horowitz, P. (1985) A fluorometric assay for the biotin-avidin interaction based on the displacement of the fluorescent probe 2-anilinonaphthalene-6-sulfonic acid. *Anal. Biochem.* 151:178-181.

Murthy, C.V.R. and Adiga, P.R. (1984) Purification of biotin-binding protein from chicken egg yolk and comparison with avidin. *Biochim. Biophys. Acta* 786:222-230.

Muzzarelli, R.A.A., Barontini, G., and Rochetti, R. (1978) Isolation of lysozyme on chitosan. *Biotechnol. Bioeng.* 20:87-94.

Piskarev, V.E., Shuster, A.M., Gabibov, A.G., and Babinkov, A.G. (1990) A novel preparative method for the isolation of avidin and riboflavin binding glycoprotein from chicken egg-white by use of high performance liquid chromatography. *Biochem. J.* 265:301-304.

Reid, L.S. (1987) Separation of lysozyme from egg white with recovery of albumen by reversibly binding the lysozyme fraction to an affinity resin. US Patent 1221046.

Rhodes, M.B., Azari, P.R., and Feeney, R.E. (1958) Analysis, fractionation and purification of egg white proteins with cellulose cation exchanger. *J. Biol. Chem.* 230:399-408.

Sternberg, M. and Hershberger, D. (1974) Separation of proteins with polyacrylic acids. *Biochim. Biophys. Acta* 342:195-206.

Szewczyk, B., Kur, J., and Taylor, A. (1982) Affinity purification of bacteriophage T4 lysozyme free of nuclease. *FEBS Lett.* 139:97-100.

Thomas, M.J., Russo, A., Craswell, P., Ward, M., and Steinhardt, I. (1981) Radioimmunoassay for serum and urinary lysozyme. *Clin. Chem.* 27:1223-1225.

Vakil, J.R., Chandan, R.C., Parry, R.M., and Shahani, K.M. (1969) Susceptibility of several micro-organisms to milk lysozyme. *J. Dairy Sci.* 52:1192-1197.

Weaver, G.L., Kroger, M., and Katz, F. (1977) Deaminated chitin affinity chromatography: a method for the isolation, purification and concentration of lysozyme. *J. Food Sci.* 42:1084-1087.

Wilkinson, B.R. and Dorrington, R.E. (1975) Lysozyme from waste egg white. *Process Biochem.* March 1975:24-25.

Yuzuriha, T., Katayama, K., and Tsutsumi, J. (1978) Studies on biotransformation of lysozyme. IV. Radioimmunoassay of lysozyme and its evaluation. *Chem. Pharm. Bull.* (Tokyo) 26:908-914.

Chapter Eight

Separation of Immunoglobulin from Egg Yolk

S. Nakai, E. Li-Chan, and K.V. Lo

Departments of Food Science and BioResource Engineering, University of British Columbia, Vancouver, British Columbia, Canada V6T 1Z4

Abstract Immunoglobulins against specific antigens can be readily produced in egg yolk by immunising hens. Immunoglobulin separated from yolk (IgY) may have broad applications in immunoassay and prophylaxis and possibly for separat-ing bioactive compounds using immunoaffinity chromatography. To facilitate the expansion of future markets for this value-added product, the production costs of IgY are most crucial. A variety of techniques which have been used for separating and purifying IgY from yolk are compared, and a strategy for developing an automated, continuous process under mild conditions without using high concentrations of chemicals or organic solvents is discussed. (*Key words:* immunoglobulins, egg, yolk, separation, purification)

Introduction

It is only within the past 10 years that immunoglobulins (Ig) extracted from egg yolk have been discovered to be more convenient to use than Ig extracted from mammalian blood. The obvious advantage is the ease of collecting eggs from laying hens rather than collecting blood from animals. While the bleeding of animals requires employment of skilled technicians, eggs can be collected by untrained workers. Furthermore, hens can be more economically fed and housed than laboratory animals and show less susceptibility to diseases. Another advantage of Ig from eggs is that a hen naturally lays an average of 240 eggs a year (Canadian Egg Marketing Agency, 1986), while bleeding can only be done periodically with a maximum volume of 50 ml taken from a rabbit without ill effects. The levels of IgG in yolk (IgY) are high, ranging from 9 mg/ml (Wang *et al.*, 1986) to 25 mg/ml (Rose *et al.*, 1974), almost equalling the Ig levels in blood. Assuming the average yolk volume in eggs to be 15 ml, the annual production of IgY by a hen would be 30-90 g, although the latter value may be a slight overestimation, since many other researchers have reported values on the lower side of the above range.

Comparison of Yolk and Blood for Antibody Recovery

According to Jensenius *et al.* (1981), amounts of IgY corresponding to almost half a litre of serum may be recovered from a chicken in one month. This amount is 5-10 times the amount that can be taken from the blood of a rabbit as mentioned above. A comparison of the amount of Ig isolated from egg yolk of a hen immunised with *Echinococcus granulosus* and from blood of a rabbit immunised with the same organism demonstrated the superiority of egg yolk as an Ig source (Gottstein and Hemmeler, 1985). Over a period of slightly less than 6 weeks, 298 mg of specific antibodies were obtained from the eggs compared with only 16.6 mg from the rabbit's blood, i.e. 18 times more from yolk.

Gardner and Kaye (1982) discussed advantages of using IgY in virus diagnosis in comparison with mouse monoclonal antibodies for large-scale production of antibody. They stated that production of mouse monoclonal antibodies requires sophisticated technology and the antibodies so produced are frequently too specific for routine diagnosis. Furthermore, high costs of culture media do not allow production of monoclonals at reasonable prices. However, Hassl *et al.* (1987) reported that the precipitation patterns of IgY and rabbit antiserum were not identical, indicating differences in the specificity between IgY and rabbit IgG. Therefore, IgY may not be an immediate replacement for rabbit IgG in immunoassay.

Specificity of Antigens

An advantage of IgY is that its specificity against mammalian antigens is higher than that of Ig from other laboratory animals, e.g. rabbit or sheep, due to the evolutionary distance existing between birds and mammals (Larsson and Sjoquist, 1990). A comparative study was carried out by Hassl *et al.* (1987) between Ig obtained from egg yolk and rabbit blood after immunising with antigen prepared from the parasite *Toxoplasma gondii*; it was found that the chicken egg IgG precipitated different components in the antigen mixture when compared with the rabbit. Antibodies against RNA polymerase II, which are difficult to prepare in rabbit and guinea pig, have been prepared in egg yolk (Carroll and Stollar, 1983).

Characteristic Properties of Yolk Antibody

In eggs, the white contains IgA and IgM but at relatively low concentrations of 0.7 and 0.15 mg/ml, respectively, while yolk is characteristic in containing a considerably higher concentration of 25 mg IgG/ml (Rose *et al.*, 1974).

Furthermore, intramuscular injection with an antigen such as *Escherichia coli* cells or their lipopolysaccharide (LPS) resulted in elevated levels of antigen-specific serum IgG and IgY, while no effect was detected in egg white Ig activity (Shimizu *et al.*, 1989).

The main components of the water-soluble fraction of egg yolk are livetins, with three varieties (α-, β-, and γ-livetins) in addition to many endogenous enzymes. Since γ-livetin (IgY) is the largest molecule, it is relatively easy to separate from other proteins in the water-soluble fraction. The IgY molecule is larger than that of human IgG, 170,000 vs. 150,000 (Kobayashi and Hirai, 1980), with a larger molecular weight of 60,000-70,000 for the heavy chain compared with 50,000 for human IgG heavy chain. The molecular weights of the light chains of IgY and human IgG are similar, with a value of 22,000. According to Parvari *et al.* (1988), chicken Ig has four CH domains, unlike mammalian Ig which has three CH domains. They proposed that the unique chicken C_γ is the ancestor of the mammalian C_ε and C_γ subclasses. Shimizu *et al.* (1992) compared the molecular stability of IgG derived from chicken and rabbit based on molecular structure. The molecular rigidity of IgY was lower than that of rabbit IgG as demonstrated in intrinsic fluorescence and circular dichroism patterns. The isoelectric point of IgY was slightly more acidic, due to a higher carbohydrate content, than that of IgG of mammals (Higgins, 1975).

IgY was reported to be fairly resistant to trypsin or chymotrypsin digestion, but more sensitive to pepsin digestion than rabbit IgG (Otani *et al.*, 1991) or bovine IgG (Shimizu *et al.*, 1988). Stability characteristics of IgY at temperatures below 65°C and pH above 4 were similar to those reported for rabbit and bovine IgG.

Unlike mammalian IgG, chicken IgG does not fix mammalian complement, and binds with neither protein A nor Fc receptors (Jensenius *et al.*, 1981). Also unlike mammalian IgG, IgY does not bind protein G (Akerstrom *et al.*, 1985). Therefore, affinity chromatography with immobilised protein A or protein G, which is commonly used for purification of mammalian IgG, cannot be applied for IgY purification.

Utilisation of Yolk Antibody

Because of the above advantages and characteristic properties of IgY, there are possibilities for extensive utilistion of IgY. A variety of IgY with specificity against bacteria, viruses, parasites, and toxins has been prepared (see Shimizu and Hatta, 1991 for a short review). IgY reactive to different proteins has also been reported. These IgY are intended for use in immunoassay or in biosensors for diagnosis as well as for detection of microbial contamination and adulteration of foods. Potential large-scale utilisation of IgY in prophylaxis and therapy was suggested by Losch *et al.* (1986). Yolken *et al.* (1988) discussed a possibility of using antiviral IgY, especially IgY against rotavirus which is the major cause of diarrhoea, for consumption by infants and young children.

Also, application of IgY in immunoaffinity chromatography is likely in the future for separation of bioactive compounds from natural resources or fermentation culture media, because of the simplicity of the separation procedures such as a single pass operation (Sada *et al.*, 1986). Using anti-lactoferrin monoclonal antibody immobilised on Affi-Gel (Bio-Rad, 1988), it was possible to extract 97% pure lactoferrin directly from unpasteurised skim milk in a single chromatographic step with 97% recovery (Kawakami *et al.*, 1987). A synthetic polymer which can withstand pressure up to 1000 p.s.i., thus being suitable for continuous use in large-scale affinity chromatography, is commercially available (Bio-Rad, 1987). The most critical points for practical, extensive application of IgY are costs of the antibodies as well as supporting gel materials for immobilising IgY for constructing chromatography columns.

The objectives of this paper are to compare the techniques used for separation of IgY from egg yolk and to discuss the possibilities of automation for large-scale preparation of IgY for industrial production.

Separation Technology

The first step for separation of IgY from egg yolk is the extraction of a water-soluble fraction containing the antibodies. A variety of techniques for the next step of separation of IgY from the water-soluble fraction has been reported. However, for automated, continuous separation which is appropriate for industrial production, membrane process and chromatography are preferable because in addition to the ease of semi-continuous operation they eliminate the need for use of high concentrations of chemicals or organic solvents.

Extraction of Water-Soluble Fraction

For extraction of a water-soluble fraction from egg yolk with water, two factors may be critical, i.e. pH and the extent of dilution of egg yolk. It was found that pH was most important for obtaining the highest recovery of IgY. An optimum recovery of 93-96% for IgY in the water-soluble fraction was obtained by six-fold water dilution of yolk at pH 5.0 after incubation at 4°C for 6 h (Akita and Nakai, 1992). Lipid contamination in this fraction was only 1-2%. For removal of the residual lipids in the water-soluble fraction recovered from yolk, food grade gums have been used. Subsequent to the work on delipidation of the water-soluble fraction using sodium alginate (Hatta *et al.*, 1988), Hatta *et al.* (1990) reported that λ-carrageenan was the most effective among 12 food gums, successfully achieving a delipidation efficiency of 99.6%.

The theory of the Signer tank, which was originally developed for a continuous countercurrent distribution device, was utilised to explain chromatographic separation with two phases, i.e. stationary phase and mobile phase (Brenner et al., 1965). We successfully applied this theory to the washing of cheese whey concentrated on the membrane in an ultrafiltration tank in order to obtain a high level of protein with low contents of lactose and ash by diafiltration. The whey protein is continuously washed with water with stirring on the membrane in the tank, extracting water-soluble lactose and ash which were filtered out of the tank through the membrane. Extraction of water-soluble proteins from egg yolk is similar in principle to this cheese whey protein concentrate preparation. The Signer tank is a series of identical tanks placed side by side; the mobile phase continually extracts solute from each tank. When n tanks are used to form the set, the concentration of water-soluble proteins left from the nth tank $C_n(V)$ after pumping a volume V of water into the first tank which originally contained solute at a concentration of C_0 can be computed using the following equation (Brenner et al., 1965):

$$C_n(V) = C_0.e^{-V/V_m} [1 + 1/1!.V/V_m + 1/2!.(V/V_m)^2 + ...$$

$$+ 1/(n - 1)!.(V/V_m)^{n-1}] \tag{1}$$

where V_m is the volume of each tank.

The number of tanks in the set to achieve adequate washing on a practical basis in the case of whey protein concentrate preparation was computed to be six or seven using equation 1. Assume that six tanks are placed in a circle, each with a filter through which the aqueous extract but not lipoproteins can pass. A set of five tanks is used for serial extraction using water as the mobile phase, meaning that filtered water extract from each tank is used for extraction in the subsequent tank. When the level of residual water-soluble proteins in the first tank in the set has reached the preset level, e.g. 5% of C_0 with 95% extraction efficiency, the water pump is switched from the first tank to the second tank which now becomes the first tank in the new set, to which the extra tank in the circle containing fresh yolk is linked, thus becoming the last tank in the new set, and then extraction is immediately continued. While the extraction is performed in the new set, the first tank in the previous set is emptied and then filled with a new batch of fresh yolk. The water extraction is thus carried out semi-continuously. The amount of water for extraction that is saved by using this system increases as the set extraction efficiency level increases.

Precipitation and Extraction

Polson and von Wechmar (1980) found that polyethylene glycol (PEG) with a molecular weight of 6000 was a better precipitant for IgY than were organic solvents, such as ether and toluene, and ammonium sulphate. After separating a water-soluble fraction from egg yolk with 3.5% (w/v) PEG, IgY was precipitated by increasing the PEG concentration to 12%. Polson *et al.* (1985) improved their PEG method using an extra purification treatment with 25% ethanol at a temperature of -20°C to reprecipitate IgY recovered by the 12% PEG method, thereby removing the PEG efficiently from IgY preparation. Traces of alcohol were removed by evaporation or dialysis. Polson (1990) recently further modified his method by replacing 3.5% PEG for separation of water-soluble proteins from yolk with chloroform and claimed a higher recovery of IgY without affecting the antibody activity.

Jensenius *et al.* (1981) used dextran sulphate to precipitate IgY from a water-soluble fraction separated from egg yolk. Akita and Nakai (1992) precipitated and purified IgY from the water-soluble fraction separated from yolk as described above using ammonium sulphate followed by precipitation with sodium sulphate or ethanol. IgY thus separated had purity above 93% with a recovery rate over 98%.

Ultrafiltration

Ultrafiltration for separating IgY from a yolk water-soluble fraction based on molecular weight differences is difficult due to clogging of the membrane with lipid-containing mucous materials. Akita and Nakai (1992) successfully used ultrafiltration, but only after preliminary separation of crude IgY by ammonium sulphate precipitation. This two-step operation improved the purity of IgY from 30% to >93%. However, recently we successfully applied ultrafiltration directly to the water-soluble fraction by adjusting the pH of the fraction to 9.0. Purity of 80% was readily achieved for IgY with a recovery of 95%.

Ion Exchange Chromatography

Although both anion and cation exchange celluloses have been successfully used for Ig separation, the most common choice has been anion exchangers, especially diethylaminoethyl (DEAE) cellulose (Fahey and Terry, 1978). A beaded form of cellulose (Sephacel, manufactured by Pharmacia, Uppsala, Sweden) improved column capacity and flow properties. Anion exchange chromatography was reported as the final step in purification of IgY in many papers (e.g. Carroll and Stollar, 1983; Akita and Nakai, 1992).

A DEAE-Sephacel column was used to separate IgY from a water-soluble fraction of yolk derived from hens immunised with LPS of *E. coli* and β-lactoglobulin (McCannel and Nakai, 1990). Of the total IgY, 32.3% did not bind to the column; this fraction had low specific antibody activity. A smaller portion of 20.2% was recovered in the first gradient elution, with a purity of 40.1%. ELISA indicated relatively high levels of specific antibody activity in this fraction. The first portion of the second gradient resulted in the elution of 15.8%

of the applied IgY, with a higher purity of 59.8% but with lower specific antibody activity, especially to LPS. Therefore, it may be important that DEAE-Sephacel chromatography can separate immunologically active IgY subpopulations from inactive subpopulations.

Recently, use of cation exchange chromatography was proposed for separation of IgY from a water-soluble fraction of egg yolk (Fichtali *et al.*, 1992). A high protein capacity, cross-linked carboxymethyl cellulose (Indion HC-2, manufactured by Phoenix Chemicals, New Zealand) was used for constructing a column which was equilibrated with pH 5.0 phosphate buffer. The pH of column equilibration and the water-soluble fraction to be applied is critical, as it lies between the isoelectric points of IgY and other livetins. While IgY is adsorbed, other livetins which are anions at pH 5.2 therefore pass through the column unadsorbed. This pH is convenient as it matches the best pH for separating the water-soluble fraction from egg yolk (Akita and Nakai, 1992). IgY with a purity of 60% was obtained at a recovery of about 70% by the method of Fichtali *et al.* (1992). The behaviour of IgY on such cation exchange columns may be the same in principle as the starch gel electrophoretic behaviour of livetins observed by McIndoe and Culbert (1979). At pH 5.4, nine out of a total of ten bands migrated toward the anode; the remaining, diffuse band corresponded to plasma γ-globulin.

However, considerable difference in isoelectric focusing (IEF) pattern was observed between IgY and rabbit IgG by Hassl *et al.* (1987) with more bands at pH 4.7-4.8 in IgY than in IgG. They stated that fractionation of water-soluble yolk proteins according to their isoelectric points was not likely to lead to the development of a satisfactory purification step as the yolk protein bands are very close and linked in IEF. Separation based on molecular weight difference, e.g. gel filtration, may be preferable for purification of IgY based on this IEF result.

Metal Chelate Interaction Chromatography

Immobilised metal affinity chromatography (IMAC) was proposed as a new approach to protein fractionation by Porath *et al.* (1975) and applied for separation of serum proteins (Porath and Olin, 1983). This method, also called metal chelate interaction chromatography (MCIC), was applied for separating immunoglobulins from waste blood (Lee *et al.*, 1988) and from cheese whey (Al-Mashikhi *et al.*, 1988). The column capacity of MCIC for IgG, which was already reported by Al-Mashikhi *et al.* (1988) to be high when applied to whey, was further improved by using competitive displacement (Li-Chan *et al.*, 1990). About 100 mg of IgG with 75-95% purity could be recovered per millilitre of MCIC gel, which is 5-10 times greater than the capacity of DEAE-cellulose columns.

Markedly different behaviour of IgY from that of mammalian IgG was, however, observed when MCIC was used to separate IgY from the water-soluble fraction of egg yolk (McCannel and Nakai, 1989). It was found that the recovery of IgY in unbound fractions was more efficient than that in bound fractions unlike the IgG recovery from blood and whey. This different chromatographic behaviour of IgY may be due to the differences in the molecular structure of IgY from that of mammalian IgG, thereby affecting the exposure of histidine residues in the molecules. The working principle of MCIC was reported to be a strong affinity of immobilised transition metal ions to the imidazole ring through coordination complex formation (Porath *et al.*, 1975). Egg yolk water-soluble fraction containing approximately 60 mg of protein could be treated per ml of MCIC gel, yielding 10 mg IgY with a relatively high purity of 75% but at a low recovery of 27% (McCannel and Nakai, 1989). However, this result was still superior to a low capacity of hydrophobic interaction chromatography column (2 mg of protein applied per ml gel) reported by Hassl and Aspock (1988).

Important Strategy in IgY Separation

An ideal system of separating IgY from egg yolk would be an automated, continuous system capable of high processing capacity, without generating any waste. To achieve this goal, column processes are most suitable, especially those that combine different separation principles. Due to greater selectivity, affinity chromatography will be more frequently utilised in the future. According to the review work of Higgins (1975), several groups have used affinity chromato-graphy to produce relatively homogeneous populations of chicken immuno-globulin, and the dinitrophenyl hapten has often been used. However, due to high costs of antibody products, it may be some time before this technique can be used for separation of IgY from eggs. Instead, IgY may be used to separate mammalian immunoglobulins from animal sources, e.g. blood and milk, when it becomes available at low costs as a result of large-scale industrial production.

For semi-continuous chromatography, two or three identical columns are used so as to complete a cycle of regeneration, sample application and washing, and elution (sometimes elution and regeneration are performed simultaneously). We have recently completed designing a computer-controlled system for separating IgY from the water-soluble fraction of egg yolk using a cation exchange column (Fichtali *et al.*, 1992).

As repeatedly stated, combinations of different separation principles are recommended for improving the purity of separated compounds. Hollow-fibre cartridge ultrafiltration for crude separation followed by continuous cation exchange chromatography for further purification is under investigation in our laboratory. MCIC, which is a very efficient method for separation of IgG from blood and milk, may not be recommended for separation of IgY, except for specific purification purposes. The high price of MCIC gel could be another obstacle for the wide use of MCIC. Gel filtration chromatography may be useful, but only for the final stage of purification of the separated IgY, mainly due to its low column capacity and difficulty in designing a system for continuous operation (Nakai *et al.*, 1991). Hydrophobic interaction chromatography and

DEAE-cellulose chromatography are both useful also, but only at the final purification stage; however, they may have particular advantages for separation and purification of specific subpopulations of IgY. Affinity chromatography using immobilised antigens may also be useful for purification of IgY subpopulations having specific antigen binding capacity.

Full utilisation of the residue after separating IgY from egg yolk is important for the industry. Mayonnaise with improved consistency was prepared from egg yolk after separating the water-soluble fraction from the original yolk (Kwan *et al.*, 1991). From the yolk pellet remaining after separation of water-soluble proteins, yolk oil, undenatured apoproteins, lecithin, and cholesterol were separated with the minimum use of organic solvents (Kwan *et al.*, 1991).

Comparing all advantages and disadvantages, the best possible combination of techniques for separating IgY from egg yolk would be the semi-continuous serial filtration system for extracting the water-soluble proteins from yolk, from which crude IgY is extracted by ultrafiltration, followed by purification using ion exchange chromatography. Gel filtration chromatography may be needed for further purification depending on the purity required for the IgY products.

Acknowledgements

The authors are grateful for a grant for the research partnership support programme provided by the Natural Sciences and Engineering Research Council of Canada, Agriculture Canada, and the Canadian Egg Marketing Agency.

References

Akerstrom, G., Brodin, T., Reis, K., and Borg, L. (1985) Protein G: a powerful tool for binding and detection of monoclonal and polyclonal antibodies. *J. Immunol.* 135:2589-2592.

Akita, E.M. and Nakai, S. (1992) Immunoglobulins from egg yolk: isolation and purification. *J. Food Sci.* 57:629-634.

Al-Mashikhi, S.A., Li-Chan, E., and Nakai, S. (1988) Separation of immunoglobulins and lactoferrin from cheese whey by chelating chromatography. *J. Dairy Sci.* 71:1747-1755.

Bio-Rad (1987) Affi-Prep high performance affinity media. *Bulletin 1298.* Bio-Rad Laboratories, Richmond, CA.

Bio-Rad (1988) Affi-gel Hz immunoaffinity kit. *Bulletin 1424.* Bio-Rad Laboratories, Richmond, CA.

Brenner, M., Niederwieser, A., Pataki, G., and Webber, R. (1965) Theoretical aspects of thin-layer chromatography. In: Stahl, E. (ed.), *Thin-Layer Chromatography.* Academic Press, New York, pp. 75-133.

Canadian Egg Marketing Agency (1986) *The Amazing Egg*. Canadian Egg Marketing Agency, Ottawa, Ontario.

Carroll, S.B. and Stollar, B.D. (1983) Antibodies to calf thymus RNA polymerase II from egg yolks of immunized hens. *J. Biol. Chem.* 258:24-26.

Fahey, J.L. and Terry, E.W. (1978) Ion exchange chromatography and gel filtration. In: Weir, D.M. (ed.), *Handbook of Experimental Immunology. Vol. 1. Immunochemistry*. 3rd edn. Blackwell Scientific Publications, London, pp. 8.1-8.16.

Fichtali, J., Charter, E.A., Lo, K.V., and Nakai, S. (1992) Separation of egg yolk immunoglobulins using an automated liquid chromatography system. *Biotechnol. Bioeng.* 40:1388-1394.

Gardner, P.S. and Kaye, S. (1982) Egg globulins in rapid virus diagnosis. *J. Virol. Methods* 4:257-262.

Gottstein, B. and Hemmeler, E. (1985) Egg yolk immunoglobulin Y as an alternative antibody in the serology of echinococcosis. *Z. Parasitenkunde* 71:273-276.

Hassl, A. and Aspock, H. (1988) Purification of egg yolk immunoglobulins: a two-step procedure using hydrophobic interaction chromatography and gel filtration. *J. Immunol. Methods* 110:225-228.

Hassl, A., Aspock, H., and Flamm, H. (1987) Comparative studies on the purity and specificity of yolk immunoglobulin Y isolated from eggs laid by hens immunized with *Toxoplasma gondii* antigen. *Zentralbl. Bakt. Mikrobiol. Hyg.* A267:247-253.

Hatta, H., Sim, J.S., and Nakai, S. (1988) Separation of phospholipids from egg yolk and recovery of water-soluble proteins. *J. Food Sci.* 53:425-427,432.

Hatta, H., Kim, M., and Yamamoto, T. (1990) A novel isolation method for hen egg yolk antibody "IgY". *Agric. Biol. Chem.* 54:2531-2535.

Higgins, D.A. (1975) Physical and chemical properties of fowl immunoglobulins. *Vet. Bull.* 45:139-154.

Jensenius, J.C., Andersen, I., Hau, J., Crone, M., and Koch, C. (1981) Eggs: conveniently packaged antibodies. Methods for purification of yolk IgG. *J. Immunol. Methods* 46:63-68.

Kawakami, H., Shinmoto, H., Dosako, S., and Sogo, Y. (1987) One-step isolation of lactoferrin using immobilized monoclonal antibodies. *J. Dairy Sci.* 70:752-759.

Kobayashi, K. and Hirai, H. (1980) Studies on subunit components of chicken polymeric immunoglobulins. *J. Immunol.* 124:1695-1704.

Kwan, L., Li-Chan, E., Helbig, N., and Nakai, S. (1991) Fractionation of water-soluble and insoluble components from egg yolk with minimum use of organic solvents. *J. Food Sci.* 56:1537-1541.

Larsson, A. and Sjoquist, J. (1990) Chicken IgY: utilizing the evolutionary difference. *Comp. Immunol. Microbiol. Infect. Dis.* 13:199-201.

Lee, Y.Z., Sim, J.S., Al-Mashikhi, S., and Nakai, S. (1988) Separation of immunoglobulins from bovine blood by polyphosphate precipitation and chromatography. *J. Agric. Food Chem.* 36:922-928.

Li-Chan, E., Kwan, L., and Nakai, S. (1990) Isolation of immunoglobulins by competitive displacement of cheese whey proteins during metal chelate interaction chromatography. *J. Dairy Sci.* 73:2075-2086.

Losch, Y., Schranner, I., Wanke, R., and Jurgens, L. (1986) The chicken egg, an antibody source. *J. Vet. Med.* B33:609-619.

McCannel, A.A. and Nakai, S. (1989) Isolation of egg yolk immunoglobulin-rich fractions using copper-loaded metal chelate interaction chromatography. *Can. Inst. Food Sci. Technol. J.* 22:487-490.

McCannel, A.A. and Nakai, S. (1990) Separation of egg yolk immunoglobulins into subpopulations using DEAE-ion exchange chromatography. *Can. Inst. Food Sci. Technol. J.* 23:42-46.

McIndoe, W.M. and Culbert, J. (1979) The plasma albumins and other livetin proteins in egg yolk of the domestic fowl. *Int. J. Biochem.* 10:659-663.

Nakai, S., Kitts, D.D., Durance, T., and Li-Chan, E. (1991) Bioactive substances from animal resources for potential food uses. 8th World Congress of Food Science and Technology. Toronto.

Otani, H., Matsumoto, K., Saeki, A., and Hosono, A. (1991) Comparative studies on properties of hen egg yolk IgY and rabbit serum IgG antibodies. *Lebensm. -Wiss. u. -Technol.* 24:152-158.

Parvari, R., Avivi, A., Lentner, F., Ziv, E., Tel-Or, S., Burstein, Y., and Schechter, I. (1988) Chicken immunoglobulin γ-heavy chains: limited VH gene repertoire, combinational diversification by D gene segments and evolution of the heavy chain locus. *EMBO J.* 7:739-744.

Polson, A. (1990) Isolation of IgY from the yolk of eggs by a chloroform polyethylene glycol procedure. *Immunol. Invest.* 19:253-258.

Polson, A. and von Wechmar, M.B. (1980) Isolation of viral IgY antibodies from yolks of immunized hens. *Immunol. Comm.* 9:476-493.

Polson, A., Coetzer, T., Kruger, J., Von Maltzahn, E., and Van der Merwe, K.J. (1985) Improvements in the isolation of IgY from the yolks of eggs laid by immunized hens. *Immunol. Invest.* 14:323-327.

Porath, J. and Olin, B. (1983) Immobilized metal ion affinity adsorption and immobilized metal ion affinity chromatography of biomaterials. Serum protein affinities for gel-immobilized iron and nickel ions. *Biochemistry* 22:1621-1630.

Porath, J., Carlsson, J., Olsson, I., and Belfrage, G. (1975) Metal chelate affinity chromatography, a new approach to protein fractionation. *Nature* 258:598-599.

Rose, M.E., Orlans, E., and Buttress, N. (1974) Immunoglobulin classes in the hen's egg: their segregation in yolk and white. *Eur. J. Immunol.* 4:521-523.

Sada, E., Katoh, S., Sukai, K., Tohma, M., and Kondo, A. (1986) Adsorption equilibrium in immuno-affinity chromatography with polyclonal and monoclonal antibodies. *Biotechnol. Bioeng.* 28:1497-1502.

Shimizu, M. and Hatta, H. (1991) Passive immunization using IgY. *Saibokogaku (Cell Engineering)* 7:557-560.

Shimizu, M., Fitzsimmons, R.C., and Nakai, S. (1988) Anti-*E. coli* immunoglobulin Y isolated from egg yolk of immunized chickens as a potential food ingredient. *J. Food Sci.* 53:1360-1366.

Shimizu, M., Fitzsimmons, R.C., and Nakai, S. (1989) Serum and egg antibody responses in chickens to *Escherichia coli*. *Agric. Biol. Chem.* 53:3233-3238.

Shimizu, M., Nagashima, H., Sano, K., Hashimoto, K., Ozeki, M., Tsuda, K., and Hatta, H. (1992) Molecular stability of chicken and rabbit immunoglobulin G. *Biosci. Biotechnol. Biochem.* 56:270-274.

Wang, K., Hoppe, C.A., Datta, P.K., Fogelstrom, A., and Lee, Y.C. (1986) Identification of the major mannose-binding proteins from chicken egg yolk and chicken serum as immunoglobulins. *Proc. Natl Acad. Sci. USA* 83:9670-9674.

Yolken, R.H., Leister, F., Wee, S.B., Miskuff, R., and Vonderfecht, S. (1988) Antibodies to rotaviruses in chickens' eggs: a potential source of antiviral immunoglobulins suitable for human consumption. *Pediatrics* 81:291-295.

Chapter Nine

Egg Cholesterol Removal by Supercritical Fluid Extraction Technology

G.W. Froning

Department of Food Science and Technology, University of Nebraska Lincoln, 143 H.C. Filley Hall, East Campus, PO Box 830919, Lincoln, NE 68583-20919, USA

Abstract Research has indicated that supercritical carbon dioxide extraction can effectively remove two-thirds of the cholesterol and 35% of the fat from dried egg yolk without hindering functionality. Generally, higher temperatures and pressures are more efficient. Phospholipids are not removed during supercritical carbon dioxide extraction which may help maintain emulsification properties. The fatty acid composition of neutral lipids appears to be affected by supercritical carbon dioxide extraction. Supercritical carbon dioxide extraction has been found to be less effective on liquid egg yolk, although use of an ethanol entrainer shows some promise. The future application of supercritical carbon dioxide extraction technology by the egg industry relates to its cost-effectiveness. If the costs can be reduced, this technology may have excellent potential.
(*Key words:* cholesterol, egg yolk, extraction)

Introduction

Although the egg is a source of high quality protein, vitamins, and minerals, the lipid and high cholesterol content has often been emphasised as a negative dietary factor. Today's egg has recently been found to have considerably less cholesterol than that reported 20 years ago. The cholesterol content of the egg yolk is now approximately 213 mg in a 50 g egg (Table 9.1).

Table 9.1. Proximate composition of edible portion (50 g) of one large egg.

Nutrient	Handbook no. 8-1 (1989)	Handbook no. 8-1 (1976)
Water	37.66	37.28
Food energy (k cal)	75	79
Protein (g)	6.25	6.07
Total lipid (g)	5.01	5.58
Crude fibre (g)	0	0
Ash (g)	0.47	0.47
Cholesterol (mg)	213	247

Data taken from USDA (1976) and USDA (1979).

It is postulated that this change may be due to production factors (e.g. genetics and nutrition of the bird) or improved analytical methods. This difference is probably attributable to both of these factors.

Production factors have been reviewed and found to have some modest influence on the composition of the egg (Hargis, 1989; Hargis and Van Elswyk, 1991). The cholesterol content of the egg yolk is altered very little by diet. However, the fatty acid content can be substantially changed through the hen's diet. If the hen is fed diets containing oils high in unsaturated fats, the egg yolk can be made more unsaturated. Recently, there has been increased effort to enrich eggs in ω3 fatty acids. There is some evidence that eating diets high in ω3 fatty acids will lower serum cholesterol.

Future advancements in removing cholesterol and lipids from egg yolk will probably be made through processing technology. Cholesterol-free egg substitutes are being marketed in the United States (Leutzinger *et al.*, 1977). These egg substitutes are usually made with egg white with added vegetable oil and seasonings.

Other than supercritical fluid extraction there have been various attempts to alter the lipid composition of eggs through processing procedures. Utilisation of organic solvents has been studied (Larsen and Froning, 1981; Warren *et al.*, 1988). Although organic solvents effectively remove lipid components, proteins are denatured and phospholipids are also extracted thereby altering functional properties. Other approaches have used enzymatic modifications for cholesterol reduction (Dehal *et al.*, 1991) or adsorbents to remove cholesterol (Morris, 1991). Supercritical fluid extraction has shown promise recently as a potential approach to modifying the lipid and cholesterol content of egg yolk.

Supercritical Fluid Extraction

The Process and Previous Work

Although supercritical fluid extraction has received much emphasis lately, the process was reported as early as 1879 by Hannay and Hogarth. When a fluid is at its supercritical level, it has a density similar to that of a liquid while having a viscosity comparable to that of a gas. Since carbon dioxide is non-toxic, inert, and nonflammable, it is the fluid choice for supercritical fluid for foods. The low critical temperature (31°C) for carbon dioxide is particularly applicable to food systems. When compared with other gases and liquids, carbon dioxide appears to be the best supercritical fluid (Table 9.2).

Table 9.2. Critical temperatures and pressures of selected supercritical fluid extractants.

Fluid	Pressure (atm)	Temperature (°C)
Carbon dioxide	73	31
Ethylene	50	10
Ethane	48	32
Propylene	46	92
Propane	42	97
Sulphur hexafluoride	37	46
Ammonia	111	132
Nitrous oxide	71	36
Nitrogen	34	-147
Water	218	374

Morris (1982) reviewed the various potential applications for supercritical fluid technology. The food industry is currently using this technology to decaffeinate coffee and extract flavour components from foods. Supercritical fluids have been used by several researchers to extract lipid components and flavours from foods. Friedrich et al. (1982) extracted a high quality soy oil from full-fat soy flakes. Cholesterol has been extracted successfully from butterfat (Kauffman et al., 1982; Arul et al., 1987). Wehling (1991) observed that 80-90% of the lipids and cholesterol could be extracted from dried chicken or beef using supercritical carbon dioxide. Wong and Johnston (1986) studied the solubility of cholesterol in supercritical carbon dioxide and observed that ethanol as a co-solvent substantially improved the solubility of cholesterol. Leiner (1986) has studied the extraction of cholesterol and lipids from eggs using supercritical carbon dioxide. Zosel (1976) used supercritical ethylene to extract fat from egg yolk.

Supercritical Extraction of Egg Yolk

Research at the University of Nebraska has emphasised supercritical extraction of dried egg yolk (Froning *et al.*, 1991). Initial work indicated problems in extraction of cholesterol and lipids from liquid egg yolk. Some other research, which is reported later in this paper, has emphasised liquid egg yolk.

The schematic diagram of the supercritical fluid extraction apparatus used at Nebraska is shown in Fig. 9.1. This unit has a compressor capable of delivering carbon dioxide at pressures up to 680 atm. The capacity of the stainless steel extraction vessel was 300 ml and the stainless steel separation vessel holds 103 ml. The temperature of the extraction vessel was monitored with a thermocouple. The vessel temperature was digitally controlled with flexible electrical heating elements attached to the exterior of the extraction vessel. The pressure in both vessels was controlled manually by back-pressure regulators.

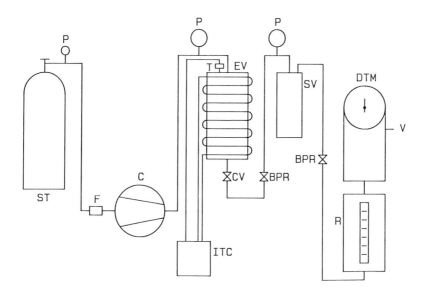

Fig. 9.1. Supercritical fluid extraction unit (ST) carbon dioxide supply tank, (P) pressure gauge, (F) in-line filter, (C) compressor, (EV) extraction vessel, (T) thermocouple, (ITC) indicating temperature controller, (CV) check value, (BPR) back-pressure regulator, (SV) separation vessel, (R) rotameter, (DTM) dry test meter, (V) vent (Froning *et al.*, 1991).

Extraction conditions included 163 atm/140°C, 238 atm/40°C, 306 atm/45°C, and 374 atm/55°C. Flow rate of carbon dioxide was maintained between five and ten standard l/min. Flow was continued until 45 g carbon dioxide ± 1 g had passed through the extractor.

Analyses and measurements included proximate composition, cholesterol, phospholipids, fatty acid composition, colour and functional properties of control and extracted samples.

Proximate composition and cholesterol content are shown in Table 9.3 and Table 9.4, respectively. Total lipids were significantly reduced at higher temperatures and pressures of supercritical carbon dioxide extraction. As much as 36% of the lipids were removed as compared with the control. Conversely, protein was concentrated at higher extraction conditions. At highest extraction conditions (306 atm/45°C; 374 atm/55°C) approximately two-thirds of the cholesterol was removed.

Table 9.3. Effect of supercritical carbon dioxide extraction of spray-dried egg yolk at various temperatures and pressures on proximate composition (Froning *et al.*, 1991).

Treatment	Total lipids[a] (%)	Protein[a] (%)	Moisture[a] (%)
Control	61.12a	33.01a	5.72a
163 atm/40°C	59.51b	35.43b	3.73a
238 atm/45°C	53.54c	41.26c	4.01a
306 atm/45°C	40.43d	53.04d	6.68b
374 atm/55°C	38.96d	54.06e	5.95b

[a] Means with different letters in the same column are significantly different ($P < 0.05$).

Table 9.4. Effect of supercritical carbon dioxide extraction of spray dried egg yolk at various temperatures and pressures on cholesterol and phospholipid content (mg/g sample) (Froning *et al.*, 1991).

Treatment	Cholesterol[a] (mg/g)	Phosphatidyl ethanolamine[a] (mg/g)	Phosphatidyl choline[a] (mg/g)
Control	18.52a	27.77d	182.32d
163 atm/40°C	15.54b	30.79c	204.46c
283 atm/45°C	13.33c	35.20b	238.87b
306 atm/45°C	6.38d	46.45a	326.40a
374 atm/55°C	6.34d	44.06a	328.17a

[a] Means with different letters in the same column are significantly different ($P < 0.05$).

Phospholipids were not removed by supercritical carbon dioxide extraction (Table 9.4). Thus, upon extraction phospholipids were concentrated. This probably explains the good functional properties of supercritical extracted egg yolk (Table 9.5). Emulsion stability and sponge cake volume were not adversely affected except at the highest extraction conditions (374 atm/55°C). Apparently, the highest extraction conditions may have denatured some proteins thereby reducing functional properties.

As extraction conditions increased, Hunter Lab L values significantly increased while a_L values and b_L values significantly decreased (Table 9.6). Therefore, supercritical carbon dioxide extraction produced a lighter coloured egg yolk with less redness and yellowness. This indicates that xanthophyll pigments may have been extracted. Other researchers have also shown that some carotenoids are removed by the supercritical carbon dioxide.

Table 9.5. Effect of supercritical carbon dioxide extraction of spray-dried egg yolk at various temperatures and pressures on emulsion stability and sponge cake volume (Froning *et al.*, 1991).

Treatment	Emulsion stability[a] (g oil/5 g mayonnaise)	Sponge cake volume[a] (cm^3)
Control	0.0072b	237.33b
163 atm/40°C	0.0077b	259.00a
238 atm/45°C	0.0148b	245.33a
306 atm/45°C	0.4850b	256.50a
374 atm/55°C	1.3455a	234.57b

[a] Means with different letters in the same column are significantly different ($P < 0.05$).

Table 9.6. Effect of supercritical carbon dioxide extraction of spray-dried egg yolk at various temperatures and pressures on Hunter Lab colour values (Froning *et al.*, 1991).

Treatment	L[a]	a_L[a]	b_L[a]
Control	74.80c	1.08a	37.45a
163 atm/40°C	78.97b	-0.33c	34.95b
238 atm/45°C	80.72a	-0.46c	34.07b
306 atm/45°C	80.39a	0.06b	30.53c
374 atm/55°C	80.17a	0.27b	26.38d

[a] Means with different letters in the same column are significantly different ($P < 0.05$).

Warren *et al.* (1991) found that supercritical carbon dioxide extraction did not significantly affect the fatty acid composition of phospholipids (Table 9.7). However, both the unsaturated and saturated fatty acids (Table 9.8) of the neutral lipids were decreased by supercritical extraction which is similar to that of Froning *et al.* (1991). Concentration of phospholipids may increase the possibility of oxidative rancidity although recent work at the University of Nebraska indicated no increase in thiobarbituric acid (TBA) values during storage.

Table 9.7. Distribution of phospholipid fatty acids by degree of saturation on a dry matter basis[a] (Warren *et al.*, 1991).

Treatment	Saturated (g/100 g)	Monounsaturated (g/100 g)	Diunsaturated (g/100 g)	Polyunsaturated (g/100 g)
Control	6.2	2.9	1.9	0.7
45°C, 238 atm	5.4	2.8	1.8	1.7
45°C, 306 atm	5.7	3.1	2.2	0.9
55°C, 374 atm	4.9	2.2	1.5	0.7

[a] Data not significantly different among treatments ($P > 0.05$).

Supercritical carbon dioxide extraction of liquid egg yolk has been less successful (Novak *et al.*, 1991). Novak *et al.* extracted dried egg yolk using a pressure of 4000 p.s.i.g. and a temperature of 40°C while extracting liquid egg yolk at 3500 p.s.i.g. and at a temperature of 35°C. They removed 76% of the cholesterol from dried egg yolk and 34% of the cholesterol from liquid egg yolk. The supercritical carbon dioxide extraction process partially coagulated the egg yolk. These workers also used a 5% ethanol co-solvent in one phase of their study which improved the solubility of cholesterol.

Table 9.8. Distribution of neutral lipid fatty acids by degree of saturation on a dry matter basis (Warren *et al.*, 1991).

Treatment	Saturated (g/100 g)	Monounsaturated (g/100 g)	Diunsaturated (g/100 g)	Polyunsaturated (g/100 g)
Control	11.6a	15.4a	5.9a	0.5a
45°C, 238 atm	8.0a	10.5a	5.2a	1.0a
45°C, 306 atm	6.2a	6.9a	2.9ab	0.5a
55°C, 374 atm	4.7a	3.5a	1.8b	0.7a

Values in the same column with no common letters differ significantly ($P < 0.05$).

Future Application

Supercritical carbon dioxide offers a safe process of effectively removing cholesterol from egg yolk. Since this technology is now being used in the food industry for other purposes (e.g. for extraction of flavours and decaffeinating of coffee), the supercritical carbon dioxide extraction process has shown proven success. The process for removal of cholesterol may find future application providing the firm can utilise the extraction process for multiple uses. For example, extraction of natural flavours using supercritical fluid extraction offers a market for ingredients with a high profit margin.

References

Arul, J., Boudreau, A., Makhlouf, J., Tardif, R., and Sahasrabudhe, M.R. (1987) Fractionation of anhydrous milk by supercritical carbon dioxide. *J. Food Sci.* 52:1231-1236.

Dehal, S.S., Freier, T.A., Young, J.W., Hartman, P.A., and Beitz, D.C. (1991) A novel method to decrease the cholesterol content of foods. In: Haberstroh, C. and Morris, C.E. (eds), *Fat and Cholesterol Reduced Foods - Technologies and Strategies.* Gulf Publishing Co., Houston, TX, pp. 203-220.

Friedrich, J.P., List, G.R., and Heakin, A.J. (1982) Petroleum-free extraction of oil from soybeans with supercritical CO_2. *J. Am. Oil Chem. Soc.* 59:288-292.

Froning, G.W., Wehling, R.L., Cuppett, S.L., Pierce, M.M., Niemann L., and Siekman, D.K. (1991) Extraction of cholesterol and other lipids from dried egg yolk using supercritical carbon dioxide. *J. Food Sci.* 55:95-98.

Hannay, J.B. and Hogarth, J. (1879) On the solubility of solids in gases. *Proc. Royal Soc. London* 29:324.

Hargis, P.S. (1989) Modifying egg yolk cholesterol in the domestic fowl: a review. *World Poultry Sci. J.* 44(1):17-29.

Hargis, P.S. and Van Elswyk, M.E. (1991) Modifying yolk fatty acid composition to improve the health quality of shell eggs. In: Haberstroh C. and Morris C.E. (eds), *Fat and Cholesterol Reduced Foods - Technologies and Strategies.* Gulf Publishing Co., Houston, TX, pp. 249-260.

Kauffman, V.W., Biernoth, G., Frede, E., Merk, W., Precht, D., and Timmen, H. (1982) Frakzionierung von butterfet durch extraktion mit uberkritischem CO_2. *Milchwissenschaft* 37:92-96.

Larsen, J.E. and Froning, G.W. (1981) Extraction and processing of various components from egg yolk. *Poultry Sci.* 60:160-167.

Leiner, S. (1986) Application of dense gases to extraction and refining. In: Stahl, E., Quin, K.W., and Gerald, D. (eds), *Dense Gases for Extraction and Refining.* Springer Verlag, New York, NY, p. 101.

Leutzinger, R.L., Baldwin, R.E. and Cotterill, O.J. (1977) Sensory attributes of commercial egg substitutes. *J. Food Sci.* 42:1124.

Morris, C.E. (1982) New process tool supercritical CO_2. *Food Eng.* (April), pp. 89-91.

Morris, C.E. (1991) Dairy science removes cholesterol, replaces fat. In: Haberstroh C. and Morris, C.E. (eds), *Fat and Cholesterol Reduced Foods-Technologies and Strategies*. Gulf Publishing Co., Houston, TX, pp. 201-202.

Novak, R.A., Reihtler, W.J., Pasin, G., King, A.J., and Zeidler, G. (1991) Supercritical fluid extraction of cholesterol from liquid egg. In: Haberstroh, C. and Morris, C.E. (eds), *Fat and Cholesterol Reduced Foods - Technologies and Strategies*. Gulf Publishing Co., Houston, TX, pp. 289-298.

USDA (1976) Composition of foods, dairy and egg products - raw, processed, prepared. *Agriculture Handbook* No. 8-1. USDA, ARS, Washington, DC 20250.

USDA (1989) Composition of foods, dairy and egg products - raw, processed, prepared. *Agriculture Handbook* No. 8-1, USDA, ARS, Washington, DC 20250.

Warren, M.H., Brown, H.G., and Davis, D.R. (1988) Solvent extraction of lipid components from egg yolk solids. *J. Am. Oil Chem. Soc.* 65:1136-1139.

Warren, M.W., Ball, H.R., Jr, Froning, G.W., and Davis, D.R. (1991) Lipid composition of hexane and supercritical carbon dioxide reduced cholesterol dried egg yolk. *Poultry Sci.* 70:1991-1997.

Wehling, R.L. (1991) Supercritical fluid extraction of cholesterol from meat products. In: Haberstroh, C. and Morris, C.E. (eds), *Fat and Cholesterol Reduced Foods - Technology and Strategies*. Gulf Publishing Co., Houston, TX, pp. 133-139.

Wong, J.M. and Johnston, K.P. (1986) Solubilization of biomolecules in carbon dioxide based supercritical fluids. *Biotechnol. Progr.* 2(1):30-39.

Zosel, K. (1976) Process for the separation of mixtures of substances. US Patent 3,969,196.

Chapter Ten

Removing Cholesterol from Liquid Egg Yolk by Carbon Dioxide-Supercritical Fluid Extraction

G. Zeidler, G. Pasin, and A. King[1]

Department of Food Science and Technology and [1]Department of Avian Sciences, University of California, Davis, CA 95616, USA

Abstract Carbon dioxide-supercritical fluid extraction (CSF) technique was known to extract cholesterol successfully from egg yolk powder at levels of about 70%, but was inefficient in removing cholesterol from liquid yolk. However, as the liquid yolk market is the larger and more lucrative segment of the egg market, effort was made to develop a method to extract cholesterol from liquid egg by CSF. Vapour-liquid equilibrium (VLE) study of the solubility of cholesterol in carbon dioxide-supercritical fluid demonstrated that optimal conditions are at pressure ranges between 2500 and 4000 p.s.i.g. and temperature ranges between 25 and 45°C. Using 5% ethanol as co-solvent increased solubility twofold, where 5% methanol increased solubility 10- to 100-fold. Moisture reduction from 50 to 35% substantially increases cholesterol solubility. Using optimised conditions and ethanol as co-solvent increased extraction of cholesterol from liquid yolk from 11% to 46%. Another feature that increased extractability of cholesterol was a preparation of yolk-oil emulsion by adding oil to yolk at a ratio of 2:1. Egg yolk is a complicated oil in water emulsion where the lipoprotein complex is dispersed in water. Under these conditions phospholipid extraction is predominant to cholesterol extraction. Preferable conditions for extraction of cholesterol over phospholipid could be obtained by modifying the emulsion characteristics of the egg yolk into water in oil emulsion. These conditions could be practically achieved through reduction of moisture or by incorporating oil to the egg yolk which is then treated by the supercritical extraction process. Using this method improved cholesterol extraction to 71.4% and 83% when ethanol was used as co-solvent.
(Key words: cholesterol, egg yolk, extraction, supercritical)

Introduction

For generations, eggs were known world-wide to be a wholesome food as far as protein quality, digestibility, cost of protein unit, flavour, and the ease of preparation of a variety of popular dishes are concerned. However, in the last decade high levels of cholesterol remain one of the major factors which constantly pull consumption figures down. In the US egg consumption constantly dropped from over 400 eggs per capita per year immediately after World War II to 232.4 eggs per capita per year in 1991. The USDA corrected its statement about the amount of cholesterol in large eggs from 274 mg per large egg to 213 mg in 1989. However, the correction was not meaningful enough to have an impact on consumption patterns although the American Heart Association increased its recommendations on the weekly consumption from 3 eggs to 4. One large egg still provides 71% of the 300 mg daily recommendation for cholesterol, where a standard egg dish which routinely contains two eggs exceeds by 41% the daily recommendation. Eggs are still the food which contributes more cholesterol than any other food to the daily diet (46.5%) and therefore are negatively branded by the medical and nutritional professions. In recent years, nutritional and physiological trials indicated that dietary saturated fats contribute more than dietary cholesterol to elevated serum cholesterol levels (Moore, 1989). However, the medical and the nutrition professions would still prefer to see a major reduction of these two components in the diet rather than increase in the limit on cholesterol consumption. Yaffee *et al.* (1991) demonstrated that egg consumption in California varies with age group. Advanced age groups of 50 years and older named cholesterol as the major factor in reducing or ceasing egg consumption whereas college students believed that cholesterol was not a problem if moderation in consumption was practised routinely and regular exercise taken. The strong demand for low cholesterol eggs and egg products has triggered the development of many methods for cholesterol reduction as well as numerous egg substitute products where the yolk is replaced by a soybean oil based component.

Recently, cholesterol reductase from cucumber leaves was used to convert cholesterol into coprostanol, a non-absorbed sterol found naturally in the human intestine, or trying to transplant cholesterol reductase genetic information into lactic acid bacteria for future fermentation of cholesterol out of foodstuff (Anon, 1989; Lamb, 1989).

However, extraction methods were found to be the most promising in providing short-term commercial solutions. Solvent extraction of egg lipids and cholesterol was used by many researchers. Melnick (1969), Tokarska and Clandinin (1985), Warren *et al.* (1988) and others used various solvent combinations to remove up to 98% of the cholesterol as well as the egg lipids. A specifically designed steam stripping system (Marchner and Fine, 1989, 1991) could remove up to 95% cholesterol from dairy butterfat and it is the first system to be in commercial use. Adsorption techniques have also been developed whereby specific compounds such as saponins (University of California method), activated carbon (New Zealand Dairy Research Institute method), or β-cyclodextrin (Strattan, 1991; SKW (Germany) method, see Cully and Vollbrech, 1990) as well as other compounds specifically adhere to the egg cholesterol and the complex is then removed from the liquid egg by centrifugation or other methods.

Fioriti and Stahl (1973) and Fioriti *et al.* (1978) developed an oil extraction method where yolk and highly unsaturated vegetable oil are mixed together in a high shear blender followed by separation of the cholesterol-containing oil by centrifugation. Bracco and Viret (1982) improved the method by reducing the pH from 6.0, the natural yolk pH, to 5.3 which resulted in improved cholesterol extraction from 50-70% to 90%.

Carbon dioxide-supercritical fluid extraction (CSF) offers significant advantages over many conventional separation techniques especially when applied to foods and pharmaceuticals (Mitchell, 1986; Perrut, 1988; Stahl *et al.*, 1988; Warren *et al.*, 1988; Johnston and Penninger, 1989; Parkinson and Johnson, 1989). Major advantages of the process are that the processing is at low temperature (around 30-40°C), there are no toxic or undesired residues, the specificity is high and it often produces separations not easily achieved by other methods. It is commercially used in decaffeination of coffee and tea, and in extraction of hops, and commercial interest has been demonstrated in extracting β-carotene, xanthine, flavours, cholesterol, spices, and natural pesticides (Margolis *et al.*, 1982; Zosel, 1982; Patel, 1985; Cygnarowicz and Seider, 1990). Cholesterol extraction was experimented using meats, dairy products, and dry egg powder where about 70% was removed (Froning *et al.*, 1990; Froning, 1991). On the other hand, cholesterol removal from liquid eggs was found to be very low (11%) under uncontrolled conditions (Rossi *et al.*, 1989; Novak *et al.*, 1991). In this project process parameters and conditions which allow removal of up to 85% of the cholesterol from liquid egg yolk were established and the stage for pilot plant scale-up was set.

Materials and Methods

Grade A, large shell eggs from the University of California, Davis, hatchery were separated, freeze-dried, and then further sieved through a 14-mesh screen, in order that all tests would be conducted on the same material and the same level of cholesterol. The final dried yolk was reconstituted to 50% moisture with deionised water to imitate liquid yolk or with the appropriate amount of water to create reduced moisture yolk with desired moisture levels. The supercritical solvent was carbon dioxide (Liquid Carbonic, commercial grade, at least 99.5% purity, obtained in dip-tube cylinders). Reagent grade ethanol and methanol (Fisher Scientific) were used as co-solvents. Safflower oil was chosen as the vegetable oil used for preparing the oil-yolk preparate. The yolk to oil ratio used was 1:2 and the pH used was 6.0 and 5.0. The preparate was prepared by blending the liquid yolk and the oil at 2000 r.p.m. in a high shear blender (Polytron). Cholesterol was determined by supercritical fluid chromatography as well as by the official AOAC method, California Department of Food and Agriculture method (1984), and by the cholesterol oxidase method (Beutler and Michael, 1975) which was modified to analyse cholesterol in eggs in order to improve cholesterol analysis methodology (R. Feeny, personal communication).

Solubility Apparatus

SCP's Screener (Fig. 10.1), a supercritical screening unit supplied by Supercritical Processing Inc., was used to measure the solubility of egg yolk cholesterol in the supercritical solvent.

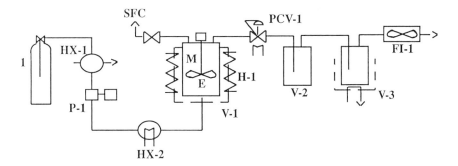

Fig. 10.1. SCP screener for measure of cholesterol solubility. 1: liquid CO_2 cylinder; H-1: electrical heating mantle; HX-1: heat exchanger to sub-cool CO_2; E: extracting vessel; P-1: reciprocal pump to pressurise CO_2; M: magnetic stirrer; HX-2: heat exchange for heating CO_2; SFC: supercritical fluid chromatograph.

For each isothermal experiment, about 100 g of liquid yolk were placed in the 300 ml extractor, which was then sealed and purged with carbon dioxide to remove air. Liquid carbon dioxide from cylinders was further cooled with glycol in a heat exchanger (HX-1) to prevent cavitation in the pump. The extractor was pressurised using a packaged plunger, reciprocating pump (P-1), which was stopped when the lowest desired pressure was reached. Solvent flowing into the extractor from the pump was preheated (HX-2), and an electrical heating mantle (H-1) maintained the extractor temperature. The extractor was held at these conditions without flow while the contents were stirred with a magnetic mixer at 20 r.p.m. for 2 min, then allowed to settle for 45 min. A multiport valve was used to introduce a 10 μl sample of the supercritical fluid phase directly to a supercritical fluid chromatograph (Hewlett Packard 1082 LC with SFC hardware). Cholesterol concentration was then determined with a 25.0 cm × 4.6 mm ID Supelcosil LC-CN column (Supelco) and UV detection, based on an external standard; the supercritical phase was sampled three times for each solubility measurement.

The pump was then restarted and the extractor pressure increased by 500 p.s.i.g. (3.4 MPa) to a new pressure level, whereupon the pump was stopped. The extractor contents were stirred and sampled, as described. The process was repeated at 500 p.s.i.g. (3.4 MPa) increments until the upper pressure of 4500 p.s.i.g. (31 MPa) was reached. The entire process was repeated with fresh feed charges at different temperature levels.

Batch Extraction

Isothermal batch extractions of liquid egg yolk were conducted on SCP's Feasibility Unit (Fig. 10.2). Feed samples of about 50 g were placed in a 300 ml extractor, which was sealed and purged. Liquid solvent from a dip-tube cylinder was subcooled with glycol in a heat exchanger (HX-1) and brought to extraction pressure with a reciprocating, packed-plunger pump (P-1).

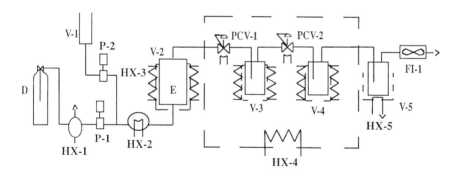

Fig. 10.2. Feasibility unit. D: solvent's dip tube cylinder; HX-3: heat exchanger for temperature control of extractor; HX-1: heat exchange for solvent subcooling; PCV-1: pressure control valve; P-1: reciprocating pump to pressurise solvent; V-2: electric heat exchanger; V-3: separator vessel; V5: dry ice cold trap for solvent; HX-2: heat exchanger for controlled heating of solvent; FI-1: dry test meter; E: Extractor.

The high pressure solvent was preheated in an electric heat exchanger (V-2). Electric heaters (HX-3) maintained the extractor temperature, and an electrically heated pressure control valve (PCV-1) was used to control extraction pressure. Solvent flowed from the extractor through the control valve, which dropped the steam pressure to near atmospheric before entering the separation vessel (V-3). Separator vessel temperature was maintained with electric heaters.

Materials that were dissolved in the solvent collected in the separator. Solvent from the separator flowed through a dry ice cold trap (V-5) and a dry test meter (FI-1) to measure flow, and was vented. The extraction was continued until the desired solvent:feed ratio was achieved; at that point, the pump was stopped and the system was depressurised through the separator and vent.

Results and Discussion

The typical composition of egg yolk is given in Table 10.1 for comparison with the experimental results in Table 10.2.

Table 10.1. Chemical composition of hen's egg yolk.

Constituent	% by weight
Water	47.5
Lipids	33.0
Protein	17.4
Carbohydrate (free)	0.2
Inorganic elements	1.1
Others	0.8
Macromolecules and complexes	
Low density lipoprotein	30.0
Granules	12.0
Levitins	8.0
Lipid composition (from total lipids)	
Triglycerides	71-73
Cholesterol	4-6
Phospholipids	23-25
Lecithin (in phospholipids)	70-77
Fatty acids C_{16}-C_{18}	99.5
Saturated fatty acids	44
Monounsaturated fatty acids	44
Polyunsaturated fatty acids	10.2

From Burlee and Vadehra (1989).

Vapour liquid equilibrium experiments (VLE) on the solubility of cholesterol from liquid egg yolk in supercritical carbon dioxide indicates that solubility of cholesterol increases with increasing extraction pressure and temperature (Fig. 10.3).

Table 10.2. Experimentally determined egg yolk composition.

Nutrient	Amount (g) per 100 g
Moisture	52.0
Proteins	17.0
Lipids	28.0
Carbohydrates	1.0
Ash	2.5
Saturated fatty acids	8.6
Monounsaturated fatty acids	7.0
Polyunsaturated fatty acids	3.8
Cholesterol	1.28

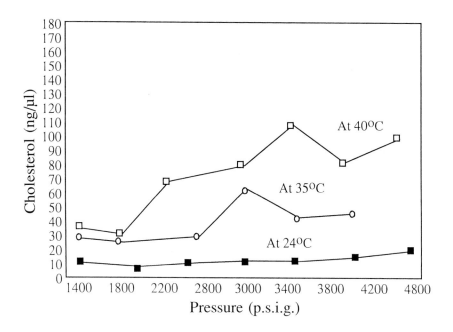

Fig. 10.3. Cholesterol solubility in supercritical carbon dioxide in the absence of a co-solvent.

A sharp increase in solubility at 2500 p.s.i.g. was found. However, solubility increases only slightly at pressures above 3500 p.s.i.g. Solubility was low at subcritical temperature but increased up to 40°C. Addition of ethanol at 5% level increased cholesterol solubility by a factor of two. Use of ethanol as co-solvent in batch extraction improved cholesterol extraction by a factor of 3. Addition of 5% methanol increased cholesterol solubility by a factor of 10-100 (Figs 10.4-10.8).

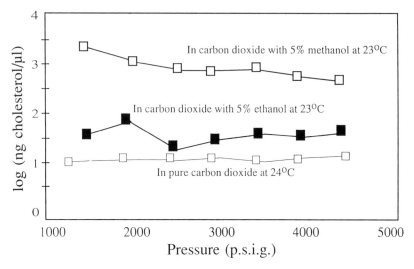

Fig. 10.4. Cholesterol solubility in supercritical carbon dioxide at 23-24°C.

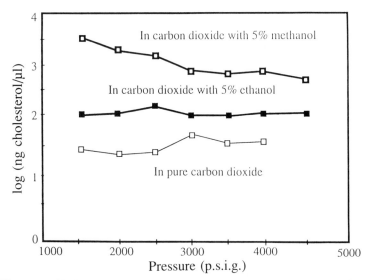

Fig. 10.5. Cholesterol solubility in supercritical carbon dioxide at 35°C.

However, although methanol has limited uses in extracting spices and hops, it is unlikely that it will be accepted as a commercial co-solvent in eggs due to its toxicity. Therefore, it is of interest to find that the improved cholesterol extraction of methanol could be achieved by moisture reduction in the yolk (Fig. 10.9).

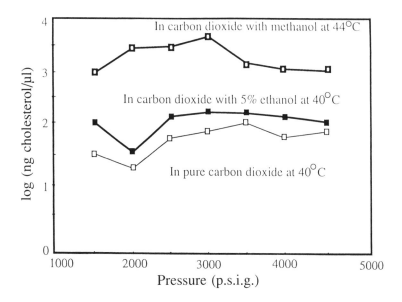

Fig. 10.6. Cholesterol solubility in supercritical carbon dioxide at 40-44°C.

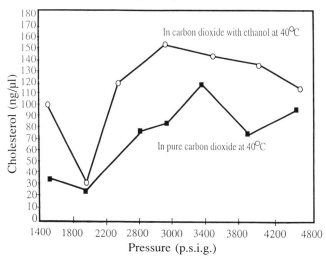

Fig. 10.7. Cholesterol solubility in supercritical carbon dioxide in the presence and absence of ethanol.

The preparation of yolk-oil emulsion in a ratio of 1:2 improved the extractability of cholesterol to 57% (Table 10.3) which is more than 3 times higher than the 18% achieved in pure CO_2 extraction. Adding 5% ethanol to the yolk-oil preparation resulted in only a small improvement (60%).

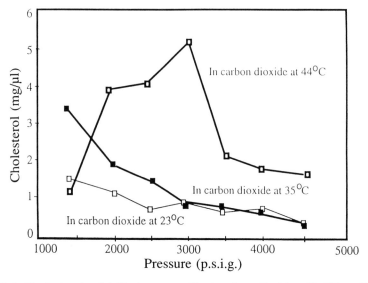

Fig. 10.8. Cholesterol solubility in supercritical carbon dioxide with 5% methanol at 23-44°C.

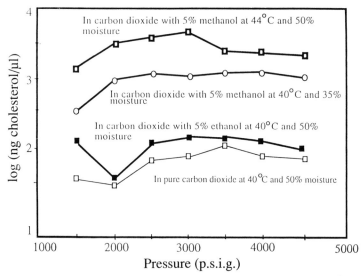

Fig. 10.9. Cholesterol solubility in supercritical carbon dioxide at 40-44°C and 35-50% moisture.

pH modification from 6.0 to 5.0 did not have any significant effect (57.7% versus 56.7% cholesterol, respectively) when supercritical extraction was employed. This is in contrast to the major effect which was seen in the oil extraction method (Bracco and Viret, 1982).

Table 10.3. Extraction of cholesterol from eggs (batch extracting).

Group feed type	Liquid yolk	Recon-stituted yolk	Dry yolk	Dry yolk	Liquid yolk	Yolk:oil 1:2	Yolk:oil 1:2	Yolk:oil 1:2	Yolk:oil 1:2
% Moisture	50	50	4	4	50	17	17	17	17
T (°C)	50	35	60	40	35	35	35	35	35
P (p.s.i.g.)	4000	3500	3000	4000	3500	3500	3500	3500	3500
Solv/feed	50	50	100	100	50	50	50	50	50
Time (h)	3.25	4.75	4.25	4.25	3.0	2.5	4.5	3.0	3.0
Co-solvent	-	-	-	-	5% ethanol	-	5% ethanol	-	5% ethanol
Cholesterol (% removal)	11.0	18.0	36	76	46	57	60	(1) 48.8 (2) 55.8 ‾‾‾‾ 71.4	(1) 57.0 (2) 60.0 ‾‾‾‾ 82.8

Combining oil extraction and supercritical extraction methods further increases cholesterol extraction to 71.4%, and to 83% when ethanol is used as co-solvent. These cholesterol extraction levels are comparable to extraction results achieved by other methods or to extraction of cholesterol from egg powder (76%). In initial trials, it was noticed that solidification of liquid yolk was related to loss of moisture and oil from the yolk. When a yolk-oil emulsion was prepared, favourable conditions for cholesterol extraction were formed where the moisture level of the emulsion dropped from 50% to about 17%, and lipid content increased from 33% to 77.7%. As a result, less moisture and egg oil were lost during the process and the product kept its fluidity.

The process is now ready for pilot plant scale-up experimentation where the system operation could be refined, and enough product could be produced to address the issue of product quality, palatability, and composition.

Acknowledgements

This work has been supported by the California Egg Commission, Upland, California and conducted with collaboration of Liquid Carbonic, Supercritical Processing Group, Allentown, Pennsylvania, the Department of Food, Science & Technology, and the Department of Avian Sciences, University of California,

Davis, California. The authors wish to thank these organisations which enabled the conduct of this research.

References

Anon (1989) New processes for removing cholesterol from dairy products. *Dairy Res. Rev.* 4(1).

Beutler, H.D. and Michael, G. (1975) Eine Eigehalts Bestimmung fuer die Routine-Analytik: Enzymatische Bestimmung von Cholesterin. *Getreide Mehl Brut* 30:116-118.

Bracco, U. and Viret, J.L. (1982) Decolorization of egg yolk. US Patent 4,333,959.

Burlee, R.W. and Vadehra, D.V. (1989) *The Avian Egg: Chemistry and Biology.* John Wiley and Sons, New York, NY.

California Department of Food and Agriculture (1984) Feed and Fertilizer Section, Chemistry Laboratory Services. Sacramento, CA. *Cholesterol in Eggs. Direct Saponification Method.* File: CHLESTRL MET:5D

Cully, J. and Vollbrech, H.R. (1990) Process for the removal of β-cyclodextrin from egg yolk or egg yolk plasma. US Patent 4,980,180.

Cygnarowicz, M.S. and Seider, W.D. (1990) Design and control of a process to extract beta carotene with supercritical carbon dioxide. *Biotechnol. Prog.* 6(1):82-91.

Fioriti, J.A. and Stahl, H.D. (1973) Low cholesterol egg process. US Patent 3,717,474.

Fioriti, J.A., Stahl, H.D., Sims, R.J., and Clifford, H. S. (1978) Low cholesterol egg product and process. US Patent 4,103,040.

Froning, G.W. (1991) Supercritical fluid extraction of cholesterol from dried eggs. In: Habestrok, C. and Morris, C.E. (eds), *Fat and Cholesterol Reduced Foods: Technologies and Strategies.* Gulf Publishing Company, Houston, TX, pp. 277-288.

Froning, G.W., Wehling, R.L., Cuppett, L.L., Pierce, M.M., Niemann, L., and Siekman, D.K. (1990) Extraction of cholesterol and other lipids from dried egg yolk using supercritical carbon dioxide. *J. Food Sci.* 55:95-98.

Johnston, K.P. and Penninger, J.M.L. (eds) (1989) Supercritical fluid. *Science and Technology ACS Symposium Series 406.* American Chemical Society, Washington, DC.

Lamb, L.F. (1989) Cholesterol reduction technologies overview - special report. *Wisconsin Milk Marketing Board Research Review* no. 2, pp. 1-4.

Margolis, G., Pagliaro, F.A., and Chiovini, J. (1982) Removal of xanthine stimulants from cocoa, UK Patent GB 2 095 091 A.

Marchner, S.S., and Fine, J.B. (1989) Physical process for simultaneous deodorization and cholesterol reduction of fats and oils. US Patent 4,804,555.

Marchner, S.S. and Fine, J.B. (1991) Physical process for the deodorization and/or cholesterol reduction of fats and oils. US Patent 4,996,072.

Melnick, D. (1969) Low cholesterol dried egg yolk and process. US Patent 3,563,765.

Mitchell, C. (1986) *Food Manufacture* 61(12):58.

Moore, T.J. (1989) The cholesterol myth. *Atlantic* September 1989, pp. 37-60.

Novak, P.A., Reightler, W.J., Pasin, G., King, A.J., and Zeidler, G. (1991) In: Habestrok, C. and Morris, C.E. (eds), *Fat and Cholesterol Reduced Foods: Technologies and Strategies.* Gulf Publishing Company, Houston, TX, pp. 289-298.

Parkinson, G.D. and Johnson, E. (1989) Supercritical processes wins CPI acceptance. *Chem. Eng.* 96(7):35-39.

Patel, C. (1985) Hops extracts. *The Brewer* 71(2):43-45.

Perrut, M. (ed.) (1988) *Proceedings of International Symposium on Supercritical Fluids.* Institut National Polytechnique de Lorraine, France.

Rossi, M., Schiraldi, A., and Spedicato, E. (1989) Supercritical fluid extraction of cholesterol and its oxidized products from eggs: preliminary experiments, flavors and off-flavors. *Proceedings of the 6th International Flavor Conference.* Rethymnon, Crete, Greece, July 5-7, 1989.

Stahl, E., Quirin, K.D., and Gerard, D. (1988) *Dense Gases for Extraction and Refining.* Springer-Verlag, New York.

Strattan, C.E. (1991) Cyclodextrins and biological macromolecules. *Biopharmacology* 4(10):44-51.

Tokarska, B. and Clandinin, M.T. (1985) Extraction of egg yolk oil of reduced cholesterol content. *Canad. Inst. Food Sci. Technol. J.* 18:256-258.

Warren, M.W., Brown, H.G., and Davis, D.R. (1988) Solvent extraction of lipid components from egg yolk solids, *J. Amer. Oil Chem. Soc.* 65(7):1136-1139.

Yaffee, M., Schutz, H.G., Stone, J., Bokhari, S., and Zeidler, G. (1991) Consumer attitudes toward eggs differ by age. *Misset-World Poultry* 7(10): 14-15.

Zosel, K. (1982) Process for the direct decaffeination of aqueous coffee extract solutions. US Patent 4,348,422.

Chapter Eleven

New Extraction and Fractionation Method for Lecithin and Neutral Oil from Egg Yolk

J.S. Sim

Department of Animal Science, University of Alberta, Edmonton, Alberta, Canada T6G 2P5

Abstract The egg yolk consists of 66% lipids on a dry matter basis. Thus eggs may be regarded as an ideal oil crop. The purpose of this report is to describe a series of attempts made in the author's laboratory that led to a patented technology for separating yolk lipids directly from fresh or dried egg yolks by using a simple ethanolic aqueous solvent system. The bench-top egg oil extractor consists of a temperature-controlled chamber extracting and filtering system which forces a pressurised solution of egg yolk in water and alcohol through a 50 mesh nylon filter. The ethanol filtrate was further fractionated into neutral (97% pure triglycerides) and polar lipids (89% crude lecithin) by cold temperature crystallisation. Thus, it is now possible simultaneously to extract and fractionate egg yolk into neutral oil, lecithin, and lipid-free egg yolk protein without employing toxic organic solvents. This technology is patented and available for commercial exploitation.

(Key words: egg yolk, oil source, aqueous ethanol solution, triglyceride, lecithin)

Introduction

An average egg provides about 6 g of lipids which are contained exclusively in the yolk. More than 66% of the total dry yolk mass is fats, thus the yolk can be regarded as a potentially important oil source. The egg yolk consists of lipids and protein, 66 and 33%, respectively, on a dry matter basis. Egg yolk oil has been regarded as an essential oil base for infant formula because it resembles the fatty acid composition of human milk. Egg lecithins are preferentially adopted by the pharmaceutical and liposome industries because of their high entrapment efficiency, stability, and low cost. The increasing demand for ovolecithin and essential egg oil by food, pharmaceutical, and biotechnology industries needs a rapid and economical method to produce lecithin and egg oil directly from fresh egg yolk without using toxic solvents.

The present industrial methods are based on a solvent extraction system requiring diethylether, hexane, and chloroform. Hexane is the major solvent in use; however, recent price increases, safety, environmental, and health concerns have generated a need for alternative methods. Furthermore, the hexane extraction alters the nutritional and functional integrity of lecithin removed from egg yolk. The objective of this report is to describe a series of studies conducted in the author's laboratory which produced a patented technology to simultaneously extract and fractionate egg yolk into neutral oil, lecithin, and lipid-free egg yolk protein without employing toxic organic solvents.

Centrifugal Behaviour of Fresh Egg Yolk in an Aqueous Solution

To understand the fresh yolk in an aqueous solution, a wide range of pH, ionic strength of solution, temperature, and centrifugal forces was investigated by the Computer-Aided Mapping Simplex Optimisation technique. Results from a series of experiments show that pH is the most important determinant in separating fresh egg yolk into two distinct fractions, lipid-rich yolk plasma and protein-rich yolk granules, without altering nutritional and functional properties (Fig. 11.1).

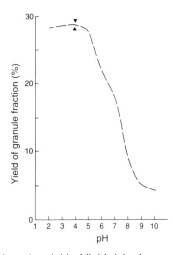

Fig. 11.1. Effect of pH on the yield of lipid-rich plasma and low-lipid granular egg yolk fractions by centrifugation in 0.29 M salt solution.

From this experiment a standardised procedure was devised as follows: fresh egg yolk is diluted in four volumes of 0.29 M NaCl solution at pH 4 at room temperature (RT) and spun at 3000 *g* for 15 min. The plasma fraction (supernatant) and granule fraction (bottom layer) are collected separately. The two fractions are then lyophilised and subjected to chemical and functional analysis. Results suggest that an egg yolk product containing low lipids (43% less than normal) and high protein (50% more than normal) can be produced without damaging its good food quality (Table 11.1). The granule fraction, in particular, may have an important nutritional and potentially significant commercial merit for human food as a low-lipid and high-protein egg yolk product.

Table 11.1. Total lipids and lipid classes of plasma, granule, and control egg yolk fractions.

	Lyophilised egg yolk fractions		
	Whole	Plasma	Granule
Yield (%)	100.00	71.60	28.40
Dry matter (%)	98.90	95.90	99.70
Protein (%)	33.10	23.00	65.90
Lipids (%)	62.88	35.82	35.05
Lipid classes (%)			
Triglyceride	67.09 ± 0.38	68.91 ± 5.80	53.05 ± 0.91
	(42.18)	(43.35)	(19.12)
Phospholipids	28.75 ± 1.11	28.21 ± 4.73	41.28 ± 0.40
	(18.08)	(18.56)	(14.88)
Cholesterol	2.94 ± 0.48	2.40 ± 0.4	3.86 ± 0.13
	(1.85)	(1.58)	(1.39)
Lipid/protein ratio	1.89	2.86	0.54

Values in brackets represent grams of liquid per 100 g lyophilised egg yolk fraction.

Differential Solubility by Non-Polar and Polar Solvents

Spray-dried yolk powder, granule, and plasma yolk fractions were subjected to sequential extraction by a polar solvent (ethanol) and non-polar solvent (diethylether). Phospholipids preferentially extracted by the initial polar solvent and neutral lipids, mostly triglycerides, were extracted by non-polar solvent (diethylether) (Table 11.2).

Water:Alcohol Ratios

Various ratios of water:alcohol were employed for maximum extractability of lecithin from egg yolk. There was a clear relationship between alcohol content and extractability and purity of lecithin extraction (Fig. 11.2).

Table 11.2. Phospholipid (lecithin) and neutral oil production by sequential mono-phasic solvent system from lyophilised egg yolk fractions.

	Yolk fractions		
	Plasma	Granule	Whole
Ethanol extraction			
Total lipids (%)	18.88	15.96	15.59
Phospholipid (%)	89.16	88.58	88.32
Phospholipids (g/100 g)	16.83	14.14	13.77
Diethylether extraction			
Total lipids (%)	31.82	16.52	33.45
Triglycerides (%)	95.38	92.02	97.19
Triglyceride (g/100 g)	30.35	15.20	32.51
Total lipids extracted (g/100 g)	47.16	29.38	46.28

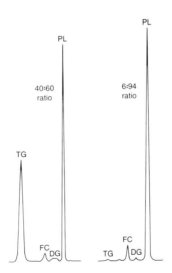

Fig. 11.2. Iatroscan chromatographs of lipid extracts form plasma yolk fraction by water/ethanol solvent system. Phospholipid (PL) extractability and purity increased linearly with ethanol volume. DG: diglyceride; FC: free cholesterol; TG: triglyceride.

Extraction Temperature

Egg yolk was repeatedly extracted by an aqueous alcohol solution at different extraction temperatures, RT, 60°C, and boiling temperature of solvent (BT) (79°C). Solubility characteristics of egg yolk lipids in an aqueous salt solution with varying ethanol concentration were repeatedly investigated. Initially the lipid-rich yolk plasma fraction containing 65% lipids (dry matter basis) was sequentially extracted with three volumes of 95% ethanol and one volume of 0.29 M NaCl solution for 15 min at RT and consecutively at BT below 79°C. The filtrate of the RT and BT extraction are designated as 'crude lecithin' and 'crude egg oil' fractions, respectively (Table 11.3).

Table 11.3. Extraction behaviour of egg yolk lipid fraction at different temperatures by alcoholic aqueous solution.

Step	Phospholipids (g/100 g yolk)	Cholesterol (g/100 g yolk)	Triglycerides (g/100 g yolk)
1st extraction (RT)	9.1	0.9	0.3
2nd extraction (60°C)	1.3	-	9.9
3rd extraction (60°C)	-	-	7.1
4th extraction (60°C)	-	-	0.8
5th extraction (79°C)	-	-	0.2
Total lipids	10.4	0.9	18.4
Extractability (%)	93.6	112.5	95.0

Cold Temperature Crystallisation

The sequential two-step extraction system at RT and BT for 'crude lecithin' and 'neutral egg oil' production was further tested. The crude fractions were subjected to cold temperature crystallisation at 2-5°C to allow all the neutral lipids (mainly triglycerides) to crystallise. The supernatant containing mostly phospholipids and the crystallised precipitate containing almost exclusively triglycerides were separated by filtering at 2-5°C and dried via rotary evaporation under low temperature and vacuum. Results suggest that this process can yield more than 90% pure lecithins and egg oil (Fig. 11.3).

Design of Egg Oil Extractor

A bench top model of a unit for extracting egg oil by an alcoholic aqueous system was designed (Fig. 11.4) and constructed (Fig. 11.5) and has been repeatedly tested using dry and fresh liquid egg yolk. The following examples were drawn from a series of laboratory trials as an optimised procedure for the production of crude lecithin and neutral egg oil from dry yolk and fresh liquid yolk samples, respectively.

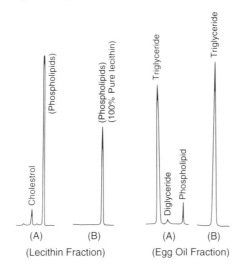

Fig. 11.3. Iatroscan of crude lecithins and egg oils extracted at (A); further purification was made for the lecithins by ZnCl$_2$ precipitation and for the egg oil by cold-temperature crystallisation (B).

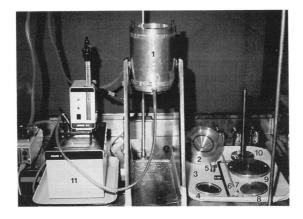

Fig. 11.4. Bench-top model of a unit for extracting egg oil by an alcoholic aqueous system (disassembled): 1. temperature-controlled extraction chamber; 2. removable bottom; 3. triple-layered circular polyethylene 300 mesh fitter (Fyntex); 4. perforated filter holder; 5. metal brushing; 6. steel arm; 7. stirrer; 8. bottom lid; 9. top lid; 10. extracting piston; 11. circulating water-bath with heating element.

Fig. 11.5. Egg oil extractor unit in extraction and filtration mode.

Extraction Time

Extraction times of 15, 30, and 60 min at RT and 60°C were tested. One hundred grams of fresh egg yolk (48.1% water, 35.0% lipids, 15.1% protein, and 1.8% ash) were added to 400 ml of undenatured 95% aqueous ethanol. The slurry was homogenised in a blender and charged into the chamber of an extraction/filtration (EF) unit set for the extraction mode. Extraction with stirring was continued for 15, 30, and 60 min, after which the EF unit was reset for the filtration mode. A triple layer of Fyntex 300 mesh (pore size 79 μm) nylon filter was used. The slurry was isothermally filtered with gradually increasing pressure to prevent the clogging of the filter.

The filter cake was re-suspended in 100 ml of 95% aqueous ethanol and filtered as described above. The filtrate was collected in a 500 ml Erlenmeyer flask and stored overnight at 2-5°C. Two phases were formed: a crystal phase at the bottom of the flask and a clear isotropic liquid phase. Two phases were separated by filtration at 5°C using a Whatman #2 ashless filter paper. Solvent removal from the liquid phase was achieved by evaporation at RT in a vacuum rotary evaporator. The residue was found to contain lipids only and was further analysed for the lipid class composition using the TLC/flame ionisation detector (FID) Iatroscan technique. This fraction was designated as 'crude lecithin'. The bottom crystal phase was melted at room temperature and transferred into a round bottom evaporation flask with several rinsings with hot ethanol. Complete solvent removal was carried out as described above. The residue was found to contain 99.8% lipids and 0.2% protein contaminant. Lipids were analysed by the TLC/FID Iatroscan technique (Table 11.4).

Table 11.4. Effect of varying extraction time on the egg yolk lipid-extraction with aqueous with aqueous ethanol solution.

	Time (min)		
	15	30	60
[Room temperature]			
Crude lecithin			
Yield (g)	11.16 ± 1.10	10.92 ± 0.44	12.15 ± 1.37
PL	10.17 ± 0.97	9.88 ± 0.48	10.85 ± 1.37
FC	0.71 ± 0.07	0.77 ± 0.03	0.75 ± 0.11
TG	0.25 ± 0.10	0.23 ± 0.02	0.40 ± 0.31
Crude oil			
Yield (g)	18,20 ± 1.10	17.57 ± 0.91	17.31 ± 1.08
PL	1.02 ± 0.49	0.92 ± 0.15	0.66 ± 0.18
FC	0.06 ± 0.02	0.06 ± 0.01	0.04 ± 0.04
TG	17.13 ± 0.90	16.58 ± 0.77	16.60 ± 1.28
Total lipid			
Extracted (g)	29.35 ± 1.19	28.49 ± 0.82	29.45 ± 2.44
Extractability (%)	83.90 ± 3.41	81.44 ± 2.34	84.14 ± 6.99
[Boiling temperature]			
Crude lecithin			
Yield (g)	11.04 ± 0.36	11.76 ± 0.23	11.39 ± 0.63
PL	9.79 ± 0.24	10.42 ± 0.29	10.16 ± 0.52
FC	0.88 ± 0.09	0.82 ± 0.25	0.71 ± 0.19
TG	0.33 ± 0.05	0.36 ± 0.08	0.44 ± 0.20
Crude oil			
Yield (g)	22.68 ± 0.32	22.72 ± 1.11	21.77 ± 0.41
PL	1.34 ± 0.10	1.19 ± 0.08	0.90 ± 0.18
FC	0.11 ± 0.10	0.12 ± 0.09	0.04 ± 0.03
TG	21.22 ± 0.43	21.40 ± 1.18	20.85 ± 0.31
Total lipid			
Extracted (g)	33.72 ± 0.68	34.48 ± 1.19	33.16 ± 0.43
Extractability (%)	96.36 ± 1.93	98.54 ± 3.39	94.77 ± 1.24

PL, phospholipid; FC, free cholesterol; TG, triglyceride.

This fraction was designated as 'crude oil'. Extractability at 60°C was higher than that at 40°C. Neither the extractability nor the purity of oil products was influenced by the length of extraction time. Therefore, 15 min was adopted.

Solvent:Yolk Ratio

One hundred grams of fresh egg yolk was extracted with 200 ml, 300 ml, 400 ml, and 500 ml of ethanol at 60°C for 15 min. Extraction, filtration, fractionation, and lipid analysis were performed as described earlier. Four hundred ml/100 g fresh egg yolk feed stock gave 96.36% extractability. Eleven milligrams of lecithin and 22 mg neutral egg oil were produced per 100 g fresh egg yolk with little cross-contamination (Table 11.5).

Table 11.5. Fresh liquid yolk which contains almost 50% water by weight requires only four volumes of ethanol, resulting in 84% aqueous ethanol solution.

| Ethanol (ml/100 g yolk) | 200 | 300 | 400 | 500 |
Resulting Solution	(75%)	(80%)	(84%)	(85%)
Crude Lecithin(g)	6.50 ± 0.06	9.84 ± 0.31	11.04 ± 0.36	12.73 ± 0.07
Crude Oil(g)	24.45 ± 0.86	21.95 ± 0.76	22.68 ± 0.32	21.49 ± 0.19
Total Lipids(g)	30.95 ± 0.84	31.74 ± 1.04	33.72 ± 0.68	34.22 ± 0.23
Extractability(%)	88.45 ± 2.40	90.71 ± 2.98	96.36 ± 1.93	97.78 ± 0.64

Optimised Procedures

Based upon the data obtained from this research, the following conclusions are drawn:

1. Fresh egg yolk can be physically fractionated into 'plasma', lipid-rich egg yolk, and 'granule' low lipid egg yolk powder by centrifugation.
2. Egg yolk lipids largely comprise 'alcohol extractable' and 'ether extractable' fractions.
3. Lecithin extractability and purity increases when alcohol content in an aqueous solvent system increases. Higher alcohol content in the solvent system decreases triglyceride contamination in the extracted lecithin fraction.
4. More than 82% phospholipids and 100% cholesterol are extractable by an aqueous alcohol solution at RT. The rest of the neutral lipids require a higher extraction temperature up to BT (79°C).

5. Extractability at 60°C is higher than that at 40°C. Neither the extractability nor the purity of oil products was influenced by the length of extraction time. Therefore, 15 min was adopted.

6. Fresh liquid yolk which contains almost 50% water by weight requires only four volumes of ethanol, resulting in 84% aqueous ethanol solution.

Results led us to design and fabricate a bench top egg oil extractor. An operating procedure with standardised conditions for the bench-top scale model has been tested with varying parameters of extraction time, temperature, and solvent volume ratios. Optimum conditions for the egg oil extractor were devised by varying the solvent ratio (ethanol:salt solution), extracting temperature, feed stock and solvent ratio, and extraction time (Fig. 11.6).

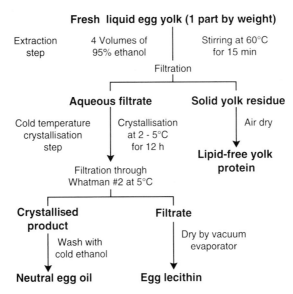

Fig. 11.6. Schematic flow chart (A) of sequential extraction, fractionation, and purification procedure for lecithin and neutral oil from fresh egg yolk by an ethanolic aqueous solvent system and Iatroscan chromatograms (B) showing cholesterol contamination in the phospholipid fraction.

Further Purification of Crude Lecithin

The predominant contaminant in the crude lecithin fraction produced from either dry yolk or fresh liquid egg yolk is free cholesterol. The strong affinity of cholesterol for the polar lecithin molecules in an aqueous solution due to its hydrophobic nature creates formidable difficulty in the purification of the final lecithin product. Interestingly, the lecithin fraction is contaminated chiefly with the free form of cholesterol (Figs 11.6 and 11.7).

ZnCl$_2$ or MgCl$_2$ salt precipitation method was planned for the purification of lecithin from neutral lipids. This method, however, requires organic solvents, like chloroform and hexane, to free the metal residues and to resolve the purified final lecithin product. Since the mandate of this project is exclusively to avoid toxic/flammable organic solvent, an attempt to adopt a supercritical carbon dioxide fluid extraction (SCFE) technology (Milton-Roy Model X10) (Fig. 11.7) was made, and the selective solubility and partition behaviour of egg yolk lipids were repeatedly investigated. As a tentative finding, it was possible to remove more than 90% of cholesterol from egg yolk lipids including crude lecithins by supercritical carbon dioxide fluid at 3500 p.s.i. and 60°C.

Fig. 11.7. Laboratory scale supercritical carbon dioxide fluid extractor (modified model X10, Milton-Roy Co.) in operational mode. Dry egg yolk powder or crude lecithin fraction (3 g) was subjected to SCFE at 3500 p.s.i. and 60°C and more than 90% cholesterol was removed, but a small amount of triglycerides was also co-eluted in the case of egg yolk powder.

Acknowledgements

The author wishes to thank both the Natural Sciences and Engineering Council of Canada (NSERC) and the Canadian Egg Marketing Agency (CEMA) for their financial support and for patenting the technology derived from this research work.

Chapter Twelve

Preparation of Pure Phospholipids from Egg Yolk

L.R. Juneja, H. Sugino, M. Fujiki, M. Kim, and T. Yamamoto[1]

Central Research Laboratories, Taiyo Kagaku Co., Ltd, 1-3 Takaramachi, Yokkaichi, Mie 510 and [1]Department of Biotechnology, Fukuyama University, Fukuyama, Hiroshima 729-02, Japan

Abstract A method was established for large-scale preparation of chromatographically homogeneous high purity phospholipids (PLs) using an ion exchange column. After extracting with acetone three times, the residue obtained was treated with ethanol twice. The ethanol extract was evaporated and treated again with cold acetone to remove remaining neutral lipids. Egg yolk lecithin (>95% PLs containing >80% phosphatidylcholine (PC)) was obtained giving 7.5% (90% based on PLs) yield. The lecithin extract was applied to a silica column (Si 10-15 mm) and PLs were separated with methanol:water (98:2) to separate PC from phosphatidylethanolamine (PE), lysophosphatidylcholine (LPC) and sphingomyelin (SM). In one of the fractions, the choline-containing PLs, e.g. PC, LPC and SM, were further purified on ion exchange cellulose using dichloromethane:methanol (9:1). The purity of PC, PE, and LPC was >98% (yield 70-80%) and the purity of SM was 92.7% (yield 76%). The fatty acid: phosphorus molar ratio of the purified PLs was 2.0. Various other PLs not found in egg yolk, e.g. phosphatidylserine (L or D isomer) or phosphatidylglycerol, were prepared enzymatically by using phospholipase D from pure PC. The yield of *trans*-phosphatidylation was approximately 100%.

(*Key words:* lecithin, phospholipids, phosphatidylcholine, phosphatidylethanolamine, phosphatidylserine, phosphatidylglycerol, lysophosphatidylcholine, sphingomyelin, egg yolk)

Introduction

Egg yolk lecithin has been popularly used in the manufacture of pharmaceutical emulsions. Research into the structure and function of biological membranes and the development and application of liposomes has led to increasing use of phospholipids (PLs) in recent years (Martin, 1990; Dearden *el al.*, 1982). Highly purified PLs are necessary for the solution of pressing biochemical and medical problems, for example, the creation of highly specific methods to diagnose some infectious diseases, e.g. malaria. More and more PLs of high purity with a certain fatty acid composition and certain polar head groups are needed. At present, available methods do not yield highly pure PLs.

PLs are natural bisurfactants which have many applications in the cosmetic, food, and pharmaceutical industries (Juneja *et al.*, 1987a,b, 1988, 1989b). In living microorganisms, PLs act in flow and transport processes through biological membranes. Thus they are mediators between the surrounding medium and the cytoplasm as well as between cytoplasm and the cell compartments.

If one starts with liquid egg yolk, the solvent of choice is acetone which removes first the water and then the triglycerides of the yolk, leaving acetone-insoluble PLs and proteins. This mixture is then split by alcohols again to extract the phosphorus-containing lipids (Lundberg, 1973). A somewhat exceptional process for obtaining egg lecithin is by extraction with dimethyl ether under medium pressure in closed equipment (Yano *et al.*, 1979). Numerous extraction processes for biological materials to obtain lipid mixtures similar to the lecithin composition have been mentioned in the literature (Singleton *et al.*, 1965; Aneja *et al.*, 1971; Fager *et al.*, 1977; Porter *et al.*, 1978; Ramesh *et al.*, 1978; Do and Ramachandran, 1980; Nielsein, 1980; Geurts van Kessel *et al.*, 1981; Hanson *et al.*, 1981; Chen and Kou, 1982; Primes *et al.*, 1982; Szuhaj, 1983; Farag *et al.*, 1984; Christie, 1985; Gunther, 1985; Kolarovic and Fournier, 1986; Baharami *et al.*, 1987; Amari *et al.*, 1990; Glass, 1990; Kiyashchitsky *et al.*, 1991; Hanras and Perrin, 1991; Takamura and Kito, 1991). However, most of them have some drawbacks for economical large-scale preparation of pure PLs. In the past few years, a new technology has been discussed in the literature (Froning *et al.*, 1990). This is the treatment of lipid mixtures with supercritical gases but it is not economical because of the high equipment cost. The drawback of available techniques is that it is very difficult to remove lysophosphatidyl-choline (LPC) and sphingomyelin (SM) from phosphatidylcholine (PC). It is possible to separate PLs on the basis of degree of unsaturation.

In the present study, a method was established for large-scale preparation for high purity PLs using silica gel followed by ion exchange column chromatography.

Materials and Methods

Materials

Authentic PLs, PC, phosphatidylethanolamine (PE), phosphatidic acid (PA), phosphatidylserine (PS), phosphatidylglycerol (PG) were obtained from Avanti Polar Lipids (Birmingham, AL, USA), and LPC and SM of egg origin were purchased from Sigma (St Louis, MO, USA). Phospholipase D (PLD) of *Streptomyces* origin was obtained from Honen Co. (Yokohama, Japan).

Fractionation of Egg Phospholipids

Laboratory scale. The method used for the extraction of egg PLs from fresh yolk was a modification of the method described by Pangborn (1950).

Large scale. Fresh egg yolk (500 kg) was blended thoroughly with acetone (1 litre) at 25°C, then the mixture was allowed to stand for 1 h and then filtered. The acetone extract contained most of the neutral lipids and pigment. The solids were washed thrice with 1500 l of acetone and then were suspended in 1250 l ethanol, mixed, and allowed to stand for 1 h; the mixture was then filtered. The extraction was repeated with 1250 l of ethanol and the combined ethanol extracts were concentrated to dryness on a steam bath with nitrogen under reduced pressure (Fig. 12.1).

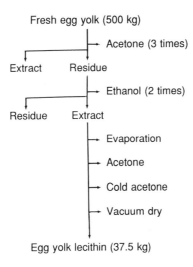

Fig. 12.1. Preparation of egg yolk lecithin.

Column Chromatographic Fractionation of Egg Lecithin

Silica gel chromatography. The column chromatographic fractionation of the above extracted lecithin was carried out on a silica gel column (10-20 mm × 250 mm; Soken Kagaku Co., Japan). The solvent flow was immediately adjusted to 10 ml/min. Two grams of the extracted crude PLs were dissolved in sufficient $CHCl_3$ to give a 20% solution and loaded on the column. The PLs were eluted with varying ratios of ethanol or methanol and water. The chromatographic system (Japan Analytical Industry Co.) used for the method development studies consisted of a pump, injector, and detector set at 205 nm. All separations were performed at room temperature.

Ion exchange cellulose chromatography. The fraction of PC mixed with LPC and SM obtained from the silica gel chromatography was further fractionated on the ion exchange cellulose column (20 × 500 mm) using CH_2Cl_2/CH_3OH stepwise elution at the flow rate of 10 ml/min.

Transphosphatidylation of PC to Other PLs

The transphosphatidylation reaction was performed for PS or PG synthesis from PC in the biphasic reaction mixture. PC (17.8 mM) was added to 2 g of ethyl acetate and the mixture was sonicated (Sonicator, Ohtake Works, Japan; continuous pulse for 3 min) under ice-cold conditions in a sonication vessel. One gram of a mixture of PLD from *Streptomyces*/aqueous buffer (desired pH) containing glycerol and L-serine or D-serine (3.4 M) was introduced into the reactor to which PC/ethyl acetate had been added. The reaction was carried out at 30°C under continuous stirring (Juneja *et al.*, 1989).

Measurement of PLs

High performance liquid chromatography (HPLC) method. PLs were analysed by HPLC as reported previously (Juneja *et al.*, 1989a).

Thin-layer chromatography/flame ionisation detection (TLC/FID) method. The PL samples were analysed by TLC/FID (Iatroscanner TH10, Iatron Laboratories, Inc., Tokyo) using chloroform:methanol:water (65:25:3 by volume) as an eluent, according to the procedure of Juneja *et al.* (1987a).

TLC method. PL samples dissolved in chloroform were applied to silica gel G (Merck) plates and chromatographs of the PLs were developed using the solvent system $CHCl_3:CH_3OH:NH_3$ (65:25:4 by volume), in a saturated chamber. Dittmer and Dragendorff reagents were used for confirmation of phosphorus and choline containing compounds, respectively.

Results and Discussion

Preparation of Lecithin from Egg Yolk

Fresh egg yolks (500 kg) treated with acetone and alcohol yielded 37.5 kg of lecithin containing 95% PLs based on the total PLs present in the egg yolk (Fig. 12.1). The egg yolk lecithin had a high PC content (80.8%) (Table 12.1). The average yield of extracted PLs from the above extraction procedure was 7.2% based on the weight of fresh egg yolk.

Table 12.1. Composition of egg yolk lecithin.

Phospholipid	Concentration (%)
Phosphatidylcholine	80.8
Phosphatidylethanolamine	11.7
Lysophosphatidylcholine	1.9
Sphingomyelin	1.9
Neutral lipids + others	3.7

Fractionation of Egg Lecithin on Silica Gel

PLs typically eluted from a silica column using ethanol or methanol in the order PE, PC mixed with LPC and SM, and pure PC. PLs were eluted with various solvent compositions. The elution time of all of the fractions containing PLs was decreased from 300 to 100 min, when ethanol:water (90:10) was replaced by methanol:water (98:2). The elution time was 250 min using industrial denatured alcohol (Solmix; containing ethanol:methanol = 87:13):water (90:10).

Effect of Silica Particle Size on the Separation of PC from Lecithin

Silica particles of various sizes were investigated for separation of PC from lecithin using methanol:water (98:2) eluent. The recovery of PC of approximately 95% purity was not very different with varying particle size, but the recovery of PC of >99% purity was increased with fine particle size. The maximum recovery of PC (>99% purity) was obtained with silica (Si) particles of 10-20 mm (Fig. 12.2), while the recovery with larger (40-64 mm) particles was 30%.

The PC-a fraction, containing PC, SM and LPC, was further fractionated on an ion exchange cellulose column using $CH_2Cl_2:CH_3OH$ stepwise elution at a flow rate of 10 ml/min (Fig. 12.3). The recovery of LPC was 97.2%. The purity of the SM fraction refractionated on the ion exchange cellulose column under the same conditions was 93.5% with 76% yield.

Separation of PC from LPC and SM

The separation of egg yolk lecithin (2 g) on the silica gel (10-20 mm) eluted 225 mg of PE (>99% purity) and 810 mg of PC (>99% purity) by using methanol: water (98:2) as an eluent, at a flow rate of 10 ml/min (Fig. 12.4). The fraction was divided into two major fractions. The former fraction (PC-a) contained SM and LPC; the purity of the PC was approximately 95% and the recovery was 700 mg (Fig. 12.4).

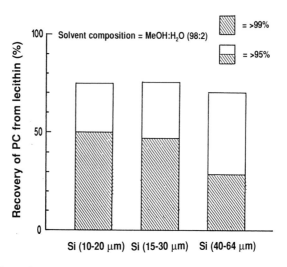

Fig. 12.2. Effect of silica particle size on the separation of PC from egg yolk. Egg yolk lecithin, 2 g; column size, 20 x 250 mm; eluent, methanol:H_2O = 98:2; flow rate, 10 ml/min.

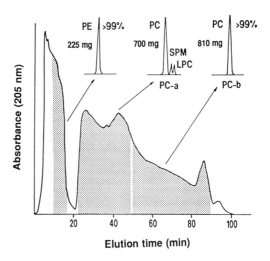

Fig. 12.3. Separation of PC from LPC and SM on an ion exchange cellulose column. PC-a fraction, 2 g; silica particle size, 10-20 mm; column size, 20 x 250 mm; eluent, methanol:H_2O = 98:2; flow rate, 10 ml/min.

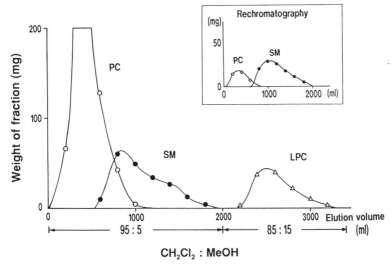

Fig. 12.4. Fractionation of egg yolk lecithin on silica gel column. Egg yolk lecithin, 2 g; silica particle size, 10-20 mm; column size, 20 x 250 mm; eluent, methanol: H_2O = 98:2; flow rate, 10 ml/min.

Fatty Acid Composition of the PC Fractions

A unique UV (205 nm) detection response was obtained during the elution of the PC fraction. The fraction of PC mixed with SM and LPC (PC-a) had high absorbance and the >99% pure PC fraction (PC-b) had low absorbance (205 nm). The fatty acid composition of both fractions was examined (Table 12.2). In the latter fraction the amount of saturated fatty acid, e.g. stearic acid (18:0), increased from 2.3 to 13.0% and oleic acid (18:1) increased from 19.0 to 39.1% while linoleic acid (18:2) decreased from 31.3 to 10.7%. However, palmitic acid (16:0) decreased from 41.3 to 32.8% (Table 12.2).

Table 12.2. Fatty acid composition of egg yolk PC.

Fatty acid	PC-a[1] (%)	PC-b[2] (%)
Palmitic acid (16:0)	41.3	32.8
Palmitoleic acid (16:1)	4.7	0.5
Stearic acid (18:0)	2.3	13.0
Oleic acid (18:1)	19.0	39.1
Linoleic acid (18:2)	31.3	10.7
Others	1.4	3.9

[1] PC (>99%) purified from PC+SM+LPC fraction.
[2] >99% PC

Recovery of PLs from Egg Yolk Lecithin

The overall recovery of egg yolk PC of 99.2% purity was 81.8% (Table 12.3). The purity of PE was 99.3% with the recovery of 79.8%. The purity of LPC and SM was 98.9% and 92.7%, respectively.

Table 12.3. Recovery of phospholipids from egg yolk lecithin.

Phospholipid	Purity (%)	Recovery (%)
Phosphatidylcholine	99.2	81.8
Phosphatidylethanolamine	99.3	79.8
Lysophosphatidylcholine	98.9	70.5
Sphingomyelin	92.7	76.1

Enzymatic Conversion of PLs

The conversion of PC to PS by PLD preparation was studied in the presence of L-serine or D-serine. At a concentration of 3.4 M serine, which is the solubility of serine in buffer of pH 5.6 at 30°C, the PLD yielded 99.8% and 99.5% of PS using 17.8 mM PC with L-serine and D-serine, respectively (Table 12.4).

Almost no by-product (PA) was formed (Fig. 12.5) giving a selectivity of approximately 100%. Similar yields were obtained with PG formation from PC.

Table 12.4. Enzymatic conversion of phospholipids.

Acceptor	Y-OH[a]	Product	Recovery (%)	Selectivity[b] (%)
PC	L-Serine	(L-) PS	99.8	99.2
PC	D-Serine	(D-) PS	99.5	99.7
PC	Glycerol	PG	99.4	99.3

[a] Undissociable -OH groups
[b] Selectivity = 100 × (Product) / [(Products) + (By-product)]
(Product) = phosphatidylserine or phosphatidylglycerol
(By-product) = phosphatidic acid

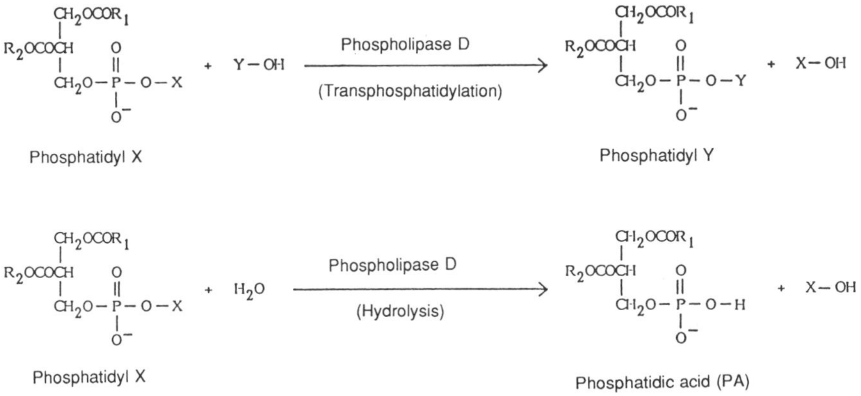

$$X = -CH_2CH_2N^+(CH_3)_3$$

Phosphatidyl X = Phosphatidylcholine (PC)

Fig. 12.5. Enzymatic conversion of phospholipids.

Conclusion

Egg yolk lecithin (37.5 kg) was obtained from 500 kg of fresh egg yolk. The PC content of the lecithin was 80.8%. The separation of PC from LPC and SM was accomplished in the present study. The results demonstrated that it was possible to obtain PLs of widely different degrees of unsaturation. PC was efficiently converted to PS and PG using PLD. The yield of transphosphatidylation was almost 100%.

References

Amari, J.V, Brown, P.R., Grill, C.M., and Turcotte, J.G. (1990) Isolation and purification of lecithin by preparative high-performance liquid chromatography. *J. Chromatogr.* 517:219-228.

Aneja, R., Chadha, J.S., and Yoell, R.W. (1971) A process for the separation of phosphatide mixtures: the preparation of phosphatidylethanolamine-free phosphatides from soy lecithin. *Fette-Seifen-Anstrichmittel.* 73:643-651.

Baharami, S., Gasser, H., and Redl, H. (1987) A preparative high performance liquid chromatography method for the separation of lecithin: comparison to thin layer chromatography. *J. Lipid Res.* 28:596-598.

Chen, S.S. and Kou, A.Y. (1982) Improved procedure for the separation of phospholipids by high performance chromatography. *J. Chromatogr.* 227: 25-31.

Christie, W.W. (1985) Chromatographic analysis of phospholipids. *Z. Lebensm. Ginters Forsch.* 181:171-182.

Dearden, S.J., Hunter, T.F., and Philip, J. (1982) A rapid method for the preparation of microvesicles of egg yolk lecithin. *Biochim. Biophys. Acta* 689:415-418.

Do, U.H. and Ramachandran, S. (1980) Mild alkali-stable phospholipids in chicken egg yolk: characterization of 1-alkenyl and 1-alkyl-sn-glycero-3-phosphoethanolamine, sphingomyelin, and 1-alkyl-sn-glycero-3-phosphocholine. *J. Lipid Res.* 21:888-894.

Fager, R.S., Shapiro, S., and Litman, B.J. (1977) A large scale preparation of phosphatidylethanolamine, lysophosphatidylethanolamine and phosphatidylcholine high performance liquid chromatography: a partial purification of molecular species. *J. Lipid Res.* 18:704-709.

Farag, R.S., El-Sharabassy, A.A.M., Abdel Rahim, G.A., Hewedy, E.M., and Ragab, A.A. (1984) Biochemical studies on phospholipids of hen's egg during incubation. *Steifen-ote-Fette-wachse.* 110:122-124.

Froning, G.W., Wehling, R.L., Cuppett, S.L., Pierce, M.M., Niemann, L., and Siekman, D.K. (1990) Extraction of cholesterol and other lipids from dried egg yolk using supercritical carbon dioxide. *J. Food Sci.* 55:95-98.

Geurts van Kessel, W.S.M., Tieman, M., and Demel, R.A. (1981) Purification of phospholipids by preparative high pressure liquid chromatography. *Lipids* 16:58-63.

Glass, R.L. (1990) Separation of phospholipid and its molecular species by high performance liquid chromatography. *J. Agric. Food Chem.* 38:1684-1686.

Gunther, B.R. (1985) Method for the preparation of phosphatidylcholine of low oil content. US Patent 4,496,486.

Hanras, C. and Perrin, J.L. (1991) Gram-scale preparative HPLC of phospholipids from soybean lecithins. *J. Am. Oil Chem. Soc.* 68:804-808.

Hanson, V.L., Park, J.Y., Osborn, T.W., and Kiral, R.M. (1981) High performance liquid chromatographic analysis of egg yolk phospholipids. *J. Chromatogr.* 205:393-400.

Juneja, L.R., Hibi, N., Inagaki, N., Yamane, T., and Shimizu, S. (1987a) Comparative study on conversion of phosphatidylcholine to phosphatidylglycerol by cabbage phospholipase D in micelle and emulsion systems. *Enzyme Microb. Technol.* 350:350-354.

Juneja, L.R., Hibi, N., Yamane, T., and Shimizu, S. (1987b) Repeated batch and continuous operations for phosphatidylglycerol synthesis from phosphatidylcholine with immobilized phospholipase D. *Appl. Microbiol. Biotechnol.* 27: 146-151.

Juneja, L.R., Kazuoka, T., Yamane, T., and Shimizu, S. (1988) Kinetic evaluation of conversion of phosphatidylcholine to phosphatidylethanolamine by phospholipases D from different sources. *Biochim. Biophys. Acta* 960:334-341.

Juneja, L.R., Yamane, T., and Shimizu, S. (1989a) Enzymatic method of increasing phosphatidylcholine. *J. Am. Oil Chem. Soc.* 66:714-717.

Juneja, L.R., Kazuoka, T., Goto, N., Yamane, T., and Shimizu, S. (1989b) Conversion of phosphatidylcholine to phosphatidylserine by various phospholipases D in the presence of L- or D-serine. *Biochim. Biophys. Acta* 1003:277-283.

Kiyashchitsky, B.A., Mezhova, I.V., Krasnopoisky, Yu.-M. and Shvets, V.I. (1991) Preparative isolation of polyphosphoinositides and other anionic phospholipids from natural sources using chromatography on adsorbents containing primary amino groups. *Biotechnol. Appl. Biochem.* 14:284-295.

Kolarovic, L. and Fournier, N.C. (1986) A comparison of extraction methods for the isolation of phospholipids from biological sources. *Anal. Biochem.* 156: 244-250.

Lundberg, B. (1973) Isolation and characterization of egg lecithin. *Acta Chem. Scand.* 27:2515-2549.

Martin, F.J. (1990) Pharmaceutical manufacturing of liposomes In P. Tyle (ed.), *Specialized Drug Delivery Systems.* Marcel Dekker, Inc. New York, pp. 266-315.

Nielsein, J.R. (1980) A simple chromatographic method for purification of egg lecithin. *Lipids* 15:481-484.

Pangborn, M.C. (1950) A simplified purification of lecithin. *J. Biol. Chem.* 188: 471-476.

Porter, N.A., Wolf, R.A., and Nixon, J.R. (1978) Separation and purification of lecithin by high performance liquid chromatography. *Lipids* 14:20-24.

Primes, K.J., Sanchez, R.A., Metzner, E.K., and Patel, K.M. (1982) Large scale purification of PC from egg yolk phospholipids by column chromatography. *J. Chromatogr.* 236:519-522.

Ramesh, B., Prabhudesai, A.U., and Vishwanathan, C.V. (1978) Simultaneous extraction and preparative fractionation of egg yolk lipids using the principle of adsorption. *J. Am. Oil Chem. Soc.* 55:501-502.

Singleton, W.S., Gray, M.S., Brown, M.L., and White, J.L. (1965) Chromatographically homogeneous lecithin from egg phospholipids. *J. Am. Oil Chem. Soc.* 42:53-56.

Szuhaj, B.F. (1983) Lecithin production and utilization. *J. Am. Oil Chem. Soc.* 60:306-309.

Takamura, H. and Kito, M. (1991) A highly sensitive method for quantitative analysis of phospholipid molecular species by high performance liquid chromatography. *J. Biochem.* 109:436-439.

Yano, N., Fukinbara, I., and Takano, M. (1979) A process for obtaining yolk lecithin from raw egg yolk. US Patent 4,157,404.

Chapter Thirteen

A Simple Procedure for the Isolation of Phosvitin from Chicken Egg Yolk

J.N. Losso and S. Nakai

Department of Food Science, University of British Columbia, Vancouver, British Columbia, Canada V6T 1Z4

Abstract A simplified method was developed to isolate phosvitin from chicken egg yolk, which involves the removal of water-soluble proteins, except phosvitin, with acidified water, the removal of lipids, and the isolation of phosvitin from the granules with 1.74 M NaCl at pH 7.0. A yield of 100-113 mg phosvitin per egg was obtained. Amino acid profiles and nitrogen and phosphorus contents of different isolation procedures appear to yield similar if not identical phosvitin. Polyacrylamide gel electrophoresis of native phosvitin isolated by four different procedures revealed the presence of similar polypeptide subunits.
(*Key words:* phosvitin, chicken egg yolk, isolation)

Introduction

Egg is a complete food rich in bioactive compounds such as livetins, lysozyme, ovotransferrin, avidin, ovalbumin, etc. In times when egg consumption is in decline, value added products from eggs offer the egg industry tremendous opportunities for growth.

Phosvitin, the most highly phosphorylated protein known, represents 1% of the egg yolk (Burley and Vadehra, 1989), accommodates nearly all the egg's metal (Grognan and Taborsky, 1987), and binds iron more strongly than the other cations (Taborsky, 1980). Iron-phosvitin strength exceeds the chelating strength of citrate and nitrilotriacetic and may exceed that of transferrin (Taborsky, 1980). The levels of circulating phosvitin and circulating iron in chicken blood vary in roughly parallel fashion (Greengard *et al.*, 1964). Lu and Baker (1986, 1987) and Yamamoto *et al.* (1990) have reported the potential of phosvitin as a protein with antioxidant activity, in egg and non-egg products, that can be put to commercial use. Food proteins with antioxidant activity such as phosvitin have the advantages of contributing to safety and nutrition, and endow functional properties as foaming, emulsification capacity, and heat stability. Phosvitin seems very resistant to heat treatment (Mecham and Olcott, 1949; Itoh *et al.*, 1983). But cost and availability of phosvitin makes its use uncertain. Existing methods for the isolation of phosvitin from chicken egg yolk (Mecham and Olcott, 1949; Joubert and Cook, 1958; Sundararajan, 1960; Burley and Cook, 1961; Wallace *et al.*, 1966; Tsutsui and Obara, 1984; Wallace and Morgan, 1986, etc.) are laborious, time consuming, involve the use of solvents that are not food grade, render other yolk proteins unsuitable for further use, and recover only small concentrations of phosvitin. A simple method, based on the work of Tsutsui and Obara (1984), Yamamoto *et al.* (1990), and Akita and Nakai (1992), was developed for the preparation of large concentrations of phosvitin from egg yolk using food grade reagents.

Materials and Methods

Materials

Eggs, from White Leghorn chickens, were obtained from the University of British Columbia animal farm and used within a week. Commercial phosvitin was purchased from Sigma (St Louis, MO). All chemicals were analytical grade.

Phosvitin Isolation

Phosvitin was isolated from egg yolk as outlined in Figure 13.1. Egg yolk was washed with distilled water and rolled on filter paper to remove the adhering albumen. The yolk membrane was punctured with a needle and the content was collected in a container. The egg yolk was diluted 1:10 with cold water acidified with HCl to pH 2.85, stirred to bring the final pH to 5.0-5.3 (Akita and Nakai, 1992) and allowed to stand for 6 h at 4°C. Then the precipitate was collected by centrifugation at 10,000 g for 25 min at 4°C. The supernatant, which contained all of the water-soluble proteins except phosvitin, was used for the isolation of livetins (Akita and Nakai, 1992). The pellet was extracted with 20 volumes of hexane:ethanol (3:1) at 4°C for at least 3 h, and filtered through Whatman #4 filter paper. The solvent was removed from the filtrate by rotary evaporation and the lipid stored under nitrogen at -20°C. The resulting cake was air dried and extracted with 10 times its volume of 1.74 M NaCl, pH 7.0, and filtered through Whatman #4 filter paper and then through 0.45 µm membrane, then the filtrate was dialysed (Microacilyzer, Asahi Co., Japan) for 6 h at room temperature and lyophilised.

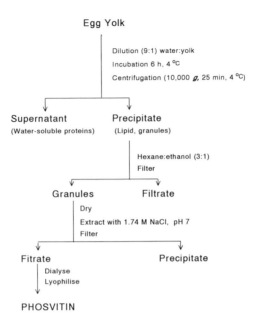

Fig. 13.1. Isolation of phosvitin from chicken egg yolk.

Phosvitin was also isolated following the procedures of Tsutsui and Obara (1984), and Wallace and Morgan (1986) as reported by Yamamoto *et al.* (1990).

Amino Acid Analysis

Analyses were performed following the procedure of Spackman *et al.* (1958) using a Beckman 121 MB Automatic Amino Acid Analyzer (Beckman Instrument Co., Fullerton, CA) equipped with a Hewlett Packard integrator Model 3396A. Tryptophan was determined by the method of Penke *et al.* (1974).

Nitrogen and Phosphorus Determinations

Nitrogen concentrations were determined on a Technicon AutoAnalyzer II system (Technicon, Elmsford, NY) following micro-Kjeldahl digestion (Concon and Soltes, 1973).

Phosphorus contents of the different preparations were determined by the method of Morrison (1964).

Electrophoresis

Polyacrylamide gel electrophoresis under denaturing and native conditions was carried out on a Phast system (Pharmacia, Uppsala, Sweden) with 8-25%

gradient gels. Gels were stained with 0.05% Coomassie Brilliant Blue R 250 in a solution containing 25% isopropanol, 10% acetic acid, 1% Triton X-100, and 0.1 M aluminium nitrate and destained in 7% aqueous acetic acid (Cutting, 1984).

Molecular weight standards were rabbit muscle phosphorylase B (97.4 kDa) bovine serum albumin (66.2 kDa), hen egg white ovalbumin (45 kDa), bovine carbonic anhydrase (31 kDa), soybean trypsin inhibitor (21.5 kDa), and hen egg white lysozyme (14.4 kDa).

Results

Amino Acid Composition

The amino acid composition of phosvitin isolated by the different procedures is shown in Table 13.1.

Table 13.1. Amino acid composition (mol %) of phosvitin isolated by different procedures.

Residue	Isolation procedure			
	Yamamoto[a]	Sigma[b]	Tsutsui[c]	This report
Asp	7.18	6.66	7.10	7.07
Thr	2.10	1.90	3.30	2.16
Ser	54.23	54.90	55.10	51.12
Glu	6.98	5.90	4.75	6.90
Pro	0.00	1.40	1.65	1.57
Gly	2.96	2.56	3.85	3.07
Ala	4.11	3.67	3.40	4.16
Cys	0.00	0.00	0.00	0.00
Val	1.97	1.92	1.00	1.90
Met	0.23	0.17	0.05	0.65
Ile	0.91	0.93	0.50	1.03
Leu	1.30	1.32	1.15	1.91
Tyr	0.47	0.42	0.15	0.58
Phe	1.02	0.81	0.50	1.07
Lys	6.11	7.03	6.80	6.73
His	4.63	4.87	4.40	4.53
Arg	5.46	4.97	5.60	5.19
Trp	0.32	0.58	0.31	0.00

[a] Yamamoto *et al.*, 1990.
[b] Sundrarajan *et al.*, 1960.
[c] Tsutsui and Obara, 1984.

Notable features are the absence of cysteine (or cystine) and the presence of a relatively high concentration of serine in all preparations. Phosvitin has been reported to contain as low as 30% serine (Allerton and Perlmann, 1965; Mok *et al.*, 1966; Itoh *et al.*, 1983), to lack sulphur-containing amino acids and to contain a low concentration of aromatic amino acids (Mano and Lipman, 1966).

Nitrogen and Phosphorus Concentration

Nitrogen and phosphorus concentrations of the isolated phosvitin are shown in Table 13.2. Nitrogen content is the best method of determining phosvitin concentration because the protein contains very small concentrations of aromatic amino acids and absorbs poorly or not at all at 280 nm (Wallace and Morgan, 1986). The molar ratio of N:P is often used as an index of phosvitin purity, a high value being related to high purity. Values reported here compare well with the N:P ratio reported by Wallace and Morgan (1986).

Table 13.2. Nitrogen and phosphorus contents and yield of phosvitin.

Procedure[a]	Nitrogen (%)	Phosphorus (%)	N:P	Yield (mg/egg)
Sigma	10.12	9.00	2.49	n.d[*]
Yamamoto	14.60	8.74	3.70	50-70[*]
Tsutsui	12.00	9.40	2.83	n.d[*]
This report	15.20	9.34	3.60	100-113

[*] Wallace and Morgan (1986), n.d[*] = not determined.
[a] See Table 13.1 for the references.

Electrophoresis

Under native conditions, all preparations were composed of two major subunits, the α- and β-phosvitins (Abe *et al.*, 1982; Itoh *et al.*, 1983) (Fig. 13.2). Abe *et al.* (1982) reported the molecular weight of α- and β-phosvitin to be 160,000 and 190,000 respectively. Under denaturing conditions, phosvitin separated into several subunits (Fig. 13.3). Shantz and Dawson (1972) reported that phosvitin separates into 18 subunits of molecular weight ranging from 12,400 to 119,000 under SDS-PAGE. Phosvitin is composed of a cluster of minor bands called 'phosvettes' (Wallace *et al.*, 1990) which are difficult to visualise even with overloaded protein. Considerable smearing of Sigma phosvitin has been reported by Wallace and Morgan (1986).

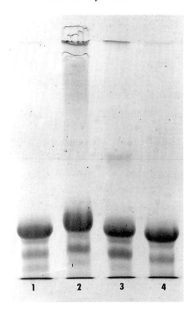

Fig. 13.2. Electrophoretic pattern of native phosvitin isolated by four different procedures: Sigma (lane 1), our procedure (lane 2), Wallace and Morgan (lane 3), and Tsutsui and Obara (lane 4).

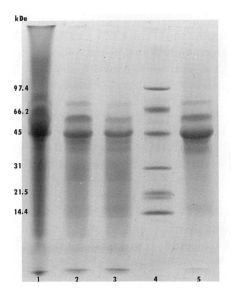

Fig. 13.3. Electrophoretic pattern of SDS-PAGE of phosvitin isolated by four different procedures: Sigma (lane 1), our procedure (lane 2), Wallace and Morgan (lane 3), and Tsutsui and Obara (lane 5). Lane 4 represents the molecular weight standards.

Discussion

The procedure described here uses the ability of pH to remove water-soluble proteins and the ability of alcohols to liberate proteins bound to lipids and render vitellin insoluble, and allow the release of phosvitin.

The water-soluble proteins obtained from the first separation are used to isolate bioactive compounds livetins. The lipids removed with cold hexane/ethanol at 4°C can be recovered after solvent evaporation and can find application in food as the solvents used are food grade.

The state of purity of even the best phosvitin preparation has been a vexing problem; the nature of the apparent heterogeneity of hen phosvitin is nebulous and some of the observed heterogeneity may have its basis in the existence of genetic variants (Taborsky, 1974).

Our procedure is competitive in terms of the simplicity, safety, and non-destructiveness towards other yolk proteins, and yield of phosvitin.

References

Abe, Y., Itoh, T., and Adachi, S. (1982) Fractionation and characterization of hen's egg yolk phosvitin. *J. Food. Sci.* 47:1903-1907.

Akita, E. and Nakai, S. (1992) Immunoglobulins of chicken egg: isolation and purification. *J. Food Sci.* 57:629-634.

Allerton, S.E. and Perlmann, G.E. (1965) Chemical characterization of the phospho-protein phosvitin. *J. Biol. Chem.* 240:3892-3898.

Burley, R.W. and Cook, W.H. (1961) Isolation and composition of avian egg yolk granules and their constituents and β-lipovitellin. *Can. J. Biochem. Physiol.* 39:1295-1307.

Burley, R.W. and Vadehra, D.V. (1989) *The Avian Egg: Chemistry and Biology.* John Wiley and Sons, New York, NY.

Concon, J.M. and Soltes, D. (1973) Rapid micro-Kjeldahl digestion of cereal grains and other biological materials. *Anal. Biochem.* 53:35-41.

Cutting, J.A. (1984) Gel protein stains: phosphoproteins. *Methods Enzymol.* 104:451-455.

Greengard, O., Sentenac, A., and Mendelson, N. (1964) Phosvitin, the iron carrier of egg yolk. *Biochim. Biophys. Acta* 90:406-407.

Grognan, J. and Taborsky, G. (1987) Iron binding by phosvitins: variable mechanism of iron release by phosvitin of diverse species characterized by different degree of phosphorylation. *J. Inorg. Biochem.* 29:33-47.

Itoh, T., Abe, Y., and Adachi, S. (1983) Comparative studies on the α- and β-phosvitin from hen's egg yolk. *J. Food. Sci.* 48:1755-1757.

Joubert, F.J. and Cook, W.H. (1958) Preparation and characterization of phosvitin from hen egg yolk. *Can. J. Biochem. Physiol.* 36:399-408.

Lu, C.L. and Baker, R.C. (1986) Characteristics of egg yolk phosvitin as an antioxidant for inhibiting metal-catalyzed phospholipid oxidation. *Poultry Sci.* 65:2065-2073.

Lu, C.L. and Baker, R.C. (1987) Effect of pH and food ingredients on the stability of egg yolk phospholipids and the metal-chelator antioxidant activity of phosvitin. *J. Food. Sci.* 52:613-616.

Mano, Y. and Lipman, F. (1966) Characteristics of phosphoproteins (phosvitins) from a variety of fish roes. *J. Biol. Chem.* 241:3822-3833.

Mecham, D.K. and Olcott, H.S. (1949) Phosvitin, the principal phosphoprotein of egg yolk. *J. Am. Chem. Soc.* 71:3670-3679.

Mok, C.C., Grant, C.T., and Taborsky, G. (1966) Countercurrent distribution of phosvitin. *Biochemistry* 5:2517-2523.

Morrison, W.R. (1964) A fast, simple and reliable method for the micro-determination of phosphorus in biological materials. *Anal. Biochem.* 7:218-224.

Penke, B., Ferenczi, R., and Kovacs, K. (1974) A new acid hydrolysis method for determining tryptophan in peptides and proteins. *Anal. Biochem.* 60:45-50.

Shantz, R.C. and Dawson, L.E. (1973) Electrophoretic examination of native and phosvitin fraction of avian egg yolk. *Poultry Sci.* 53:969-974.

Spackman, D.H., Stein, W.H., and Moore, S. (1958) Automatic recording apparatus for use in the chromatography of amino acids. *Anal. Chem.* 30: 1190-1206.

Sundararajan, T.A., Sampath Kumar, K.S.V., and Sarma, P.S. (1960) A simplified procedure for the preparation of phosvitin and vitellin. *Biochim. Biophys. Acta* 38:360-362.

Taborsky, G. (1974) Phosphoproteins. *Adv. Protein Chem.* 28:1-187.

Taborsky, G. (1980) Iron binding by phosvitin and its conformational consequences. *J. Biol. Chem.* 255:2976-2985.

Tsutsui, T. and Obara, T. (1984) Preparation and characterization of phosvitin from hen's egg yolk granule. *Agric. Biol. Chem.* 48:1153-1160.

Wallace, R.A. and Morgan, J.P. (1986) Isolation of phosvitin: retention of small molecular weight species and staining characteristics on electrophoretic gels *Anal. Biochem.* 151:256-261.

Wallace, R.A., Jared, D.W., and Eisen, A.Z. (1966) A general method for the isolation and purification of phosvitin from vertebrate eggs. *Can. J. Biochem.* 44:1647-1655.

Wallace, R.A., Carnevali, O., and Hollinger, T.G. (1990) Preparation and rapid resolution of *Xenopus* phosvitin and phosvettes by high performance liquid chromatography. *J. Chromatogr.* 519:75-86.

Yamamoto, Y., Sogo, N., Iwao, R., and Miyamoto, T. (1990) Antioxidant effect of egg yolk on linoleate in emulsions. *Agric. Biol. Chem.* 54:3099-3104.

Chapter Fourteen

Modified Avidins for Application in Avidin-Biotin Technology: an Improvement on Nature

E.A. Bayer and M. Wilchek

Department of Biophysics, The Weizmann Institute of Science, Rehovot 76100, Israel

Abstract The avidin-biotin system was initially described nearly two decades ago as a versatile system destined for a spectrum of scientific, industrial, and clinical applications. It has since proved effective for a variety of biotechnological purposes. The story is one of continual development, both in breadth (number and types of application) and depth (novel reagents and general improvements in procedure and performance). In the beginning, the egg white glycoprotein, avidin, served as the sole model for application; eventually, however, the bacterial protein, streptavidin, all but replaced avidin for most applications. Indeed, the sales worldwide of avidin dropped to almost insignificant levels. In recent years, there have been attempts to return the egg white protein to its initial status using protein modification techniques. The resultant modified avidin exhibits improved molecular characteristics both over the native protein (and previous derivatives thereof) as well as over streptavidin. The new product provides hope for a return of egg white avidin to its original role in avidin-biotin technology.
(*Key words:* avidin, biotin, streptavidin)

Introduction

For millennia, the egg has been an important staple in the human diet. The egg itself can be served boiled, poached, scrambled, fried, and as an omelette. In addition, the egg, after separation of the white and yolk if necessary, can be used in the baking, meat, restaurant, and dessert industries. Hundreds of millions of eggs are consumed or processed worldwide on a daily basis, and the egg industry has thrived.

In recent years, the possible health hazards, mainly associated with the fatty components of the egg yolk, have led to a decline in total egg consumption. Warranted or not, public awareness of cholesterol and triglyceride levels in the egg has resulted in a dramatic reduction of eggs and egg products in the diets of many nutrition-conscious populations (particularly in North America and Europe). As a consequence, the egg industry has supported other areas of activity in order to improve, eventually, their profit margins. One such area is designed to reduce the cholesterol levels in the yolk through genetics, nutrition, or chemical means. Another direction is to promote new egg products.

Despite being a major hindrance to the egg industry, the high levels of cholesterol and other lipids in the egg yolk can serve as an important source for the production of biologically important steroids and phospholipids (e.g. by chemical modification of resident precursors in the egg). A second series of egg products, of immense and largely untapped potential, are the egg proteins.

To date, the egg white proteins lysozyme and avidin have been manufactured in relatively large quantities. Lysozyme is of great potential as a natural protein antibiotic. Avidin, together with the vitamin biotin, has served as a prototype for the extremely versatile avidin-biotin system, which in recent years has found extensive biotechnological application, particularly in medical diagnostics (Wilchek and Bayer, 1990).

Although the principle of avidin-biotin technology was historically based on the interaction of egg white avidin with the vitamin biotin, the use of the egg white protein has given way to that of a more expensive bacterial cognate called streptavidin. This article will deal with the intriguing story of the egg white avidin-biotin system versus that of the bacterial streptavidin-biotin system. The reason for the expected return of the egg white protein in the near future will also be described.

Egg White Avidin

Periodic reports in the literature, dating back to nearly a century ago (Steinitz, 1898; Bateman, 1916), have described a toxic factor in egg white, which caused a series of peculiar anomalies in rats when administered orally. The toxic effects could be reversed by including in the diet high levels of a protective factor (initially called vitamin H), which is present in egg yolk among many other sources (Boas, 1927; György, 1931). Today we know that the toxic factor is the egg white glycoprotein avidin, which acts in this manner by depriving the organism of its normal levels of the protective factor, now known to be the vitamin biotin (Fig. 14.1).

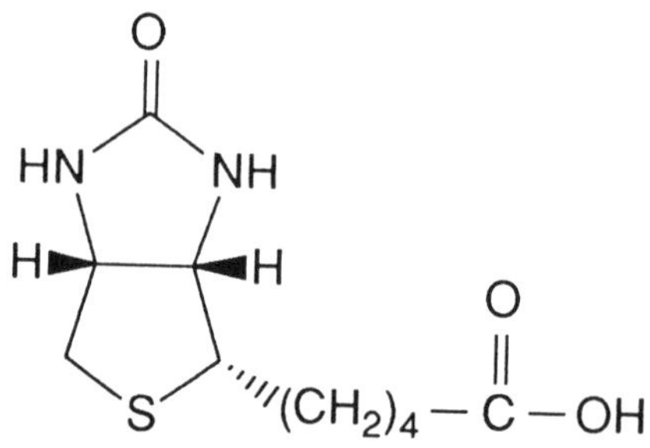

Fig. 14.1. The structure of biotin.

Avidin is produced in the oviducts of birds, reptiles, and amphibians, and deposited in the whites of their eggs (Fraps *et al.*, 1943). Avidin-like biotin-binding activity has also been shown to be induced in these animals following tissue injury (Elo, 1980).

Avidin constitutes a maximum of 0.05% of the protein in the hen egg white. Due to the strong binding characteristics between avidin and biotin, the highly stable complex formed between them has been an intriguing subject to study. Green (1975, 1990) has thoroughly characterised egg white avidin. It is a very stable tetramer consisting of four identical monomers, arranged in twofold symmetry. Each monomer has a reported molecular mass of 15,600 Da; of this value, about 10% is a single oligosaccharide moiety (comprising mainly mannose and *N*-acetylglucosamine residues), which is heterogeneous in its structure (Bruch and White, 1982). Thus, the total molecular mass of the glycosylated tetramer is about 62,400 Da, and that of the unglycosylated protein backbone is calculated (according to the primary amino acid sequence) to be 57,120 Da.

Each avidin monomer binds one molecule of biotin. The unique feature of this binding is the strength and specificity of formation of the avidin-biotin complex. The resultant affinity constant, estimated at 10^{15} M^{-1}, is the highest known for a protein and an organic ligand. It is so strong that biotin cannot be released from the binding site even when subjected to a variety of drastic conditions such as high concentrations of denaturing agents at room temperature (6 M guanidinium hydrochloride, 3 M guanidinium thiocyanate, 8 M urea, 10% β-mercaptoethanol, or 10% sodium dodecyl sulphate). Under combined treatment with guanidinium hydrochloride at low pH (1.5) or upon heating (>70°C) in the presence of denaturing agents or detergents, the protein is denatured, and biotin is dislodged from the disrupted binding site.

Avidin recognises biotin mainly at the upper ureido (urea-like) ring of the molecule. The interaction between the binding site of avidin with the sulphur-containing ring or the valeric acid chain of the vitamin is of much lower strength. The reasons for this strong binding are only now beginning to be understood (Wilchek and Bayer, 1989), mainly through modification studies (Green, 1975; Gitlin *et al.*, 1987, 1988a,b, 1990; Kurzban *et al.*, 1989, 1990, 1991) and comparative analysis of the primary (Bayer and Wilchek, 1990; DeLange and Huang, 1971; Argaraña *et al.*, 1986) and X-ray crystallographic structures (Weber *et al.*, 1989; Livnah *et al.*, 1993) of avidin and streptavidin.

The Avidin-Biotin System

The lack of interaction between the carboxy-containing side chain of biotin and avidin means that the former can be modified chemically and attached to a wide variety of biologically active material; the biotin moiety of the resultant derivative or conjugate is still available for interaction with avidin. In turn, the avidin can be derivatised with many other molecules, notably 'probes' or reporter groups of different types. This is the crux of avidin-biotin technology. Thus, a biologically active target molecule in an experimental system can be 'labelled' with its biotinylated counterpart (a binder), and the product can then be subjected to interaction with avidin, either derivatised or conjugated with an appropriate probe.

The general approach is shown schematically in the upper part of Fig. 14.2. An alternative mode is also available, as shown in the lower part of Fig. 14.2. By virtue of its four biotin-binding sites, avidin can be applied in native (unconjugated) form, together with a biotinylated probe; the two can be applied either sequentially in stepwise fashion or in a single step as preformed complexes. A list of the various biologically active pairs - the target and binder molecules - which are appropriate topics for avidin-biotin technology is presented in Table 14.1. The type of probe selected for conjunction with avidin or derivatised with biotin usually determines the specific type of application. A list of the diverse types of avidin-conjugated or biotinylated probes is presented in Table 14.2, along with their individual range of application. The rapidly expanding number of applications of avidin-biotin technology is summarised in Fig. 14.3. The reader is referred to the historical reviews and the comprehensive volume 184 in the *Methods in Enzymology* series for more detailed information about the synthesis of biotin derivatives, the biotinylation of macromolecules, avidin-containing probes and their implementation in avidin-biotin technology (Bayer and Wilchek, 1978, 1980; Bayer *et al.*, 1979; Wilchek and Bayer, 1984, 1988, 1990).

Were it not for its inherent advantages (perhaps somewhat cryptic at first glance), the interest in the avidin-biotin system would certainly have waned. These advantages eventually led to its widespread commercial exploitation. Today, dozens of companies worldwide deal with avidin-biotin products; many of these companies are centred around or solely concerned with avidin-biotin technology.

In commercial terms, the main advantages of the avidin-biotin system are its remarkable versatility and almost universal applicability. In certain applications, notably for diagnostic purposes, mediation via the avidin-biotin system affords an amplified signal which leads to increased assay sensitivity. Moreover, and not to be belittled, the various avidin-containing and biotinylated derivatives and conjugates are uncommonly stable and amenable to storage. In commercial terms, this translates as follows: avidin-derivatised and biotinylated materials can be used as shelf reagents and sold as stable components in kits.

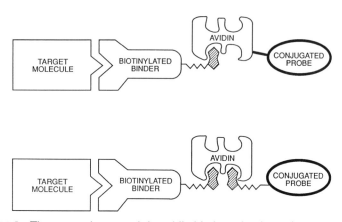

Fig. 14.2. The general approach in avidin-biotin technology. A target molecule in an experimental system is labelled with a biotinylated binder molecule. The biotinylated binder is in turn labelled with an avidin-conjugated probe (scheme at top). Alternatively, the biotinylated binder can be labelled with free avidin followed by a biotinylated probe (scheme at bottom).

Table 14.1. Target-binder pairs for application in avidin-biotin technology.

Target	Binder
Biotin	avidin, streptavidin, etc.
Avidin, streptavidin, etc.	biotin
Antigens	antibodies
Antibodies	antigens
Lectins	glycoconjugates
Glycoconjugates	lectins
Cations	anions
Anions	cations
Hydrophobic sites	hydrophobic groups
Receptors	hormones, effectors, toxins, etc.
Transport proteins	vitamins, amino acids, sugars, etc.
Membranes	liposomes
Nucleic acids, genes	DNA/RNA probes
Phages, viruses, bacteria, subcellular organelles, cells, tissues, whole organisms	all of the above

From Wilchek and Bayer (1988), with permission.

Table 14.2. Commonly used avidin-containing probes.

Probes (conjugates)	Applications
Enzymes	immunoassay, diagnostics, blotting, affinity cytochemistry, light microscopy, electron microscopy, targeted catalysis
Radiolabels	immunoassay, cytochemistry, cytological probes, blotting
Fluorescent agents	affinity cytochemistry, fluorescence microscopy, flow cytometry, immunoassay, diagnostics, blotting
Chemiluminescent agents	immunoassay, diagnostics, blotting
Colloidal gold	affinity cytochemistry, immunoassay, blotting
Ferritin, haemocyanin	affinity cytochemistry, macromolecular carrier
Phages	affinity cytochemistry, affinity targeting
Macromolecular carriers	crosslinking studies, signal amplification, affinity targeting, drug delivery, cytological probes, affinity fusion, affinity perturbation, affinity partitioning
Liposomes	affinity fusion, drug delivery, affinity targeting, signal amplification
Solid supports	affinity chromatography, immobilisation, selective retrieval, selective elimination

From Wilchek and Bayer (1990), with permission.

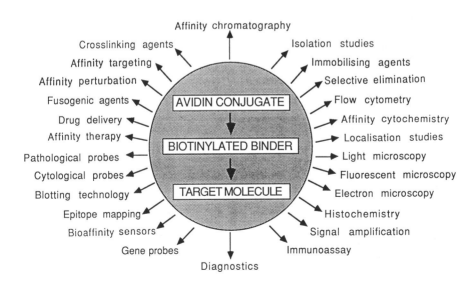

Fig. 14.3. Applications of avidin-biotin technology (from Wilchek and Bayer, 1989).

The Advent of Streptavidin

In the early 1960s, scientists at Merck, Sharp and Dohme isolated, from soil samples, a series of bacterial cultures which produced an interesting form of antibiotic, the action of which was restricted to Gram-negative bacteria and could be inhibited by the vitamin biotin (Stapely *et al.*, 1963). The antibiotic was found to be a complex which comprised two separate components - one of small molecular dimensions and the other a macromolecule - which acted together in synergistic fashion (Chaiet *et al.*, 1963). Upon further study (Chaiet and Wolf, 1964), the researchers were astonished to discover that the macromolecular component of the antibiotic complex bore striking resemblance to the egg white protein avidin, in that it bound biotin with the same unprecedented tenacity. Since the bacterial strains which produced this new antibiotic were of the genus *Streptomyces* the avidin-like protein was termed streptavidin. Most of the bacterial strains were subspecies of the known species *Streptomyces lavendulae*; one isolate exhibited unique characteristics and was dubbed *Streptomyces avidinii*. Other characteristics of the biotin-binding protein streptavidin and their comparison with those of egg white avidin will be presented in the following section.

The small molecular component was actually a series of di- and tripeptides. These were eventually identified as the 'stravidins' (Baggaley *et al.*, 1969), which contained a central unusual amino acid residue, called amiclenomycin (Acm), which is a protonated form of *p*-amino-*homo*-phenylalanine (Okami *et al.*, 1974; Kern *et al.*, 1985). Acm is a biotin anti-metabolite which inhibits one of the enzymes (DAPA-aminotransferase) of the biotin biosynthetic pathway (Hotta *et al.*, 1975; Poetsch *et al.*, 1985). One possible explanation for the mode of action of the antibiotic complex is that streptavidin binds the Acm-containing peptides, and serves to transport them to the target bacterial cell surface, possibly by binding to the outer-membranal porins (Korpela *et al.*, 1984; Kennedy *et al.*, 1989). Cell growth would be inhibited by subsequent release of the peptides from the complex, their transport intracellularly via a 'misuse' of general peptide transport systems, and intracellular hydrolysis of the peptides to provide the free Acm which would inhibit biotin synthesis in the target cell.

It is interesting to note that the precise role of egg white avidin in the egg is unknown. It is thus intriguing that the egg white protein can substitute for the bacterial streptavidin, for antibiotic action (Tausig and Wolf, 1964). One might speculate whether Acm-containing peptides are present in the egg whites or in the tissues of various birds, reptiles, and amphibians.

Avidin versus Streptavidin: Structural Considerations

Like egg white avidin, streptavidin is a very stable tetramer with similar twofold symmetry. After removal of the leader sequence from the nascent proteins, the molecular mass of the streptavidin monomer is about 16,600 Da (Table 14.3). Streptavidin is secreted very efficiently from the bacterial cell, and the secreted protein undergoes proteolytic processing which removes segments from the N- and C-termini, which account for about 20% of the total mass of streptavidin (Pähler *et al.*, 1987; Bayer *et al.*, 1989). The final molecular mass of the resultant 'core' streptavidin is calculated at 52,800 Da (monomer = 13,200). The core streptavidin is exceptionally stable, and the biotin-binding activity (like egg white avidin, one binding site per subunit) is unaffected.

Table 14.3. Some important characteristics of avidin and streptavidin.

	Avidin	Streptavidin
Source	Egg white	Bacterium
Mol. mass (Da)		
Tetramer		
unprocessed	62,700	66,500
processed	57,100	52,800
Monomer		
unprocessed	15,700	16,600
processed[a]	14,300	13,200
K_D (M)	$\sim 10^{-15}$	$\sim 10^{-15}$
Biotin/monomer	1	1
Sugar	yes	no
pI	>10	<7
Amino acid sequence[b]	1971	1986
DNA sequence[b]	1989	1986
X-ray structure[b]	1993	1989

[a] Processed avidin refers to the deglycosylated molecule, and processed streptavidin refers to the N- and C-terminally truncated form.
[b] Year of first publication.

In fact, the interaction between core streptavidin and biotinylated proteins is significantly improved over that of the intact unprocessed molecule; the lower levels of observed binding in the interaction with the latter is apparently due to steric hindrance caused by the extraneous segments. Similar to avidin, the binding of biotin to streptavidin increases the stability of the molecule; the resistance of streptavidin and the streptavidin-biotin complex to denaturing agents is even greater than that of the egg white protein (Kurzban *et al.*, 1990).

Recent advances in the X-ray structure of egg white avidin (Livnah *et al.*, 1993) and its comparison with the known structure of streptavidin (Weber *et al.*, 1989) have shown that the respective 'business' (i.e. biotin-binding) portions of the two are almost superimposable. A comparison of their sequences is especially enlightening. The two proteins show an overall sequence homology of the order of 35%; nearly all of the conserved residues are located in six domains (Bayer and Wilchek, 1990; Wilchek and Bayer, 1989), in which over 60% of the amino acids are identical (Fig. 14.4). Outside of these domains, the respective sequences are very different. Nevertheless, the important portions of the secondary, tertiary, and quaternary structures of the two proteins are very similar. It seems that the nonhomologous regions in the two proteins are functionally less important and serve simply to force the homologous domains into the correct three-dimensional orientation. The relevant reactive conserved amino acid moieties are thus emplaced identically in the structure of the corresponding protein, such that they encompass perfectly the ring structures of the biotin molecule (Livnah *et al.*, 1993).

Unlike the binding of avidin and biotin, the intact bicyclic ring system of the biotin molecule is critical to the strong interaction with streptavidin; however (and more important biotechnologically), the valeryl side chain can be subjected to unrestricted modification as in the egg white glycoprotein. Thus, based upon this point alone, both proteins are equally appropriate for use in avidin-biotin technology.

Why then, has the bacterial protein taken precedence over the egg white glycoprotein as the major component of the kits designed for use in avidin-biotin technology? The answer to this question is partially a historical one, and is clarified by reviewing the amino acid sequences of the two proteins (Fig. 14.4). As mentioned above, the egg white protein was used for early development of the avidin-biotin system. As more reports of applications in different areas began to appear in the literature, problems with nonspecific binding and high background levels began to occur. The high pI and presence of the oligo-saccharide moiety were initially presumed to account for the 'nonspecific' binding phenomena. Indeed, egg white avidin is glycosylated at the Asn-17 residue, which occurs in a typical NXT(S) or Asn-Xxx-Thr(Ser) type carbohydrate-containing consensus sequence (Fig. 14.4). Further perusal of the sequence reveals a wealth of basic amino acid residues, i.e. lysines and arginines.

The consequences of these characteristics of the avidin molecule were reflected in biotin-independent binding of various extraneous macromolecules in some target systems. For example, the presence of sugars caused an interaction of avidin with lectin-like molecules, derived, for example, from cellular systems. Moreover, the strong positive charge of avidin led to relatively strong electrostatic interactions with negatively charged molecules (such as acidic proteins and, notably, nucleic acids).

Fig. 14.4. The amino aicd sequences of avidin (left) and streptavidin (right). Homologous residues in both proteins are shown in bold type, and the homologous domains are enclosed in boxes. Positively charged residues, i.e. arginines and lysines, are emphasised by dots. The RYD-containing sequence in streptavidin, which is homologous to that of fibronectin, is underlined. The N- and C-terminal sites of postsecretory processing (via proteolytic cleavage) on streptavidin are shown by arrows. The position of the sugar residue, linked to Asn-17 of avidin, is also shown.

For this reason, the revival of streptavidin in the early 1980s served to boost the prospects for application of the avidin-biotin system, because streptavidin is nonglycosylated in the native state and has a near-neutral pI. Thus, despite the higher cost of streptavidin, its use became popular due to its reduced levels of nonspecific binding and low background staining.

Despite the advantages in using streptavidin over avidin, many laboratories observed an equivocal interaction between the bacterial protein and unidentified macromolecular components in many experimental systems. The mechanism of this interaction remained a puzzle for many years. Recent studies (Alon et al., 1990) have shown, however, that streptavidin contains an Arg-Tyr-Asp sequence, which is highly reminiscent of the universal cell surface recognition sequence (Arg-Gly-Asp) present in a variety of adhesion molecules, e.g. fibronectin, vitronectin, fibrinogen, and collagen. In consequence, streptavidin interacts strongly and in a biotin-independent manner with the integrins and related cell surface receptors.

Modified Avidins

The problems with streptavidin (high cost and biotin-independent cell binding) have prompted renewed interest in egg white avidin as the standard for avidin-biotin technology.

Numerous early attempts were devoted to eliminating the high nonspecific binding and high background associated with the avidin-mediated labelling (Duhamel and Whitehead, 1990). Some researchers promoted the use of high salt concentrations or high pH in the medium which contained the avidin-conjugated probe (Bussolati and Gugliotta, 1983; Clark et al., 1986). Others suggested the use of basic proteins (e.g. lysozyme) for general blocking of acidic groups (Bayer et al., 1987). In many cases, both approaches were employed successfully, but neither was universally applicable.

Another approach involves the modification of avidin to reduce the charge of the protein. To this end, avidin has been formylated (Guesdon et al., 1979), acetylated (Kaplan et al., 1983), and succinylated (Finn et al., 1984). The preparation of fluorescent derivatives of avidin (e.g. fluorescein or rhodamine derivatives) also reduces the pI of the protein. However, the above derivatives of avidin are all prepared via covalent attachment to the available lysines of avidin. Consequent blocking of the free amino groups precludes subsequent preparation of other types of conjugates (notably protein-protein conjugates such as avidin-labelled enzymes) which are often prepared by crosslinking via lysines using bifunctional reagents (e.g. glutaraldehyde, dimethyl suberimidate, and disuccinimidyl suberate).

A more useful and effective alternative to lysine modification is the modification via arginines. In this case, the pI of the protein is efficiently reduced and the lysines are still available for subsequent interaction. Two different derivatives of avidin which are modified in this manner are commercially available. One, ExtrAvidin®, can be obtained in various functionally derivatised or conjugated forms from Sigma Chemical Company (St Louis, MO). A second, NeutraLite Avidin™ (a product of Belovo Chemicals, Industrial Area 1, Bastogne, Belgium) is additionally modified (see below) and can be purchased in bulk quantities.

Although the reduction of the pI of egg white avidin solves one of the problems, the presence of the oligosaccharide residue remains a serious source of nonspecific (biotin-independent) interaction which restricts its applications. The return of egg white avidin as the standard for avidin-biotin technology has been contingent upon the removal of its sugars.

The possibilities for removing a sugar from a glycoprotein are quite limited; it is possible to do so either chemically or enzymatically. The chemical methods currently available, e.g. using hydrogen fluoride or periodate oxidation, are either destructive or inefficient. The well known enzymatic method, which employs *N*-glycanase (Tarentino *et al.*, 1985), is usually very expensive and not very effective for avidin when conventional methodology is used.

Nonetheless, initial progress in eliminating this problem was first achieved in studies on the biotin-binding site of avidin which were carried out using Belovo avidin. It was noticed that some of the population of avidin molecules appeared to be nonglycosylated. One unlikely explanation for this phenomenon was that a very special type of chicken produced the nonglycosylated form of avidin which somehow reached the egg white from its site of synthesis in the hen oviduct. Another more likely possibility was that the glycoprotein was being deglycosylated during the process of purification from the egg white. At any rate, nonglycosylated avidin tetramers could be isolated from Belovo avidin (Hiller *et al.*, 1987), and the nonglycosylated form was eventually termed 'Lite' avidin. Eventually, a viable enzymatic procedure for deglycosylation was established at Belovo Chemicals. The resultant product (Fig. 14.5) was subsequently modified chemically for the arginines and termed NeutraLite Avidin™. NeutraLite Avidin™ is thus a form of modified avidin which lacks the oligosaccharide moiety, exhibits a neutral pI, and bears free lysine groups for further derivatization.

For clarity, the relevant properties of the various avidin preparations, which are potential causes of their nonspecific interactions with biological material, are summarised in Table 14.4.

Nonspecific Binding Characteristics of the Avidins and Streptavidins

In recent studies in our group, we have examined the specific and nonspecific binding properties of a series of conjugates, complexes, and derivatives, prepared from the various forms of avidin and streptavidin listed in Table 14.4. For many applications, all forms are indeed suitable. For example, for electron microscopic cytochemistry, the use of conjugates prepared from unmodified egg white avidin is often more reactive with biotin-labelled cell surface molecules than the equivalent streptavidin-containing conjugates. In most cases, little background label has been encountered. Of course, if a mannose-sensitive surface lectin is present on target cells, high nonspecific background staining may be observed.

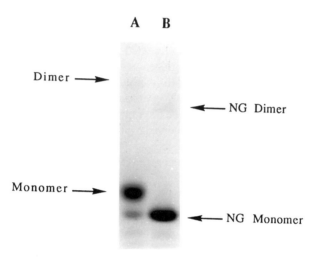

Fig. 14.5 Deglycosylation of avidin. The monomers of Belovo avidin (lane A) consist of a major glycosylated band and a minor nonglycosylated band, as demonstrated by SDS-PAGE. Following the deglycosylation procedure, the great majority of avidin monomers in 'Lite' avidin are in the nonglycosylated form (lane B).

Table 14.4 Relevant properties of avidins and streptavidin.

Protein	Charge	Sugar content	Agr-Tyr-Asp content	Free lysine groups
Egg white avidin	positive	yes	no	yes (9)[a]
Streptavidin	neutral	no	yes	yes (4)
Lite Avidin	positive	no	no	yes (9)
NeutraLite Avidin[TM]	neutral	no	no	yes (9)
ExtrAvidin®	neutral	yes	no	yes (9)
Other modified avidin	neutral	yes	no	no

[a] The number of free lysine groups in the designated protein is given in parentheses.

There is no doubt, however, that in many other systems (e.g. protein blotting and enzyme assays for nucleic acids), the use of the egg white protein is prohibitive due to its nonspecific interactions (Duhamel and Whitehead, 1990). Although the technology for countering such nonspecific binding has been available for many years, the laborious efforts necessary to set up a 'clean' system have presented an imposing barrier for the impatient scientist. In many cases, the persistence of the avidin-induced nonspecific label was too much even for the most patient scientists. For this reason, it became fashionable to substitute the more expensive bacterial streptavidin for egg white avidin. Eventually, industry entered into the avidin-biotin business, and, on the basis of proven performance, many companies embraced streptavidin as an attractive base for conjugation and derivatisation. Research and clinical laboratories alike, simply bought the resultant prestandardised kits at highly inflated prices.

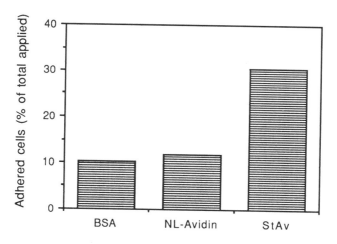

Fig. 14.6 Binding of melanoma cells to immobilised proteins. Bovine serum albumin (BSA), NeutraLite Avidin™ (NL-Avidin), and streptavidin (StAv) were bound to microtitre plates and melanoma cells were applied. The percentage of cells bound to the plates was then determined. Note the low, almost non-existent background, labelling on plates containing NeutraLite Avidin™.

In some cases, however, it is clear that streptavidin gives very high levels of nonspecific background (Alon *et al.*, 1990, 1992). For example, when streptavidin is bound to polystyrene matrices, e.g. microtitre plates, various cells will bind quite strongly to the immobilised protein. In Figure 14.6, melanoma cells are shown to bind at relatively high levels to streptavidin plates. The observed binding is about three times higher than that of plates containing NeutraLite Avidin™ or bovine serum albumin (as a negative control). In other cell types, such as bacterial cells (Fig. 14.7), the difference can be even greater. In this case, both avidin and streptavidin bind bacterial cells. Avidin presumably binds to the cells on the combined basis of sugar and charge. Indeed, if the sugar is removed from avidin (i.e. as in Lite Avidin), the binding to bacteria is markedly reduced. Further modification of the charge of Lite Avidin to produce NeutraLite Avidin™ results in nominal levels of bacteria bound to the plates. Streptavidin, on the other hand, binds bacteria strongly on the basis of the resident Arg-Tyr-Asp site.

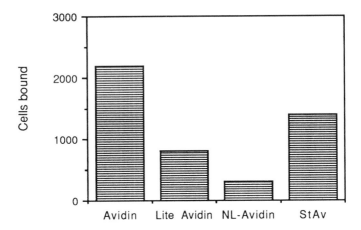

Fig. 14.7. Binding of bacterial cells to immobilised proteins. The experiment was carried out on the designated protein-adsorbed plates as described in the legend to Fig. 14.6, except that *Escherichia coli* cells were used.

In terms of positive labelling, i.e. the use of the desired avidin- or streptavidin-based probes for detecting biotinylated molecules in avidin-biotin technology, conjugates and derivatives prepared from NeutraLite Avidin™ have resulted in equivalent, and in many cases superior, labelling properties when compared with those prepared from streptavidin. Although the exact reason for the enhancement is not yet known, it may be related to the number of free lysines in NeutraLite Avidin™ relative to that in commercial streptavidin (i.e. nine versus four). At any rate, the improved properties of NeutraLite Avidin™ render its use desirable for all applications of avidin-biotin technology.

Future Trends

It is our belief that, in the next few years, NeutraLite Avidin™ will signal a return of the egg to its original role in avidin-biotin technology. NeutraLite Avidin™ is a product of protein engineering. It is one of the prime examples, and perhaps the first, in which the properties of a native protein have been altered by chemical and enzymatic means to provide a greatly improved product for applied and commercial usage. The modified avidin is thus an improvement on nature.

Acknowledgements

The authors wish to thank Dr Fabien De Meester of Belovo Chemicals (Bastogne, Belgium) for continued cooperation in the development of NeutraLite Avidin™. Large quantities (50 g and more) of avidin were provided to us by Belovo; this enabled extensive protein chemical studies of avidin which were carried out by our group over the past decade. We also thank Makor Chemicals (Jerusalem) for their support. A grant from the Fund for Basic Research, administered by the Israel Academy of Sciences and Humanities, is gratefully acknowledged.

References

Alon, R., Bayer, E.A., and Wilchek, M. (1990) Streptavidin contains an RYD sequence which mimics the RGD receptor domain of fibronectin. *Biochem. Biophys. Res. Commun.* 170:1236-1241.

Alon, R., Bayer, E.A., and Wilchek, M. (1992) Cell-adhesive properties of streptavidin are mediated by the exposure of an RGD-like RYD site. *Eur. J. Cell Biol.* 58:271-279.

Argaraña, C.E., Kuntz, I.D., Birken, S., Axel, R., and Cantor, C.R. (1986) Molecular cloning and nucleotide sequence of the streptavidin gene. *Nucleic Acids Res.* 14:1871-1882.

Baggaley, K.H., Blessington, B., Falshaw, C.P., Ollis, W.D., Chaiet, L., and Wolf, F.J. (1969) The constitution of sravidin, a novel microbiological product. *Chem. Commun.* 1969:101-102.

Bateman, W.G. (1916) The digestibility and utilization of egg proteins. *J. Biol. Chem.* 26:263-291.

Bayer, E.A. and Wilchek, M. (1978) Emerging techniques: the avidin-biotin complex as a tool in molecular biology. *Trends Biochem. Sci.* 3:N237-N259.

Bayer, E.A. and Wilchek, M. (1980) The use of the avidin-biotin complex as a tool in molecular biology. *Methods Biochem. Anal.* 26:1-45.

Bayer, E.A. and Wilchek, M. (1990) Application of avidin-biotin technology for affinity-based separations. *J. Chromatogr.* 510:3-12.

Bayer, E.A., Skutelsky, E., and Wilchek, M. (1979) The avidin-biotin complex in affinity cytochemistry. *Methods Enzymol.* 62:308-315.

Bayer, E.A., Ben-Hur, H., and Wilchek, M. (1987) Enzyme-based detection of glycoproteins on blot transfers using avidin-biotin technology. *Anal. Biochem.* 161:123-131.

Bayer, E.A., Ben-Hur, H., Hiller, Y., and Wilchek, M. (1989) Postsecretory modifications of streptavidin. *Biochem. J.* 259:369-376.

Boas, M. (1927) The effect of desiccation upon the nutritive properties of egg white. *Biochem. J.* 21:712-725.

Bruch, R.C. and White, H.B., III (1982) Compositional and structural heterogeneity of avidin glycopeptides. *Biochemistry* 21:5334-5341.

Bussolati, G. and Gugliotta, P. (1983) Nonspecific staining of mast cells by avidin-biotin-peroxidase complexes (ABC). *J. Histochem. Cytochem.* 31: 1419-1421.

Chaiet, L. and Wolf, F.J. (1964) The properties of streptavidin, a biotin-binding protein produced by *Streptomycetes*. *Arch. Biochem. Biophys.* 106:1-5.

Chaiet, L., Miller, T.W., Tausig, F., and Wolf, F.J. (1963) Antibiotic MSD-235: II. Separation and purification of synergistic components. *Antimicrob. Agents Chemother.* 3:28-32.

Clark, R.K., Tani, Y., and Damjanov, I. (1986) Suppression of nonspecific binding of avidin-biotin complex (ABC) to proteins electroblotted to nitrocellulose paper. *J. Histochem. Cytochem.* 34:1509-1512.

DeLange, R.J. and Huang, T.-S. (1971) Egg white avidin: III. Sequence of the 78-residue middle cyanogen bromide peptide complete amino acid sequence of the protein subunit. *J. Biol. Chem.* 246:698-709.

Duhamel, R.C. and Whitehead, J.S. (1990) Prevention of nonspecific binding of avidin. *Methods Enzymol.* 184:201-207.

Elo, H.A. (1980) Occurrence of avidin-like biotin-binding capacity in various vertebrate tissues and its induction. *Comp. Biochem. Physiol.* 67B:221-224.

Finn, F.M., Titus, G., and Hofmann, K. (1984) Ligands for insulin receptor isolation. *Biochemistry* 23:2554-2558.

Fraps, R.M., Hertz, R., and Sebrell, W.H. (1943) Relations between ovarian function and avidin content in the oviduct of the hen. *Proc. Soc. Exp. Biol. Med.* 52:140-142.

Gitlin, G., Bayer, E.A., and Wilchek, M. (1987) Studies on the biotin-binding site of avidin: lysine residues involved in the active site. *Biochem. J.* 242:923-926.

Gitlin, G., Bayer, E.A., and Wilchek, M. (1988a) Studies on the biotin-binding site of avidin: tryptophan residues involved in the active site. *Biochem. J.* 250:291-294.

Gitlin, G., Bayer, E.A., and Wilchek, M. (1988b) Studies on the biotin-binding site of streptavidin: tryptophan residues involved in the active site. *Biochem. J.* 256:279-282.

Gitlin, G., Bayer, E.A., and Wilchek, M. (1990) Studies on the biotin-binding site of avidin and streptavidin: tyrosine residues involved in the active site. *Biochem. J.* 269:527-530.

Green, N.M. (1975) Avidin. *Adv. Protein Chem.* 29:85-133.

Green, N.M. (1990) Avidin and streptavidin. *Methods Enzymol.* 184:51-67.

Guesdon, J.-L., Ternynck, T., and Avrameas, S. (1979) The use of avidin-biotin interaction in immunoenzymatic techniques. *J. Histochem. Cytochem.* 27: 1131-1139.

György, P. (1931) Rachitis und andere Avitaminosen. *Z. Arztl. Fortbild.* 28:377-380.

Hiller, Y., Gershoni, J.M., Bayer, E.A., and Wilchek, M. (1987) Biotin binding to avidin: oligosaccharide side chain not required for ligand association. *Biochem. J.* 248:167-171.

Hotta, M., Kitahara, T., and Okami, Y. (1975) Studies on the mode of action of amiclenomycin. *J. Antibiot.* 28:222-228.

Kaplan, M.R., Calef, E., Bercovici, T., and Gitler, C. (1983) The selective detection of cell surface determinants by means of antibodies and acetylated avidin attached to highly fluorescent polymer microspheres. *Biochim. Biophys. Acta* 728:112-120.

Kennedy, K.E., Daskalakis, S.A., Davies, L., and Zwadyk, P. (1989) Non-isotopic hybridization assays for bacterial DNA samples. *Mol. Cell. Probes* 3:167-178.

Kern, A., Kabatek, U., Jung, G., Werner, R.G., Poetsch, M., and Zähner, H. (1985) Amiclenomycin peptides - isolation and structure elucidation of new biotin antimetabolites. *Liebigs Ann. Chem.* 1985:877-892.

Korpela, J., Salonen, E.M., Kuusela, P., Sarva, M., and Vaheri, A. (1984) Binding of avidin to bacteria and to the outer membrane porin of *Escherichia coli. FEMS Microbiol. Lett.* 22:3-10.

Kurzban, G.P., Gitlin, G., Bayer, E.A., Wilchek, M., and Horowitz, P.M. (1989) Shielding of tryptophan residues of avidin by the binding of biotin. *Biochemistry* 28:8537-8542.

Kurzban, G.P., Gitlin, G., Bayer, E.A., Wilchek, M., and Horowitz, P.M. (1990) Biotin binding changes the conformation and decreases tryptophan accessibility of streptavidin. *J. Protein Chem.* 9:673-682.

Kurzban, G.P., Bayer, E.A., Wilchek, M., and Horowitz, P.M. (1991) The quaternary structure of streptavidin in urea. *J. Biol. Chem.* 266:14470-14477.

Livnah, O., Bayer, E.A., Wilchek, M., and Sussman, J. (1993) Three-dimensional structures of avidin and the avidin-biotin complex. *Proc. Natl Acad. Sci. USA* 90:5076-5080.

Okami, Y., Kitahara, T., Hamada, M., Naganawa, N., Kondo, S., Maeda, K., Takeuchi, T., and Umezawa, H. (1974) Studies on a new amino acid antibiotic, amiclenomycin. *J. Antibiot.* 27:656-664.

Pähler, A., Hendrickson, W.A., Gawinowicsz Kolks, M.A., Argaraña, C.E., and Cantor, C.R. (1987) Characterization and crystallization of core streptavidin. *J. Biol. Chem.* 262:13933-13937.

Poetsch, M., Zähner, H., Werner, R.G., Kern, A., and Jung, G. (1985) Metabolic products from microorganisms. 230. Amiclenomycin-peptides, new antimetabolites of biotin: taxonomy, fermentation and biological properties.

antimetabolites of biotin: taxonomy, fermentation and biological properties. *J. Antibiot.* 38:312-320.

Stapely, E.O., Mata, J.M., Miller, I.M., Demney, T.C., and Woodruff, H.B. (1963) Antibiotic MSD-235. I. Production by *Streptomyces avidinii and Streptomyces lavendulae. Antimicrob. Agents Chemother.* 3:20-27.

Steinitz, F. (1898) *Arch. Gesamte. Physiol.* 72:75.

Tarentino, A.L., Gomez, G.M., and Plummer, T.H., Jr (1985) Deglycosylation of asparagine-linked glycans by peptide: *N*-glycosidase F. *Biochemistry* 24: 4665-4671.

Tausig, F. and Wolf, F.J. (1964) Streptavidin - a substance with avidin-like properties produced by microorganisms. *Biochem. Biophys. Res. Commun.* 14:205-209.

Weber, P.C., Ohlendorf, D.H., Wendoloski, J.J., and Salemme, F.R. (1989) Structural origins of high-affinity biotin binding to streptavidin. *Science* 243:85-89.

Wilchek, M. and Bayer, E.A. (1984) The avidin-biotin complex in immunology. *Immunol. Today* 5:39-43.

Wilchek, M. and Bayer, E.A. (1988) The avidin-biotin complex in bioanalytical applications. *Anal. Biochem.* 171:1-32.

Wilchek, M. and Bayer, E.A. (1989) Avidin-biotin technology ten years on: has it lived up to its expectations? *Trends Biochem. Sci.* 14:408-412.

Wilchek, M. and Bayer, E.A. (eds) (1990) *Methods Enzymol.* Vol. 184. *Avidin-Biotin Technology.* Academic Press, San Diego, CA.

Chapter Fifteen

Egg White Lysozyme as a Preservative for Use in Foods

E.A. Johnson

Food Research Institute, University of Wisconsin, 1925 Willow Drive, Madison, WI 53706, USA

Abstract Egg white lysozyme is useful as a preservative against undesirable organisms in various foods. Lysozyme was demonstrated to have antibacterial activity *in vitro* against organisms of concern in food safety including *Listeria monocytogenes* and *Clostridium botulinum*. My laboratory has also demonstrated that food spoilage bacteria including *Clostridium thermosaccharolyticum, Bacillus stearothermophilus,* and *Clostridium tyrobutyricum* were highly susceptible to the lytic enzyme. In certain foods, lysozyme killed or prevented growth of *L. monocytogenes* and *C. botulinum*. Lysozyme was more active against *L. monocytogenes* in vegetables than in animal-derived foods that we tested. Lysozyme was effective in controlling toxin formation by *C. botulinum* in fish, poultry, and certain vegetables. Lysozyme activity was enhanced by certain food additives including EDTA, butylparaben, and tripolyphosphate, and also by naturally occurring antimicrobial agents including conalbumin, lactoferrin, nisin, and transferrins. Lysis of pathogens and spoilage bacteria was influenced by the nutritional composition, pH, and temperature of the growth substrates. Our results indicate that lysozyme could increase the safety and shelf-lives of several foods.
(*Key words:* lysozyme, preservative)

Introduction

In response to consumer demands for healthier and less calorific foods, the food industry is introducing products to the market place that receive minimal processing and contain reduced levels of traditional preservatives such as salt, sugar, and acids, and few or no chemical preservatives. Many of these foods rely solely on refrigeration for preserving their safety and quality, and the marketing personnel demand that they have a long shelf-life to enable distribution and maximise profits. Refrigeration, however, is a delicate method of food preservation and is prone to abuse in retail chill cabinets (Hutton et al., 1991). Failure of refrigeration can result in food spoilage or transmission of foodborne disease.

There is presently much concern within the industry and governmental agencies that minimally-processed refrigerated foods could support the growth or survival of pathogens, particularly the pathogenic bacteria *Listeria monocytogenes*, *Yersinia enterocolitica*, *Aeromonas hydrophila*, and the toxin-former *Clostridium botulinum* (Palumbo, 1986; Conner et al., 1989). These organisms can grow at refrigeration temperatures as low as 3.3°C (38°F). The safety and quality of minimally-processed refrigerated foods can be enhanced by good manufacturing practices and through the use of preservatives that eliminate or prevent the growth of undesirable bacteria and fungi (Leistner, 1978; Moberg, 1989; Scott, 1989).

Our laboratory has demonstrated that hen egg white lysozyme has anti-microbial activity against organisms of concern in food safety, including *Listeria monocytogenes* and *Clostridium botulinum*, and that lysozyme is inhibitory to certain spoilage organisms including thermophilic spore-forming bacteria and certain yeasts. Egg white lysozyme appears to have considerable potential as a harmless preservative for use in the food industry.

Properties of Lysozyme

Lysozyme was discovered and characterised in the early 1920s by Alexander Fleming, who called it 'a remarkable bacteriolytic element' (Fleming, 1922). He demonstrated that it had antibacterial activity against Gram-positive bacteria and was present in many tissues and secretions, and was particularly abundant in the white of hen's eggs. Lysozyme was the first enzyme whose three-dimensional structure was determined and whose properties were understood in atomic detail (Blake et al., 1965; Phillips, 1966). Lysozyme from hen egg white is a poly-peptide of 129 amino acids, consisting of some 1950 atoms, having a molecular weight of ~14,700. It is a basic protein and has an isoelectric point of 10.5-11.0. Crystals of hen egg white lysozyme are easily grown at pH 4.5-4.7 (Alderton and Fevold, 1946). Analysis of the crystals by a Fourier map of the electron density distribution at 2 Å resolution showed a zone of uniform density due to the poly-peptide chain and four high density zones attributable to four disulphide bridges. The lysozyme appears as roughly ellipsoidal with dimensions of about 45 × 30 × 30 Å with most of the nonpolar chains facing inwards (Blake et al., 1965).

Lysozyme (EC 3.2.1.17) acts as a mucopeptide *N*-acetylmuramyl hydrolase catalysing the hydrolysis of the ß-(1-4) linkage between *N*-acetylmuramic acid and *N*-acetylglucosamine of bacterial peptidoglycan, which is the structural component of bacterial cell walls (Salton, 1957). The activity of the enzyme on intact bacteria is inhibited through masking of the peptidoglycan by capsular polymers and by substitution of peptidoglycan with chemical groups. The enzyme can also attack other polymers and has weaker activity on the ß-1-4 linkage binding the *N*-acetylglucosamine molecules of chitin found in cell walls of certain fungi (Jollès *et al.*, 1963; Jollès and Jollès, 1984).

Lysozyme has several desirable properties as a food preservative. It is among the most stable of proteins and exhibits some remarkable resistance properties that contribute to its stability in foods. The chicken egg white lysozyme, known as type C, is one of the most durable enzymes known. It can withstand boiling for 1-2 min at pH 4-5, but it is denatured by heat at higher pHs. Although stable *in vitro*, it may be unstable to heating in various foods, particularly those of high protein content, since during heating it can form mixed disulphide-linked molecules with other proteins in the foods. Lysozyme is not inactivated by solvents and regains full activity when transferred back to an aqueous environment. The enzyme can be frozen and is stable on drying. It is active over the temperature range 1°C to near boiling. The resistance properties of egg white lysozyme make it attractive for use in foods as a preservative. The optimal pH value is between 5.3 and 6.4 which would also make it useful for preservation of low-acid foods.

Lysozyme is also economical to produce. Lysozyme in hens' egg white accounts for about 3.5% of the protein, and it is obtained from egg whites in high yield and purity by the use of cation-exchange resins. The use of lysozyme in foods began in the 1960s after efficient industrial methods were developed to extract lysozyme from egg albumen. Lysozyme currently has limited applications in the food industry. It is used to prevent gas formation or 'late blowing' by *Clostridium tyrobutyricum* during ripening of certain cheeses, particularly the varieties provolone, grana padano, Emmental, Gouda, and some others (Carini *et al.*, 1985). In the development of blowing, *C. tyrobutyricum* ferments lactate in cheese with production of carbon dioxide, hydrogen, butyric and acetic acids. In many cheeses, particularly hard-pressed varieties, the accumulation of gases can cause the cheese to 'blow' or crack the wheels forming unwanted fissures. Addition of lysozyme to the cheesemilk prevents the development of gas and cracking by *C. tyrobutyricum*. Lysozyme is usually added to cheesemilk at a concentration of 20-35 mg per litre of milk, and it associates nearly entirely with the curd giving a maximum level of 400 mg/kg of cheese. When used at this level, the enzyme does not inhibit activity of the starter organisms or beneficial microbes involved in cheese ripening, and does not affect the desired physical or organoleptic properties of the cheese (Carini *et al.*, 1985). Lysozyme has largely replaced formaldehyde and nitrate in many countries in Europe for the inhibition of spore-formers.

Lysozyme is presently approved for use in cheese in Austria, Australia, Belgium, Denmark, Finland, France, Germany, Italy, and Spain. Other countries are expected to grant approval for use soon including Canada, Norway, and the

United Kingdom. Recently, the Joint FAO/WHO Expert Committee on Food Additives (JECFA) ruled during the 39th meeting in 1992 that lysozyme is safe to use in food. It is expected that the GRAS (generally regarded as safe) status of lysozyme for use in foods will be obtained and documented in the Federal Register of the United States in 1993.

Antimicrobial Activity of Lysozyme against Spoilage Microorganisms

It is interesting that lysozyme is highly specific in its action and is effective against relatively few species of bacteria associated with foods. As mentioned earlier, lysozyme has considerable activity against the cheese contaminant *C. tyrobutyricum*, practically preventing outgrowth of spores without inhibiting desirable starter and secondary cultures required for the ripening of the cheese. This specificity for one undesirable bacterial species in a complex food containing several beneficial species emphasises lysozyme's very specific action as a food preservative. This specificity mandates that it will have highly selective roles as a food preservative, perhaps acting against only a few undesirable organisms.

Inhibition of Mesophilic and Thermophilic Spore-Formers

Our laboratory showed that lysozyme had antibacterial activity against a limited number of food pathogens and spoilage bacteria (Hughey and Johnson, 1987). In an initial screening in complex media, three bacterial species of 15 examined, *Bacillus stearothermophilus*, *Clostridium thermosaccharolyticum*, and *C. tyrobutyricum*, were found to be completely inhibited by lysozyme hydrochloride at 20 or 200 mg per litre (Table 15.1). We confirmed that two strains of *C. tyrobutyricum* isolated from blowing cheese in Italy were highly susceptible to lysozyme. A population of cells suspended in phosphate buffer was rapidly lysed by 100 mg of lysozyme per litre. Growing cultures of *C. tyrobutyricum* also lysed rapidly on introduction of lysozyme to the culture (Hughey and Johnson, 1987).

We also found during screenings that certain thermophilic spore-forming bacteria were exceptionally susceptible to lysozyme (Hughey and Johnson, 1987). These bacteria contain highly heat-resistant endospores and can cause severe spoilage of processed foods. *Bacillus stearothermophilus* can cause bland-sour spoilage, whereas *C. thermosaccharolyticum* can cause gaseous spoilage of canned vegetables (Frazier and Westhoff, 1988). Our laboratory found that cell suspensions of the vegetative cells were lysed in phosphate buffer, although a certain proportion of the population appeared more resistant. Growing cells were also inhibited on exposure of the cells to lysozyme.

Table 15.1. Antibacterial activity of egg white lysozyme.

Organisms strongly inhibited
Bacillus stearothermophilus
Clostridium thermosaccharolyticum
Clostridium tyrobutyricum

Organisms moderately inhibited
Bacillus cereus (some uninhibited)
Campylobacter jejuni
Clostridium botulinum types A, B, and E
Listeria monocytogenes
Yersinia enterocolitica

Organisms not inhibited
Clostridium butyricum
Clostridium perfringens
Escherichia coli O157:H7
Klebsiella pneumoniae
Salmonella typhimurium
Staphylococcus aureus
Vibrio cholerae

Since the thermophilic spore-formers can cause spoilage, it may be possible to decrease the numbers of these spores in foods by adding lysozyme during processing. We found in preliminary experiments that exposing spores of *B. stearothermophilus* to lysozyme at 70°C in phosphate buffer resulted in inactivation. Since lysozyme is relatively heat-resistant, particularly at low pH, it may be possible to reduce the thermal energy requirements during canning or other heat processes of foods by inclusion of lysozyme during processing.

Inhibition of Yeasts

Many yeasts grow at low pH and produce gas, ethanol, and other undesirable products in foods resulting in spoilage (Frazier and Westhoff, 1988). Certain yeasts can be pathogenic to humans and also cause uncomfortable infections, for example yeast infections of the vagina. Although yeasts do not contain peptidoglycan, the principle substrate of lysozyme, some yeasts contain chitin in their cell surface which can be acted upon by lysozyme.

We conducted preliminary screenings against several yeasts including *Saccharomyces cerevisiae, Zygosaccharomyces rouxii, Candida krusei, Candida glabrata, Candida albicans, Cryptococcus neoformans,* and *Geotrichum candidum.*

Lysozyme alone had weak activity against certain of these yeasts, and the activity was potentiated by the compounds lysolecithin or poly-L-lysine (Table 15.2). These observations demonstrate the importance of synergistic agents in lysozyme activity, and suggest that it may be possible to develop antimicrobial formulations active against yeasts that are important in foods and in medicine. Activity of lysozyme against the pathogenic yeasts *Candida albicans* and *Cryptococcus neoformans* has been noted by others (Barbara and Pellegrini, 1976).

Table 15.2. Inhibition of yeasts by lysozyme.

Condition	Growth ($A_{660\ nm}$)			
	S.c.[1]	Z.r.	C.a.	C.n.
Control	0.72	0.37	0.72	0.52
Lysozyme, 100 p.p.m.	0.75	0.38	0.72	0.51
Lysolecithin, 100 p.p.m.	0.56	0.11	1.00	0.22
Lysolecithin, 100 p.p.m. + lysozyme	0.61	0.085	0.77	0.18
Poly-L-lysine, 100 p.p.m.	0.70	0.41	0.90	0.08
Poly-L-lysine, 100 p.p.m. + lysozyme	0.73	0.43	0.90	0.015
Conalbumin, 100 p.p.m.	0.80	0.37	1.05	0.33
Conalbumin, 100 p.p.m. + lysozyme	0.85	0.30	0.87	0.25

[1] S.c., *S. cerevisiae*; Z.r., *Z. rouxii*; C.a., *C. albicans*; C.n., *C. neoformans*

Antimicrobial Activity of Lysozyme against Pathogenic Bacteria

Our laboratory has investigated the potential use of lysozyme to control pathogenic bacteria that are known to be associated with foods and to cause foodborne disease. Lysozyme could potentially serve as a useful secondary preservative, particularly in refrigerated foods, to prevent the growth or survival of pathogenic bacteria.

We initially screened several species of food-associated pathogens including *Bacillus cereus, Campylobacter jejuni, Clostridium botulinum* serotypes A, B, and E, *Clostridium perfringens* type A, *Escherichia coli* O157:H7, *Klebsiella pneumoniae, Listeria monocytogenes, Salmonella typhimurium, Staphylococcus aureus, Vibrio cholerae,* and *Yersinia enterocolitica.* Of these species, lysozyme had weak inhibitory activity against *C. jejuni, C. botulinum, L. monocytogenes,* and *Y. enterocolitica.* Most of our work has focused on *L. monocytogenes* because of the importance of this organism to human public health and its relatively common contamination of foods.

Inhibition of *Listeria monocytogenes*

Listeria monocytogenes is a Gram-positive coccobacillus found frequently in the environment that was recognised in the 1980s as a serious cause of foodborne disease in humans. Diseases due to *L. monocytogenes* usually occur in individuals who are pregnant, suffering from cancer or AIDS, or being treated with immunosuppressants, or in extremes of age (Farber and Peterkin, 1991; Schuchat *et al.*, 1991). Stillbirths and infections of the central nervous system and meningoencephalitis are among the most common manifestations. Listeriosis is associated with a distressing fatality rate of nearly 25%. Recently, a carefully conducted case-control study of 18 million persons in the United States estimated that the annual incidence is 1850 illnesses and 425 deaths annually in the United States (Schuchat *et al.*, 1992). It has been estimated that the annual costs of listeriosis cases in the United States in 1989 was $480 million or more, including $35.6 million for medical care, $221.2 million for productivity losses, and at least $225.5 million for losses due to mental distress (Food Chemical News, March 12, 1990). Because of the importance of *L. monocytogenes* in causing disease and fatalities, and our current lack of knowledge regarding its virulence properties, a zero tolerance has been emphasized in the United States for *L. monocytogenes* in ready-to-eat foods. The zero-tolerance enforced by the FDA and USDA has resulted in numerous recalls of foods and economic difficulty for some sectors of the food industry.

Contaminated foods implicated in epidemic listeriosis have included coleslaw, soft cheeses, milk, and pâté (Farber and Peterkin, 1991; Schuchat *et al.*, 1992). *Listeria monocytogenes* has been frequently isolated from food-processing plants and finished food products. Since *L. monocytogenes* can grow slowly at refrigeration temperatures, its control is of particular importance in minimally processed refrigerated foods with extended shelf-life. For control of *L. monocytogenes* in these foods, it is often necessary to use barriers such as chemical preservatives.

In 1986, our laboratory observed that hen egg lysozyme lysed cells of four strains of *L. monocytogenes* suspended in buffer. All four strains isolated during food poisoning outbreaks were susceptible. These initial results indicated the potential for specific inhibition of *L. monocytogenes* in foods. However, growth of the pathogen was delayed but not inhibited when the four strains were inoculated into media containing lysozyme at 20 or 200 mg/litre. We found that inhibition of growth by lysozyme was promoted by certain chemicals including ethylenediamine tetraacetic acid (EDTA), lactic acid, trypsin, proteinase K, conalbumin, and lactoferrin.

The susceptibility of the pathogen to lysozyme was found to depend on the physiological status of the bacterium and the medium or food in which it was suspended. Cells grown at low temperatures were found to be more sensitive than those grown at ambient temperatures (R.J. Premaratne and E.A. Johnson, unpublished; Smith *et al.*, 1991). These results were intriguing because they suggested that lysozyme may be most effective in refrigerated foods. The susceptibility of *L. monocytogenes* to lysozyme was also increased by treating cells with trypsin or proteinase K, indicating that removal of certain proteins influences the ability of lysozyme to penetrate and hydrolyse the peptidoglycan. The growth substrate is also important in determining the sensitivity of *L. monocytogenes* to lysozyme. Using a defined medium, R.J. Premaratne and E.A. Johnson (unpublished observations) have found that starvation for iron increases susceptibility. Starvation for iron changed the cell surface structure as indicated by altered protein profiles and binding of haemin. These results have indicated that *L. monocytogenes* is susceptible to hen egg lysozyme but that the degree of sensitivity depends on the growth medium and the physiological state of the pathogen.

Several investigators have found that *L. monocytogenes* survives poorly in egg albumen. Erickson and Jenkins (1992) found that *L. monocytogenes* was inactivated in unsalted pasteurised egg white and whole egg. Wang and Shelef (1991a) concluded that the antilisterial activities of egg albumen were primarily due to lysozyme and were enhanced by ovomucoid, conalbumin, and alkaline pH. These results support that lysozyme is listericidal, and suggest that other unidentified components in egg albumen may enhance its activity.

The nature of the food in which *L. monocytogenes* is present markedly influences the sensitivity of the pathogen to lysozyme. Our laboratory demonstrated that lysozyme effectively killed or prevented growth of *L. monocytogenes* Scott A in several vegetables, but it was less effective in animal-derived foods that were tested (Hughey *et al.*, 1989). The activity of lysozyme in foods was enhanced by EDTA, and lysozyme together with 1-5 mM EDTA killed inoculated populations of 10^4 cells per gram in fresh corn, fresh green beans, shredded cabbage, shredded lettuce, and carrots during incubation at 5°C. Control incubations not containing lysozyme supported growth of *L. monocytogenes* to 10^6 to 10^7 cells per gram. In carrots and cabbage, lysozyme alone was bactericidal, but in fresh corn and green beans it permitted growth. Overall, lysozyme was effective in inactivating *L. monocytogenes* in most vegetable products.

Lysozyme was less effective in controlling *L. monocytogenes* in pork sausage (bratwurst) (Hughey et al., 1989) and in pork, beef, or turkey frankfurters (V.L. Hughey and E.A. Johnson, unpublished data). In pork sausage, lysozyme + EDTA was bacteriostatic for 2-3 weeks, but the treatment did not kill significant numbers of cells and did not prevent eventual growth. Lysozyme delayed growth of *L. monocytogenes* in frankfurters, and the inclusion of lysozyme and EDTA resulted in a reduction in the final populations of *L. monocytogenes*. These results suggest that lysozyme may have preservative activity against *L. monocytogenes* in meat products but more research is needed to improve the activity.

Other researchers have attempted to control *L. monocytogenes* or other undesirable organisms in meats by use of lysozyme and synergistic agents.

A variety of substances have been suggested for use in combination with lysozyme including chelating agents such as EDTA, phytic acid, or polyphosphates, preservatives such as butyl-*p*-hydroxybenzoate, Parabens, benzoic acid, sorbic acid, and other compounds including amino acids, hydrogen peroxide, and organic acids (Proctor and Cunningham, 1988). Monticello (1989) reported that lysozyme (20-100 µg/g) together with nisin (200-300 International Units (IU) per gram) was effective in controlling *L. monocytogenes* in culture broths and in raw pork sausages, but close examination of the data indicated that the response of nisin/lysozyme was essentially the same as that of lysozyme alone. The work pointed out the need to develop synergistic factors that would enhance the activity of lysozyme in foods.

There is considerable interest in controlling *L. monocytogenes* in milk and dairy products since these foods have been responsible for large outbreaks of listeriosis. Soft-ripened cheeses have been associated with listeriosis in humans (Schuchat *et al.*, 1992), and these foods could benefit by incorporation of a preservative active against the pathogen. Hughey *et al.* (1989) prepared Camembert cheese containing *L. monocytogenes* at 10^4/g and combinations of lysozyme and EDTA. In the control cheeses without lysozyme, the numbers of *L. monocytogenes* remained relatively constant or showed a slight decline in numbers during the first 3 weeks of ripening, but thereafter *L. monocytogenes* grew steadily and eventually reached more than 10^6/g. The increase in *L. monocytogenes* was probably associated with the rise in pH that occurs during ripening of the soft cheeses. The cheeses containing *L. monocytogenes* and lysozyme showed a decline in cell numbers over the first 3 weeks of ripening (Hughey *et al.*, 1989). After about 35 days, *L. monocytogenes* started to grow and eventually reached high numbers. The initial numbers of *L. monocytogenes* in these trials was about 10^4/g, which was higher than would be found in naturally contaminated milk. In subsequent experiments, we have prepared several samples of cheeses and cheese curd containing varying levels of *L. monocytogenes*. Even with initial numbers as low as 10-50 *L. monocytogenes* per gram of milk, the pathogen persisted in Camembert cheese in the presence of lysozyme and eventually grew in the cheese. Heat-treatment of milk containing *L. monocytogenes* at 60, 62.5, or 65°C for 15 s resulted in inactivation of the pathogen in cheese curds prepared from the milk whereas *L. monocytogenes* survived better in cheese curds prepared from unheated milk. Surface-ripened cheese is a complex food substrate, and further work is needed to evaluate the effectiveness of lysozyme in controlling *L. monocytogenes*.

Listeria monocytogenes survives extremely well in milk and is also highly resistant to lysozyme in this food (Carminati and Carini, 1989; Hughey *et al.*, 1989). Carminati and Carini (1989) found that lysozyme was active against *L. monocytogenes* in milk only under certain conditions. No activity of lysozyme was found against *L. monocytogenes* strains in whole milk, whereas killing was found in acid milk (pH 5.3), but this was strain-dependent. Growth of *L. monocytogenes* was reduced by lysozyme at 4°C. The results showed that lysozyme may play a role in inhibition of *L. monocytogenes* in milk and possibly other dairy products.

These data suggest that the resistance of the organism in milk is altered, or that factors in milk protect against inactivation by lysozyme. We have carried out an extensive study of the factors influencing susceptibility of *L. monocytogenes* in milk. The Scott A strain was highly resistant in milk, but was readily inactivated by lysozyme in phosphate buffer. Some component of whole milk appeared to protect *L. monocytogenes* from lysis. We found that removal of metals from whole milk by treatment with a Chelex resin increased the susceptibility of the pathogen to lysozyme.

Replacement of calcium or magnesium to the demineralised milk restored resistance of the pathogen. Heat treatment of *L. monocytogenes* at 60°C for 15 s markedly sensitized the cells to lysozyme in demineralised milk. The results indicate that minerals or mineral-associated components protect *L. monocytogenes* from inactivation by lysozyme and heat in milk. Experiments were also done to evaluate the ability of lysozyme to inactivate *L. monocytogenes* in laboratory-prepared Cheddar-type and Camembert-type cheeses. Low numbers of *L. monocytogenes* (10-100/g) were eliminated or greatly decreased in Camembert cheese containing lysozyme, particularly if the cells were first heat-treated. *L. monocytogenes* numbers decreased markedly in Cheddar-type cheese curd when the cells were heat-stressed and lysozyme was added but lysozyme alone only slightly decreased numbers. These results indicate that *L. monocytogenes* can be reduced or eliminated in cheese-type systems by heat treatment and lysozyme.

Listeria monocytogenes has been found associated with fish and shellfish including fresh, smoked, marinated, salted, and frozen products (Farber and Peterkin, 1991; Jemmi, 1990). Lysozyme and EDTA were tested for inhibition of *L. monocytogenes* and spoilage microflora in fresh cod (Wang and Shelef, 1991b). Raw cod fillets were dipped for 10 min in a solution of lysozyme hydrochloride (3 mg/ml) and/or EDTA disodium salt (5-25 mM), patted dry, and then inoculated with *L. monocytogenes* at ~10^3 cells/g. The investigators found that lysozyme inhibited growth of *L. monocytogenes* but did not affect the natural microflora. Treatment with EDTA alone delayed slime formation but was not strongly inhibitory against *L. monocytogenes*. These results highlight the specificity lysozyme can have against target organisms, for example, *L. monocytogenes*.

Antimicrobial Activity of Lysozyme against *Clostridium botulinum*

The public health safety of low-acid foods depends on the control of *C. botulinum* to prevent botulinal toxin formation. Certain strains of 'group II' *C. botulinum* comprising serotypes B, E, and F could grow in these foods and present a botulism hazard (Conner *et al.*, 1989). Commercial pressures are also promoting the use of fish proteins in meat and poultry products, which could increase the possibility of a hazard from *C. botulinum* where it did not exist before. Furthermore, the growth potential of *C. botulinum* may be higher in newer products such as vacuum packaged and modified atmosphere foods (Farber, 1991) than in traditional food products. The food industry would benefit from the development of safety systems to control *C. botulinum* in low-acid refrigerated foods. Over the past 5 years our laboratory has studied the inhibition of *C. botulinum* by lysozyme.

We found that egg white lysozyme killed or prevented growth of spores or vegetative cells of *C. botulinum* in culture medium and in several foods. For maximum antibotulinal activity, it is necessary to use chelators or other synergistic agents. Chelating agents that increased the activity of lysozyme included diethylenetriaminepentaacetic acid (DTPA), EDTA, and cysteine. Lysis of *C. botulinum* in culture medium was enhanced at lower temperature and pH. Lysozyme plus potential synergistic agents including chelators, nisin, parabens, potassium sorbates, and avidin were tested in various foods for control of *C. botulinum*. Lysozyme showed varying degrees of antibotulinal activity in animal-derived foods including turkey suspensions in water, fresh pork sausage (bratwurst), and salmon. In turkey suspension, lysozyme hydrochloride at 100 µg/g delayed botulinal toxin formation for 11-14 days at 20°C. The delay in toxin formation was increased by an additional 14 days by synergists including cysteine (1-5 mM), proline (1 mM), and sodium lactate (2%). In cod suspensions, toxin production by *C. botulinum* type E spores was delayed for 1-3 days by lysozyme alone. The delay in toxin formation was increased by 7-21 days by inclusion of potentiating compounds including EDTA (1-5 mM), nisin (250-500 IU/g), butyl-paraben (1000 µg/g), or tripolyphosphate (1000-3000 µg/g).

Lysozyme also had antibotulinal activity in certain vegetables including asparagus, potatoes, tomatoes, and mushrooms. In potato suspensions, toxin formation was not delayed by lysozyme alone, but the combination of lysozyme and the chelators EDTA (1 mM) or DTPA (1 mM) delayed toxin formation for 3 weeks. Similar results were observed in asparagus suspensions. Our data indicate that considerable potential exists for control of *C. botulinum* in low-acid foods but further trials are needed in commercial products.

Antimicrobial Activity of Lysozyme against Other Pathogenic Bacteria and Its Use as a Preservative in Food

The use of lysozyme for the control of certain pathogenic bacteria has been reported. Teotia and Miller (1975) found that treating turkey drumsticks with 0.1% lysozyme eliminated *Salmonella senftenberg* 775W within 3 h at 22°C. *Salmonella* populations and spoilage microflora were reduced on chicken broiler parts by treatment with lysozyme and EDTA (Samuelson *et al.*, 1985; Chander and Lewis, 1980). Ng and Garibaldi (1975) showed that populations of *Staphylococcus aureus* were reduced in whole egg which they attributed to agglutination of the cells by lysozyme. Nakamura *et al.* (1990) developed an interesting lysozyme-dextran conjugate that acted on both Gram-negative and Gram-positive bacteria, and they recommended that the conjugate could have activity in foods. Intensive research has been extended to the use of lysozyme as a preservative in food systems, particularly in Japan. Lysozyme may act directly on the organisms, or it may induce immunologic factors that prevent gastro-enteritis and allergies. Lysozyme has been used as a preservative in fresh fruits and vegetables, potato salad, soy sauce, tofu, seafoods, meats, and sausages (reviewed in Proctor and Cunningham, 1988).

Discussion and Conclusions

Presently, there is considerable interest within the food industry of using natural antimicrobial preservation systems that are not toxic in the human diet and yet are inhibitory to undesirable microorganisms. The prolongation of shelf-life for quality and the control of pathogenic bacteria and spoilage organisms in refrigerated foods poses an immense challenge to the food industry. It also presents an opportunity to develop new methods of food preservation.

Egg white lysozyme has desirable properties as a food preservative. It has been consumed in the diet of humans for thousands of years and is considered a safe food ingredient. Lysozyme has effective antibacterial activity against a limited spectrum of bacteria and fungi. This specificity is an important property since it enables the destruction of undesirable organisms while enabling the growth of beneficial bacteria and yeasts. Lysozyme's activity can be enhanced by certain substances and its spectrum against target organisms broadened by using chemical synergists or physical treatments that render organisms susceptible.

Lysozyme currently is used primarily for the control of late-blowing in several varieties of cheese. We have found that lysozyme effectively lyses and kills the pathogens *Listeria monocytogenes* and *Clostridium botulinum*. These pathogens are of considerable concern in refrigerated foods with extended shelf-lives. The prodigious use of lysozyme should be valuable for increasing the shelf-life of foods while maintaining the public health safety of our food supply.

Acknowledgements

I appreciate the cooperation of several people who have contributed to this study in my laboratory including Virginia Hughey, Ann Larson, Carl Malizio, Ranga Premaratne, Alvaro Quinones, and Pamela Wilger. I also thank Societa' Prodotti

Antibiotici for support of this research and especially Drs Ettore Bosschetti and Ernani Dell'Acqua for many valuable discussions.

References

Alderton, G. and Fevold, J. (1946) Direct crystallization of lysozyme from egg white and some crystalline salts of lysozyme. *J. Biol. Chem.* 164:1-5.

Barbara, L. and Pellegrini, R. (1976) *Fleming's Lysozyme. Biological Significance and Therapeutic Applications.* [English translation]. Edizone Minerva Medica, Torino, Italy.

Blake, C.C.F., Koenig, D.F., Mair, G.A., North, A.C.T., Phillips, D.C., and Sarma, V.R. (1965) Structure of hen egg-white lysozyme. A three-dimensional fourier synthesis at 2 Å resolution. *Nature* 206:757-761.

Carini, S., Mucchetti, G., and Neviani, E. (1985) Lysozyme: activity against clostridia and use in cheese production - a review. *Microbiol. Alimen. Nutr.* 3:299-320.

Carminati, D. and Carini, S. (1989) Antimicrobial activity of lysozyme against *Listeria monocytogenes* in milk. *Microbiol. Aliment. Nutr.* 7:49-56.

Chander, R. and Lewis, N.F. (1980) Effect of lysozyme and sodium EDTA on shrimp microflora. *Appl. Microbiol. Biotechnol.* 10:253-258.

Conner, D.E., Scott, V.N., Bernard, D.T., and Kautter, D.A. (1989) Potential *Clostridium botulinum* hazards associated with extended shelf-life refrigerated foods: a review. *J. Food Safety* 10:131-153.

Erickson, J.P. and Jenkins, P. (1992) Behavior of psychrotrophic pathogens *Listeria monocytogenes, Yersinia enterocolitica,* and *Aeromonas hydrophila* in commercially pasteurized eggs held at 2, 6.7 and 12.8°C. *J. Food Prot.* 55:8-12.

Farber, J.M. (1991) Microbiological aspects of modified atmosphere packaging technology - a review. *J. Food Prot.* 54:58-70.

Farber, J.M. and Peterkin, P.I. (1991) *Listeria monocytogenes*, a food-borne pathogen. *Microbiol. Rev.* 55:476-511.

Fleming, A. (1922) On a remarkable bacteriolytic element found in tissues and secretions. *Proc. Roy. Soc. (London)*, Ser. B. 93:306-317.

Frazier, W.C. and Westhoff, D.C. (1988) *Food Microbiology*, 4th edition. McGraw-Hill Books Co., New York.

Hughey, V.L. and Johnson, E.A. (1987) Antimicrobial activity of lysozyme against bacteria involved in food spoilage and food-borne disease. *Appl. Environ. Microbiol.* 53:2165-2170.

Hughey, V.L., Wilger, P.A., and Johnson, E.A. (1989) Antibacterial activity of hen egg white lysozyme against *Listeria monocytogenes* Scott A in foods. *Appl. Environ. Microbiol.* 55:631-638.

Hutton, M.T., Chehak, P.A., and Hanlin, J.H. (1991) Inhibition of botulinum toxin production by *Pediococcus acidilacti* in temperature abused refrigerated foods. *J. Food Safety* 11:255-267.

Jemmi, T. (1990) Actual knowledge of *Listeria* in meat and fish products. *Mitt. Geb. Lebensmittelunters Hyg.* 31:144-157.

Jollès, P. and Jollès, J. (1984) What's new in lysozyme research - always a model system, today as yesterday. *Mol. Cell. Biochem.* 63:165-189.

Jollès, J., Jauregui-Adell, J., Bernier, I., and Jollès, P. (1963) La structure chimique du lysozyme de blanc d'oeuf de poule: etude detaillée. *Biochim. Biophys. Acta* 78:668-669.

Leistner, L. (1978) Microbiology of ready-to-serve foods. *Fleischwirtschaft* 58:2088-2111.

Moberg, L. (1989) Good manufacturing practices for refrigerated foods. *J. Food Prot.* 52:363-367.

Monticello, D.J. (1989) Control of microbial growth with nisin/lysozyme formulations. European Patent Application #89123445.2.

Nakamura, S., Kato, A., and Kobayashi, K. (1990) Novel bifunctional lysozyme-dextran conjugate that acts on both Gram-negative and Gram-positive bacteria. *Agric. Biol. Chem.* 54:3057-3059.

Ng, H. and Garibaldi, J.A. (1975) Death of *Staphylococcus aureus* in liquid whole egg near pH 8.0. *Appl. Microbiol.* 29:782-786.

Palumbo, S.A. (1986) Is refrigeration enough to restrain foodborne pathogens? *J. Food Prot.* 49:1003-1009.

Phillips, D.C. (1966) The three-dimensional structure of an enzyme molecule. *Sci. Amer.* 215:78-90.

Proctor, V.A. and Cunningham, F.E. (1988) The chemistry of lysozyme and its use as a food preservative and pharmaceutical. *CRC Crit. Rev. Food Sci. Nutr.* 26:359-395.

Salton, M.J.R. (1957) The properties of lysozyme and its action on microorganisms. *Bacteriol. Rev.* 21:82-98.

Samuelson, K.J., Rupnow, J.H., and Froning, G.W. (1985) The effect of lysozyme and ethylenediaminetetraacetic acid on *Salmonella* on broiler parts. *Poultry Sci.* 64:1488-1490.

Schuchat, A., Swaminathan, B., and Broome, C.V. (1991) Epidemiology of human listeriosis. *Clin. Microbiol. Rev.* 4:169-183.

Schuchat, A., Deaver, K.A., Wenger, J.D., Plikaytis, B.D., Mascola, L., Pinner, R.W., Reingold, A.L., Broome, C.V., and the Listeria Study Group. (1992) Role of foods in sporadic listeriosis. I. Case-control study of dietary risk factors. *J. Amer. Medical Assoc.* 267:2041-2045.

Scott, V.N. (1989) Interaction of factors to control microbial spoilage of refrigerated foods. *J. Food Prot.* 52:431-435.

Smith, J.L., McColgan, C., and Marmer, B.S. (1991) Growth temperature and action of lysozyme on *Listeria monocytogenes*. *J. Food Sci.* 56:1101, 1103.

Teotia, J.S. and Miller, B.F. (1975) Destruction of salmonellae on poultry meat with lysozyme, EDTA, X-ray, microwave, and chlorine. *Poultry Sci.* 54: 1388-1394.

Wang, C. and Shelef, L.A. (1991a) Factors contributing to antilisterial effects of raw egg albumen. *J. Food Sci.* 56:1251-1254.

Wang, C. and Shelef, L.A. (1991b) Behavior of *Listeria monocytogenes* and the spoilage microflora in fresh cod fish treated with lysozyme and EDTA. Poster presentation, American Society of Microbiology, Dallas, TX.

Chapter Sixteen

Large-Scale Preparation of Sialic Acid and Its Derivatives from Chalaza and Delipidated Egg Yolk

L.R. Juneja, M. Koketsu, H. Kawanami, K. Sasaki, M. Kim, and T. Yamamoto[1]

Central Research Laboratories, Taiyo Kagaku Co., Ltd, 1-3 Takaramaci, Yokkaichi 510 (Mie) and [1] Department of Biotechnology, Fukuyama University, Fukuyama 729-02 (Hirosima), Japan

Abstract Egg yolk and chalaza mixed with egg yolk membrane were studied for preparation of *N*-acetylneuraminic acid (Neu5Ac). The delipidated hen egg yolk (DEY; 500 kg containing 0.2% w/w Neu5Ac) was hydrolysed with HCl (pH 1.4) at 80°C and neutralised with NaOH. The mixture was filtered and electro-dialysed. The filtrate was applied to a column of Dowex HCR-W2 (20-50 mesh), followed by a column of Dowex 1X8 (200-400 mesh). The latter column was eluted with a linear gradient of formic acid. The eluates containing Neu5Ac were concentrated and finally rotary evaporated at 40°C. The residue was then lyophilised to yield 500 g of Neu5Ac (50% yield). The mixture of chalaza and egg yolk membrane (800 kg) processed with screw decanter (125 kg) was processed by the above method to yield 300 g of Neu5Ac (70% yield; >98% purity). TLC, HPLC, NMR, and IR spectra of the product obtained showed that Neu5Ac was the sole derivative present in egg. On the other hand, Neu5Ac-conjugated oligosaccharides (300 g) were isolated from the water-insoluble fraction of the DEY (10 kg) wherein the fraction was treated with a commercial protease preparation, the filtrate was further treated with ethanol (75%, v/v), and the precipitate was processed for the isolation of Neu5Ac conjugates.

(*Key words:* sialic acid, *N*-acetylneuraminic acid, delipidated egg yolk, chalaza, egg yolk membrane)

Introduction

Many chemical and biotechnological companies are exploring the role of sialic acid especially *N*-acetylneuraminic acid (Neu5Ac) and related compounds in cell functions to create a new class of carbohydrate-based drugs. Sialic acid as a component of glycoconjugates (mainly glycoproteins and glycolipids) is a biological material. Recently, the interest in sialic acid compounds has increased, as their role in many biological processes is being elucidated and further explored (Schauer, 1982, 1985, 1987; Reuter *et al.*, 1982). Researchers of several groups have recently identified carbohydrate molecules that may lead to development of effective oral anti-inflammatory drugs with minimum side effects (Phillips *et al.*, 1990; Lowe *et al.*, 1990). They found that cells carrying a carbohydrate ligand containing Neu5Ac, known as sialyl-Lewis X, were able to bind endothelial leucocyte adhesion molecule-1.

Sialic acids are considered to be important components for the protection of life. Sialic acids possess various biological functions based on their negative charge, acting as receptors for microorganisms, toxins, and hormones, or masking receptors and immunological recognition sites of molecules and cells (Schauer, 1987). Sialic acids have also attracted increasing attention in pathological processes e.g. sialidosis (Cantz, 1982; Renlund *et al.*, 1983) and malignancies (Erbil *et al.*, 1985). The determination of these sugars may be significant for the diagnosis and prognosis of cancer (Erbil *et al.*, 1985).

Only a few natural sources such as swallow's nest (Czarniecki and Thornton, 1977), milk protein (Deya *et al.*, 1989), and a microbial fermentation process (Tsukada *et al.*, 1990) have been explored for practical production of sialic acid. In order to fulfil the increasing demand for sialic acid, an economical large-scale method of preparing it will have to be developed.

The avian egg is considered to be a chemical storehouse because it is composed of various important chemical substances that form the basis of life. In the present investigation, delipidated egg yolk (DEY) powder, or chalaza (a ropy structure of cloudy appearance spiralling from the yolk into albumen to both ends of the egg) mixed with egg yolk membrane, a waste product of egg processing plants, was used as a raw material and a flow process for the large-scale preparation of Neu5Ac and Neu5Ac conjugates was established.

Materials and Methods

Materials

Authentic Neu5Ac of *Escherichia coli* origin was purchased from Nakalai Tesque Inc., Tokyo, Japan (Tsukada *et al.*, 1990). Authentic *N*-glycolylneuraminic acid

(Neu5Gc) from porcine submaxillary gland was obtained from Sigma Chemical Co. (St Louis, MO, USA).

Analysis of Sialic Acid

Colorimetric method. The Neu5Ac was quantified by a modification of the periodate thiobarbituric acid (TBA) method (Warren, 1959). Neu5Ac liberated by heating the material in 0.05 N H_2SO_4 at 80°C was monitored colorimetrically.

TLC method. TLC was carried out by employing 1-propanol:1 M ammonium hydroxide:water (6:2:1) as developing solvent with aluminium sheets coated with silica gel 60 (E. Merck, Germany), and detection was achieved by spraying with 5% sulphuric acid in methanol and heating for 5 min at 150°C.

HPLC method. HPLC analysis was performed with an EYELA PLC-5D (Tokyo Rikakikai Co., Tokyo, Japan) attached to a Hitachi D-2500 integrator operated at 206 nm. The column used was Bio-Rad HPX-87H (Bio-Rad Laboratories, Tokyo, Japan) and the mobile phase (0.003 N H_2SO_4) flowed at 0.65 ml/min under a pressure of 57 kg/cm² at ambient temperature. Sample (5 µl) was injected and run for 30 min.

NMR method. NMR spectra were measured in D_2O with a JEOL-GSX-400 instrument (JEOL Co., Tokyo, Japan) operated by the pulsed Fourier-transform method). Sodium 4,4-dimethyl-4-silapentanesulphonate was used as an internal standard.

IR spectroscopy. IR spectra (KBr method) of samples were examined with a Shimadzu IR 460 spectrometer (Shimadzu Corporation, Kyoto, Japan).

Measurement of sialic acid contents in various fractions of hens' egg. Fresh eggs were separated into egg white and egg yolk. Various fractions of egg, e.g. shell, shell membrane, egg white, egg yolk, yolk membrane, and chalaza were collected (Juneja *et al.*, 1991). Chalaza was carefully separated using tweezers and incubated with three volumes of 2% KCl for 2 days at 5°C and then it was washed with water to remove contaminating egg white and egg yolk and used as purified chalaza. The egg yolk fraction was squeezed through a cloth to collect egg yolk membrane and the membrane fraction remaining on the cloth was purified by repeatedly washing with water. The fractions of egg white, egg yolk, yolk membrane, chalaza, shell membrane, and egg shell were finally vacuum dried and pulverised. Egg yolk membrane and chalaza of hen, quail, pigeon, and gamecock were collected as mentioned above and their Neu5Ac contents were determined.

Measurement of sialic acid contents in several organs of the hen. A hen was sacrificed and its ovary, oviduct, and crest were separated, minced, and hydrolysed for measurement of Neu5Ac content.

Large-scale preparation of Neu5Ac from chalaza and egg yolk membrane mixture. Hen eggs were separated into egg yolk and egg white with an egg processor and passed through a stainless steel filter (mesh size = 0.8-1.0 mm). The chalaza and yolk membrane entrapped in the filter were collected and used for preparation of Neu5Ac as a model of large-scale method (Fig. 16.1).

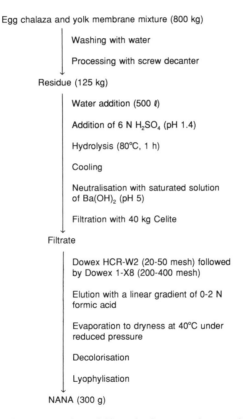

Egg chalaza and yolk membrane mixture (800 kg)

Washing with water

Processing with screw decanter

Residue (125 kg)

Water addition (500 ℓ)

Addition of 6 N H₂SO₄ (pH 1.4)

Hydrolysis (80°C, 1 h)

Cooling

Neutralisation with saturated solution of Ba(OH)₂ (pH 5)

Filtration with 40 kg Celite

Filtrate

Dowex HCR-W2 (20-50 mesh) followed by Dowex 1-X8 (200-400 mesh)

Elution with a linear gradient of 0-2 N formic acid

Evaporation to dryness at 40°C under reduced pressure

Decolorisation

Lyophylisation

NANA (300 g)

Fig.16.1. Large-scale preparation of Neu5Ac from a mixture of chalaza and egg yolk membrane.

Eight hundred kilograms of crude mixture of egg chalaza and yolk membrane was washed with water to remove contaminating egg white and egg yolk and the residue (125 kg) was homogenised with three volumes of water. The suspension was acidified to pH 1.4 with 6 N sulphuric acid and heated at 80°C for 1 h. After cooling, saturated barium hydroxide solution was added to the suspension until pH 5.0 was attained and the suspension was filtered. The filtrate was applied to a column of Dowex HCR-W2 (25 × 40 cm, H form, 20-50 mesh), followed by a column of Dowex 1-X8 (35 × 50 cm, formate form, 200-400 mesh). The latter column was washed with water, then eluted with a linear gradient of formic acid from 0 to 2 N. Neu5Ac was eluted within a narrow range of around 0.8 N formic acid. The eluates containing Neu5Ac were evaporated to dryness at 40°C under reduced pressure. The residue obtained was decolorised with activated charcoal powder and then lyophilised.

Preparation of Neu5Ac from DEY

Laboratory scale. The DEY (2 kg) was homogenised with three volumes of water. The suspension was acidified to pH 1.4 with 6 N HCl and heated for one hour at 80°C. After cooling, 6 N NaOH was added until pH 6.0 was attained and the suspension was filtered. The filtrate was electrodialysed in two steps. The final desalinate was applied to a column of Dowex HCR-W2 (1.5 × 30 cm, H form, 20-50 mesh), followed by a column of Dowex 1-X8 (1.5 × 30 cm, formate form, 200-400 mesh). The latter column was washed with water and then eluted with a linear gradient of formic acid from 0 to 2 N at a flow rate of 2 ml/min (Corfield *et al.*, 1978). The Neu5Ac fractions were collected, and then evaporated at 45°C under reduced pressure. The residue was decolorised with activated charcoal and then lyophilised.

Large scale. The milling of DEY was performed with a power mill (TYPE P-3, screen size of S16 mm; Showa Giken Co., Osaka, Japan). The finely powdered DEY (500 kg containing 0.2% w/w Neu5Ac) was hydrolysed with HCl (pH 1.4) at 80°C and neutralised with NaOH (pH 6.0). The mixture was filtered. The filtrate was electrodialysed in two steps using TS-10-360 (Tokuyama Soda Co., Tokyo, Japan) as a large-scale preparation method. The electrodialysed solution was applied to a column of Dowex HCR-W2 (25 × 40 cm, H form, 20-50 mesh), followed by a column of Dowex 1-X8 (35 × 50 cm, formate form, 200-400 mesh). The latter column was washed with water and then eluted with a linear gradient of formic acid (0-2 N). The eluates containing Neu5Ac were combined and concentrated with reverse osmosis membrane NTR-7250 (Nitto Denko Co., Tokyo, Japan). The concentrate was finally rotary-evaporated at 40°C. The residue was then lyophilised to result in 500 g Neu5Ac with a yield of 50% (Fig. 16.2). The purity of Neu5Ac was >98% (TBA method). The purity of the compound was confirmed by TLC, NMR, IR spectroscopy, and HPLC.

Analysis and Measurement of Amino Acids Interfering during Isolation of Neu5Ac

Colorimetric method. The composition of amino acids in the eluates obtained from the Dowex 1-X8 column was measured by the ninhydrin method of Horstmann (1979).

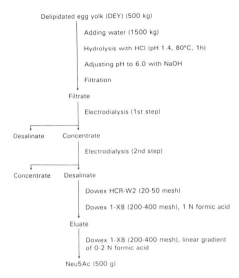

Fig. 16.2. Large-scale preparation of Neu5Ac from DEY.

Amino acid analyser. The amino acid composition of the hydrolysed DEY and the eluates of chromatography was determined (Moore and Stein, 1963); the samples were hydrolysed with 6 N HCl at 110°C for 80 h in sealed, evacuated tubes. The analyses were carried out on the resulting hydrolysate using an amino acid analyser (Hitachi 835-50; Hitachi Co., Tokyo, Japan).

Electrodialysis of the hydrolysed DEY. The hydrolysed DEY, adjusted to the desired pH, was electrodialysed with TS-2-10 and TS-10-360 (Tokuyama Soda Co., Tokyo, Japan) in both the laboratory scale and large-scale preparation of Neu5Ac. In electrodialysis, anion and cation exchange membranes are laid alternately in many gaskets as boundary layers, then NaCl solution is supplied between the gaskets and membrane, and direct current is passed through the electrodes at the both ends. The electrodialysis system consisted of ion-exchange membranes 'NEOSEPTA' CM-1 & AFN and CM-2 & ACS in the first step and second step, respectively. The system was of the fully automatic batch type. A conductivity meter was placed in the desalinate vessel of electrodialysis. The electrodialysis was carried out at room temperature (around 25°C).

Preparation of Neu5Ac conjugates. The Neu5Ac conjugates were prepared by a mild digestion of water-soluble fraction of DEY with protease (pancreatin, Amano Co., Japan) at 45°C for 4 h. The enzyme digest was centrifuged, the supernatant was treated with ethanol (75%), and sialyl-oligosaccharides were partially purified from the precipitate with ethanol using column (Dowex 1-X2, 1.5 × 30 cm, 100-200 mesh) chromatography.

Results

Neu5Ac Contents in Various Fractions of Eggs

The Neu5Ac contents of chalaza and egg yolk membrane in eggs of several avian species (hen, quail, pigeon, and gamecock) were examined (Table 16.1). The highest Neu5Ac contents in the egg yolk membrane (2.19%) and chalaza (3.21%) were found in gamecock and pigeon, respectively. Measurement of Neu5Ac contents in various fractions of hens' egg (fresh egg) showed that Neu5Ac was distributed in all parts of egg but mainly located in chalaza and egg yolk membrane (Table 16.2). Whole egg white and isolated ovomucin were found to contain 0.1% and 2% Neu5Ac, respectively, although the proportion of ovomucin in egg white was only 0.9%.

Table 16.1. Sialic acid content in eggs of various avian species.

Species	Egg yolk membrane (% dry matter)	Chalaza (% dry matter)
Hen (*Gallus domesticus*)	1.0	2.40
Quail (*Cotumix cotumix japonica*)	1.25	3.79
Pigeon (*Columba livia*)	1.85	3.21
Gamecock (Japanese bantam) (*Gallus domestica*)	2.19	2.68

Table 16.2. Sialic acid content in various fractions of hens' egg.

Fraction	Quantity (kg/tonne egg wet wt basis)	Sialic acid (g)	(% dry matter)
Egg shell	104.8	2.98	0.004
Shell membrane	6.2	1.22	0.07
Egg white	603.5	60.35	0.10
Egg yolk	281.0	267.00	0.19
Egg yolk membrane	2.3	3.52	1.80
Chalaza	2.2	3.96	2.40

Neu5Ac Contents in Various Organs of Hen

To screen for alternative Neu5Ac-rich material, some of the reproductive organs and crest of hen were examined. The Neu5Ac content of ovary was also found to be as high as 0.076% on a wet wt basis compared with 0.017% in oviduct. No Neu5Ac was detected in the crest.

Preparation of Neu5Ac from Chalaza Mixture

A mixture of hen's egg chalaza and egg yolk membrane, a waste product from an egg processing plant (Taiyo Kagaku Co.), was used as a raw material for sialic acid preparation (Fig. 16.1). Neu5Ac content (430 g in 125 kg batch of washed raw material containing chalaza and egg yolk membrane) was 0.34% (wet wt basis), although its contents in egg yolk membrane and in chalaza were 0.15% and 0.18% (wet wt basis), respectively, when fresh egg was used as the starting material in the laboratory.

Composition of DEY

The DEY, a by-product of egg processing, was analysed. The DEY contained 55.4% water, 36.4% protein, and 6.6% lipids (polar lipids 6.3%, nonpolar lipids 0.3%) and 1.6% total sugars containing 0.2% Neu5Ac.

Effect of pH on Hydrolysis of DEY

The effect of pH of the hydrolysis of DEY on the purity of Neu5Ac produced and on Neu5Ac adsorption capacity to a Dowex 1-X8 anion column was examined. The eluates obtained from the hydrolysed DEY samples adjusted to pH 6.0 showed the highest purity (55%) and adsorption (27%).

Electrodialysis of the Hydrolysed DEY prior to Column Chromatography

The filtrate of hydrolysed DEY, adjusted to pH 6.0, was electrodialysed to remove salt and other compounds to improve the resin adsorption capacity of Neu5Ac. Electrodialysis of the filtrate was performed in two steps with ion exchange membranes of different mesh size. The first electrodialysis step, using 'NOESEPTA' CM-1 and AFN membranes, was carried out down to a conductivity of 70 µS/cm. This membrane allows salts and low molecular substances to separate from macromolecules by transfer of ions. Ionic compounds having molecular weights of less than about 350, e.g. Neu5Ac, salts, amino acids, etc. were obtained in the concentrate portion, while the desalinate contained high molecular weight compounds. The concentrate containing Neu5Ac-containing salts obtained in the first step was subjected again to electrodialysis using 'NEOSEPTA' CM-2 and ACS membranes until a conductivity of 240 µS/cm (second step) was obtained. At the above conductivity value, Neu5Ac was separated from salts and almost complete recovery of Neu5Ac was obtained in the desalinate (Fig. 16.3).

Anion Exchange Chromatography of DEY after Acid Treatment and Electrodialysis

The desalinated fraction from acid-treated DEY obtained after the second step of electrodialysis, was applied to an anion exchange resin Dowex 1-X8 (1.5 × 30 cm, formate form) column. The column was washed with water and elution of Neu5Ac was carried out with 1 N formic acid. The eluates obtained were analysed for Neu5Ac (TBA method) and amino acids. The elution profile of Neu5Ac from the Dowex 1-X8 column showed that the contaminants including some amino acids were eluted before Neu5Ac, but a significant portion of other amino acid contaminant overlapped the Neu5Ac peak (Fig. 16.4).

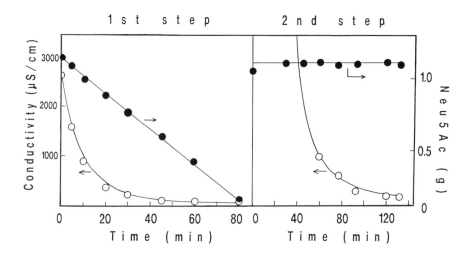

Fig. 16.3. Time-course of two step electrodialysis of hydrolysed DEY.
O-O, conductivity (µS/cm); ●-●, Neu5Ac (g).

The significant contaminants which appeared overlapping with Neu5Ac included glutamic acid, aspartic acid, leucine, serine, and threonine (Table 16.3). The amino acid composition of the concentrated eluates after the first anion chromatography showed that glutamic acid was the major contaminant left and that the other contaminants were reduced (Fig. 16.4).

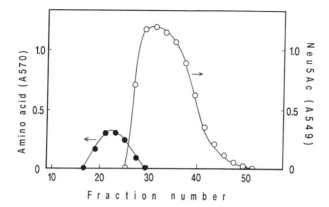

Fig. 16.4. Elution profile of Neu5Ac from Dowex 1-X8 anion exchange column by the procedure mentioned in Fig. 16.2. Neu5Ac contents were measured by the TBA method. Amino acids were measured by the ninhydrin method. ●-●, amino acids; O-O, Neu5Ac.

Table 16.3. Amino acid composition of the hydrolysed DEY and the first chromatography eluate.

Aminoacid	Hydrolysed DEY[a] (mol%)	Eluates of first chromatography[b] (mol%)
Asp	11.45	9.76
Thr	6.76	2.65
Ser	7.34	3.62
Glu	13.40	55.85
Pro	4.10	2.03
Gly	5.07	3.21
Ala	7.33	2.69
Cys	1.79	0.62
Val	7.26	1.86
Met	2.32	0.41
Ile	4.56	2.28
Leu	7.80	1.07
Tyr	3.52	0.62
Phe	3.10	2.79
Lys	6.21	5.34
His	2.40	4.34
Arg	4.10	2.03

[a] After electrodialysis.
[b] The concentration eluates obtained after first chromatography on Dowex 1-X8 (200-400 mesh).

Cation Exchange Chromatography prior to Anion Exchange Column Chromatography

Cation exchange chromatography was performed to remove compounds like glutamic acid as they interfere during anion exchange chromatography. The filtrate of hydrolysed DEY, adjusted to pH 6.0, was electrodialysed in two steps. The filtrate was applied to a column of Dowex HCR-W2 followed by Dowex 1-X8 and rechromatographed on a Dowex 1-X8 column yielding 99% pure Neu5Ac. When no cation exchange column was used, the purity of Neu5Ac was decreased to 94% (Table 16.4).

Table 16.4. Purification of Neu5AC by ion exchange chromatography.

Column resin	Dowex HCR-W2	Dowex 1-X8
First chromatography	Dowex 1-X8	
Purity of Neu5Ac (%) (1st chromatography)	84	75
Purity of Neu5Ac (%) (Rechromatography)[a]	99	94

[a] Rechromatography was carried out using Dowex 1-X8.

Purity of Sialic Acid

TLC method. TLC of sialic acid isolated from DEY and chalaza-yolk membrane mixture showed that Neu5Ac was the sole compound of the Neu5Ac type in the egg; no *N*-glycolyl was found in the compound (Fig. 16.5).

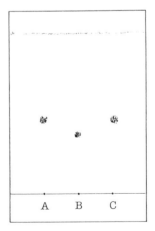

Fig. 16.5. TLC chromatogram of sialic acids. (A, authentic Neu5Ac; B, authentic Neu5Gc; C: sialic acid from large-scale process.)

NMR method. The purity of Neu5Ac was confirmed by ^{1}H- and ^{13}C-NMR spectra. The ^{1}H- and ^{13}C-NMR spectra were measured in D_2O at 400 MHz and 100 MHz, respectively (JEOL GSX-400 instrument; JEOL Co., Tokyo, Japan). The spectra matched with those of authentic Neu5Ac (Fig. 16.6).

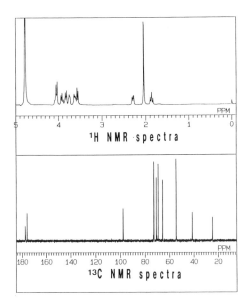

Fig. 16.6. ^{1}H- and ^{13}C-NMR spectra of sialic acid. (Solvent, D_2O; internal standard, sodium 4,4-dimethyl 4-siapentanesulphonate.)

Other methods. HPLC and IR spectra confirmed the purity of Neu5Ac.

Preparation of Neu5Ac Conjugates

Approximately 300 g of partially purified Neu5Ac conjugates were obtained from 10 kg of DEY (Fig. 16.7).

Discussion

It seems an important fact that approximately 300 g of Neu5Ac was obtained in one batch whose purity of Neu5Ac was more than 97% (TBA method). In the present experiment, the increased Neu5Ac contents of the preparation might be attributed to the processing of the raw material through a screw decanter and to the method developed in this study. The yield of Neu5Ac obtained in the final purification was 70% based on the weight of the mixture of washed egg chalaza and yolk membrane.

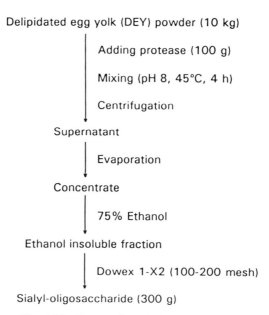

Delipidated egg yolk (DEY) powder (10 kg)

Adding protease (100 g)

Mixing (pH 8, 45°C, 4 h)

Centrifugation

Supernatant

Evaporation

Concentrate

75% Ethanol

Ethanol insoluble fraction

Dowex 1-X2 (100-200 mesh)

Sialyl-oligosaccharide (300 g)

Fig. 16.7. Preparation of Neu5Ac conjugates.

The egg yolk, after processing to obtain lipid material in an industrial scale, has until now been discarded without any further utilisation. Quantitative analysis of various egg fractions revealed that egg yolk contained the highest amount of total Neu5Ac on a per egg basis but this fact has not been noticed so far perhaps because of the presence of large amount of lipids, proteins, and other ionic compounds in the fraction. The presence of lipids leads to emulsification of yolk during processing for the isolation of Neu5Ac. The DEY used in the present study may be an attractive source for Neu5Ac, because of the very small amount of lipids present.

The mixture of chalaza and egg yolk membrane was hydrolysed with sulphuric acid and neutralised with barium hydroxide. In the case of using DEY, Neu5Ac is considered to be released by hydrolysis of glycosidic linkages with sialyl residues by acid (pH 1.4) at 80°C and pH of the hydrolysed DEY with hydrochloric acid was adjusted with sodium hydroxide. The results were as good as those using sulphuric acid hydrolysis. The effect of pH of the hydrolysed DEY was examined for the anion exchange adsorption efficiency and purity of Neu5Ac eluates. The maximum adsorption and highest purity was found to be achieved at pH 6.0. Contaminating high molecular weight compounds, salts, or remaining lipids were separated from Neu5Ac by electrodialysis.

Although two-step electrodialysis using cation and anion exchange membranes was very effective in removing large amounts of contaminants which interfere with the anion chromatographic purification of Neu5Ac, the desalinated solution finally obtained still contained charged, low molecular weight compounds. The major contaminant at this step was glutamic acid. The cation exchanger 'Dowex HCR-W2 (20-50 mesh)' column which was used prior to the chromatography by anion exchange Dowex 1-X8 (200-400 mesh) column was found to remove the major contaminants like glutamic acid. By this procedure, we achieved the purification of Neu5Ac (>98% pure) with a yield of 500 g Neu5Ac from 500 kg of the DEY (Fig. 16.2).

The analysis by TLC and NMR spectra of the compound obtained from both the DEY and chalaza mixture revealed that there was no glycolyl or O-acetyl group in the compound. In the present experiment, reverse osmosis membrane was used for concentration at room temperature. The technology resulted in very fast and efficient concentration of solution containing Neu5Ac as compared with the technique of rotary evaporation.

The chemical synthesis of sialic acid at present is complicated and expensive and thus it would be preferable to isolate sialic acid from natural sources. Various organs of hen were investigated for Neu5Ac preparation but it seemed to be difficult to collect large amounts of reproductive organs as by-products for industrial scale production of sialic acid. Although Neu5Ac contents were highest in quail's egg chalaza among the avian species examined, the hen's egg was considered best for large-scale preparation of Neu5Ac. Hens' eggs are, in a sense, mass produced and are commonly used as a food and processed globally. The DEY, chalaza, and egg yolk membrane have remained unutilised so far. Therefore, these fractions can be used as an excellent source for industrial scale preparation of Neu5Ac and Neu5Ac conjugates. Preparation of Neu5Ac from these materials is also attractive because of its relatively simple purification procedure.

References

Cantz, M. (1982) Sialidoses. *Cell. Biol. Monogr.* 10:307-320.

Corfield, A.P., Beau, J.M., and Schauer, R (1978) Desialylation of glycoconjugates using immobilized *Vibrio cholerae* neuraminidase. *Hoppe-Seyler's Z. Physiol. Chem.* 359:1335-1342.

Czarniecki, M.F. and Thornton, E.R. (1977) Carbon-13 nuclear magnetic resonance spin-lattice relaxation in the *N*-acetylneuraminic acid probes for internal dynamics and conformational analysis. *J. Am. Chem. Soc.* 99:8273-8279.

Deya, E., Ikeuchi, Y., Yoshida, H., Hiraoka, Y., and Uchida, S. (1989) Preparation of pure sialic acid. Jpn. Kokai Tokkyo Koho 40,491.

Erbil, K.M., Jones, J.D., and Klee, G.G. (1985) Use and limitation of serum total and lipid bound sialic acid concentration as markers for colorectal cancer. *Cancer* 55:404-409.

Horstmann, H.J. (1979) A precise method for the quantitation of proteins taking into account their amino acid composition. *Anal. Biochem.* 96:130-138.

Juneja, L.R., Koketsu, M., Nishimoto, K., Kim, M., Yamamoto, T., and Itoh, T.

(1991) Large-scale preparation of sialic acid from chalaza and egg yolk membrane. *Carbohydr. Res.* 214:179-189.

Lowe, J.B., Stoolman, L.M., Nair, R.P., Larsen, R.D., Berhend, T.L., and Marks, R.M. (1990) ELAM-l-dependent cell adhesion to vascular endothelium determined by transfected human fucosyltransferase cDNA. *Cell* 63:475-484.

Moore, M. and Stein, W.H. (1963) Chromatographic determination of amino acids by the use of automatic recording equipment. *Methods Enzymol.* 6: 819-831.

Phillips, M.L., Nudelman, E., Gaeta, F.C.A., Perez, M., Singhal, A.K., Hakomori, S., and Paulson, J.C. (1990) ELAM-l mediates cell adhesion by recognition of a carbohydrate ligand, sialyl-LeX. *Science* 250:1130-1132.

Renlund, M., Aula, P., Raivio, K.O, Autio, S., Sainio, K., Rapola, J., and Koskela, S.L. (1983) Salla disease: a new lysosomal storage disorder with disturbed sialic acid metabolism. *Neurology* 33:57-66.

Reutter, W., Kittgen, E., Bauer, C., and Gerok, W. (1982) Biological significance of sialic acids. *Cell. Biol. Monogr.* 10:263-305.

Schauer, R. (1982) Chemistry, metabolism and biological function of sialic acids. *Adv. Carbohydr. Chem. Biochem.* 40:131-234.

Schauer, R. (1985) Sialic acids and their role as biological masks. *Trends Biochem. Sci.* 10:357-360.

Schauer, R. (1987) Analysis of sialic acid. *Methods Enzymol.* 138:132-161.

Tsukada, Y., Ohta, Y., and Sugimori, T. (1990) Microbial production of sialic acid and related enzyme, and their application for the development of clinical diagnostics. *Nippon Nogeikagaku Kaishi* (in Japanese) 64:1437-1444.

Warren, L. (1959) The thiobarbituric acid assay of sialic acids. *J. Biol. Chem.* 234:1971-1975.

Chapter Seventeen

Isolation, Characterisation, and Application of Hen Egg Yolk Polyclonal Antibodies

J.R. Clarke, R.R. Marquardt, A.A. Frohlich, A. Oosterveld, and F.J. Madrid[1]

Department of Animal Science and [1]Department of Food Science, University of Manitoba, Winnipeg, Manitoba, Canada R3T 2N2

Abstract Antibodies directed against ochratoxin A (OA) were isolated from egg yolks of immunised hens using a yield-optimised isolation procedure. They were characterised by examining sensitivity and specificity for OA, and applied to an ELISA of swine finisher diets. A yield of 70-80% active yolk antibody (IgY) with an estimated purity of >85% could be obtained by extracting yolk with a mixture of aqueous buffer and chloroform and purified by selectively precipitating IgY with polyethylene glycol-8000. The antibody was found in an indirect competitive ELISA to be ochratoxin-group specific. A simplified, methanol-based OA extraction procedure was tested and found suitable for routine diet analysis. The ELISA developed was capable of quantitatively detecting OA at levels of 50 p.p.b. The ELISA developed was found to correlate well with the more conventional high performance liquid chromatographic techniques for OA analysis in diet.
(Key words: antibodies, egg, yolk, isolation, application, ochratoxin A, ELISA)

Introduction

Immunobased techniques such as ELISA (enzyme-linked immunosorbent assay) offer several distinct advantages over conventional chromatographic procedures, specifically the lowered requirement for sample clean-up, improved sample throughput, and reduced operating costs. In general the high limits of detection possible, the unique specificities for a particular analyte, or the potential for group detection and the general versatility of the technique have resulted in their widespread and rapid adoption for mycotoxin analysis (Pestka, 1988; Chu, 1990; Kaufman and Clower, 1991). Ochratoxin A (OA) is one such mycotoxin that is routinely monitored due to its high toxicity and widespread occurrence in foods and feeds (Kuiper-Goodman and Scott, 1989). The development of an economical OA ELISA would be propitious for rigorous human food and animal feed screening. One of the main restrictions governing the adoption of these immunobased techniques is the need for large amounts of antigen-specific antibody. Conventional collection of antigen-specific antibody has always depended on routine bleeding of animals as is the case for rabbits or expensive murine cell culture techniques. Since it is known that laying hens protect their offspring by transferring antigen-specific antibodies from their sera to the yolk (Patterson *et al.*, 1962), the limitations of routine bleeding and cell culturing can be avoided. These yolk antibodies (IgY) have been isolated, purified, and then successfully applied in immunoassays for a wide variety of antigens which have ranged from high molecular weight proteins derived from mammalian or bacterial sources to low molecular weight antigens (Gassman *et al.*, 1990).

The first part of this paper describes an optimised egg yolk IgY isolation and purification procedure. The second part describes an OA-specific IgY. The last part of this paper will illustrate the practical application of this antibody in an ELISA for OA.

Materials and Methods

Immunisation of the Laying Hens

White laying hens (Shaver's SX 288) were immunised with a bovine serum albumin-ochratoxin A (BSA-OA) conjugate prepared according to Chu *et al.* (1982). Three injections of 2 mg (protein basis) were given intramuscularly over a period of 70 days.

Indirect Competitive ELISA for OA Quantification

The amount of OA was assessed by measuring its ability to inhibit competitively the binding of anti-OA IgY to a plate coated with OA-protein conjugate in a manner similar to that described by Gendloff *et al.* (1986).

Preparation of OA-Contaminated 'Diets' for ELISA and HPLC Analysis

Swine finisher diets known to contain 89% barley, 7.7% soybean meal, 2.5% vitamin premix were mixed thoroughly with 0.72% of OA-contaminated crushed soybeans. A 'simplified' methanol-based extraction procedure was developed and involved the extraction of a 5 g sample of diet with 30 ml of methanol-water (80:20) in closed capped Nalgene centrifuge tube for 30 min. The mixture following shaking was centrifuged for 30 min at 5000 g and the supernatant was recovered. For HPLC analysis, aliquots of supernatant were removed, diluted with methanol, and filtered through a 0.5 μm filter. Samples for ELISA were applied directly to ELISA plate wells and analysed as aforementioned.

HPLC for OA

Analytes were separated on a Beckmann 5 μm Ultrasphere ODS reversed-phase C18 25 cm × 4.6 mm analytical column using an isocratic mobile phase comprised of 70% methanol and 30% distilled water adjusted to pH 2.1 using H_3PO_4. Ten to 50 μl of diet extract was injected and chromatographed using a mobile solvent flow rate of 1.6 ml/min at a column temperature of 40°C. OA peaks were detected using fluorescence emission at 333 nm and integrated.

Results and Discussion

A Simplified Procedure for Preparation of IgY

The original study systematically investigated factors affecting the yield and purity of IgY when isolated from the yolk of laying hens that had been immunised with the protein conjugated mycotoxin, OA. The method developed was an improvement of the isolation protocol of Polson (1990) yielding substantially higher yields of active and near pure IgY. The most practical ratio for the extraction of yolk with solvents was 1(10 ml):1(10 ml):4(40 ml), respectively for yolk, chloroform, and PBS. IgY activity appeared to be quantitatively precipitated with polyethylene glycol-8000 (Sigma Chemical Co., St Louis, MO) at a concentration of greater than 14% (w/w). Maximal purity was obtained after two successive precipitations with a final recovery of 75 ± 5.4%.

Gradient gel (4-20%) SDS-PAGE done according to Laemmli (1970) followed by scanning densitometry demonstrated that the final isolated product contained less than 15% non-IgY protein bands. The purity of the commercial standard (Sigma Chemical Co., St Louis, MO) was only slightly better than that obtained with this yield-optimised purification procedure.

The main disadvantage of this isolation procedure was its use of a toxic extraction solvent. Chloroform is considered toxic and requires safe handling and disposal. Although restricted in regards to feed additive applications (Yolken *et al.*, 1988; Kuhlmann *et al.*, 1988), the optimised procedure is an excellent means of preparing reagent grade antibody which should be ideally suitable for immunodiagnostic applications since the procedure yields high amounts of near pure and active IgY in relatively short amounts of time.

Characterisation of Anti-OA IgY Antibody

An indirect competitive ELISA type format was selected for the characterisation of anti-OA IgY. Assay sensitivity was found to be generally >1 g/ml. The sensitivity of this antibody preparation was generally similar to those obtained using rabbits polyclonal antibodies and mouse monoclonal antibodies for the same mycotoxin (Chu, 1990).

The anti-OA IgY was found to be group-specific as it was found to cross-react with several other OA-like molecules. The anti-OA IgY was found to cross-react 100%, 100%, 33%, 400%, 0%, and 1.9% with ochratoxins A, B, α, C, L-phenylalanine, and citrinin, respectively. The cross-reactivities are of interest as ochratoxin C and B are toxic whereas ochratoxin α and L-phenylalanine are not (Marquardt *et al.*, 1991). The presence of ochratoxin α in a sample could lead to an increased signal in the ELISA, potentially resulting in a false positive; however, it has not been reported to occur in grain. The presence of ochratoxin α is a consequence of OA proteolytic digestion or hydrolysis by microflora (Marquardt *et al.*, 1991). Although this molecule is unlikely to exist under field conditions it is a good indicator of past OA contamination. Ochratoxin α has been shown to be present in faeces, rumen, and urine of animals fed diets containing OA (Screemannarayana *et al.*, 1988; Xiao *et al.*, 1991). The low cross-reactivity observed with citrinin, a structurally and biologically related mycotoxin, is also of interest as OA and citrinin are often found to co-occur and have similar modes of inhibition (Madhyastha *et al.*, 1990). Thus there exists the possibility of discriminating between agents of toxicosis. The likely possibility of other structurally related molecules existing in the sample requires that a putative positive for OA be confirmed using techniques such as HPLC and/or TLC, since no one single technique is definitive enough to be absolutely sure of results. Thus this antibody preparation is best suited for the development of a pre-screening type assay. The practical aspects of using a pre-screening ELISA to identify putative positives will markedly decrease the sample load required for conventional types of analysis.

Application of Anti-OA IgY Antibody

Screening swine finisher diets for OA-like molecules using a quantitative ELISA is a practical application for this unique and potentially valuable antibody. Swine finisher diets contain a wide variety of components which could conceivably be encountered in routine grain testing.

The assay developed used a simple extraction procedure which minimised the amount of time required and avoided the use of extensive sample clean-up. The ELISA was evaluated by directly comparing the recoveries with a conventional HPLC analysis. The results indicate strong correlation of OA recoveries from naturally contaminated swine finisher diets (r = 0.99) (Table 17.1). Ochratoxin A in naturally fortified swine finisher diets could be detected directly at levels of 50 p.p.b. or higher using the simplified sample preparation procedure. This level of detectability is not unreasonable since OA tolerance limits in grains have been imposed by several countries and range from as high as 300 p.p.b. to levels as low as 10 p.p.b. (van Egmond, 1991). The obvious advantages in ease, time, and cost makes this procedure attractive for quality control applications were tolerance limits are specified and large sample through-put is required.

Table 17.1. Analysis of OA in naturally fortified swine finisher diets.

OA contamination (µg/kg)	Analysis of OA by ELISA (µg/kg)[a]	Analysis of OA by HPLC (µg/kg)[a]
50	55 ± 5	50 ± 2
100	115 ± 20	100 ± 10
200	215 ± 30	220 ± 10
1000	815 ± 80	850 ± 20
2000	1945 ± 150	1860 ± 110
5000	4565 ± 400	4860 ± 290

[a] Values are the mean and standard deviation of at least three replicas.

References

Chu, F.S. (1990) Immunoassays for mycotoxins: current state of the art, commercial and epidemiological applications. *Vet. Hum. Toxicol.* 32:42-50.

Chu, F.S., Lau, H.P., Fan, T.S., and Zhang, G.S. (1982) Ethylenediamine modified bovine serum albumin as protein carrier in the production of antibody against mycotoxins. *Immunol. Methods* 55:73-78.

Gassmann, M., Thömmes, P., Weiser, T., and Hübscher, U. (1990) Efficient production of chicken egg yolk antibodies against a conserved mammalian protein. *FASEB J.* 4:2528-2532.

Gendloff, E.H., Casale, W.L., Ram, B.P., Tai, J.H., Pestka, J.J., and Hart, L.P. (1986) Hapten-protein conjugates prepared by the mixed anhydride method. Cross-reactive antibodies in heterologous antisera. *J. Immunol. Methods* 92:15-20.

Kaufman, B.M. and Clower, M., Jr (1991) Immunoassay of pesticides. *J. Assoc. Off. Anal. Chem.* 74:239-247.

Kuhlmann, R., Wiedemann, V., Schmidt, P., Wanke, R., and Losch, U. (1988) Chicken egg antibodies for prophylaxis and therapy of infectious intestinal diseases. *J. Vet. Med.* B35:610-616.

Kuiper-Goodman, T. and Scott, P.M. (1989) Risk assessment of the mycotoxin ochratoxin A. *Biomed. Environm. Sci.* 2:179-248.

Laemmli, U.K. (1970) Cleaving of structural proteins during the head assembly of bacteriophage T4. *Nature* 227:680-685.

Madhyastha, S.M., Marquardt, R.R., Frohlich, A.A., Platford, G., and Abramson, D. (1990) Effects of different cereal and oilseed substrates on the growth and production of toxins by *Aspergillus allutaceus* and *Penicillium verrucosum. J. Agric. Food Chem.* 38:1506-1510.

Marquardt, R.R., Frohlich, A.A., and Abramson, D. (1990) Ochratoxin A: an important western Canadian storage mycotoxin. *Can. J. Physiol. Pharmacol.* 68:991-999.

Patterson, R., Youngner, J.S., Weigle, W.O., and Dixon, F.J. (1962) Antibody production and transfer to egg yolk in chickens. *J. Immunol.* 89:272-278.

Pestka, J.J. (1988) Enhanced surveillance of foodborne mycotoxins by immunochemical assay. *J. Assoc. Off. Anal. Chem.* 71:1075-1081.

Polson, A. (1990) Isolation of IgY from the yolks of eggs by a chloroform polyethylene glycol procedure. *Immunol. Invest.* 19:253-258.

Screemannarayana, O., Frohlich, A.A., Vitti, T.G., Marquardt, R.R., and Abramson, D. (1988) Studies of the tolerance and deposition of ochratoxin A in young calves. *J. Anim. Sci.* 66:1703-1711.

van Egmond, H.P. (1991) Worldwide regulations for ochratoxin-A. In: Castegnaro, D.M., Palestina, R., Dirherimer, G., Chermozensky, I.N., and Bartch, H. (eds), *Mycotoxins, Endemic Nephropathy and Urinary Tumours.* International Agency for Research on Cancer, Lyon, pp. 139-143.

Xiao, H., Marquardt, R.R., Frohlich, A.A., Phillips, G.D., and Vitti, T.G. (1991) Effect of a hay and grain diet on the rate of hydrolysis of ochratoxin A in the rumen of sheep. *J. Anim. Sci.* 69:3706-3714.

Yolken, R.H., Leister, F., Wee, S., Miskuff, R., and Vonderfecht, S. (1988) Antibodies to rotaviruses in chicken's eggs: a potential source of antiviral immunoglobulins suitable for human consumption. *Pediatrics* 81(2):291-295.

Chapter Eighteen

A New Process for IgY Isolation from Industrially Separated Egg Yolk Including Automation and Scale-up

J. Fichtali, E.A. Charter, K.V. Lo, and S. Nakai

Departments of BioResource Engineering and Food Science, University of British Columbia, Vancouver, British Columbia, Canada V6T 1Z4

Abstract Industrially separated egg yolk was diluted and the water-soluble proteins separated from the insoluble fraction by sedimentation. The effect of pH and dilution on the residual lipid and egg yolk immunoglobulin (IgY) recovery in the supernatant was studied. The supernatant was filtered and then applied to a column packed with a cation exchanger within an automated liquid chromatography system which was devised in our laboratory. Overall recoveries of 60-65% were obtained for 60-70% purities using a 50 ml column. Scaling up to a 1.5 litre column did not significantly affect the separation results. However, using two 1.5 litre columns in series did improve the purity of the product. The cation exchangers assessed were found to be excellent for large-scale separation of IgY due to their low cost and superior flow properties. In addition, the automated system used confers many advantages, the key elements being the time saving and the overall control of the process.
(*Key words:* separation, immunoglobulins, chromatography, egg yolk, scale-up)

Introduction

Specific antibodies have found an increasing application in diagnosis and pure research for the detection, estimation, and isolation of different molecules in food and biological systems. Another potential application is the oral administration of antibodies against bacteria or viruses through infant formula to prevent gastrointestinal infection (Otani *et al.*, 1991). Generally the specific antibodies are prepared from sera or colostrum of immunised mammals such as rabbits, goats, sheep, and cows. However, to promote new uses and secure sufficient supply of specific antibodies, sources other than animal blood may be required. It is known that serum antibodies of hyperimmunised hens are efficiently transferred and accumulated in the egg yolk (Williams, 1962; Rose *et al.*, 1974; Losch *et al.*, 1986) providing a valuable source of antibody. This antibody, equivalent to chicken serum IgG, is termed IgY to denote that it is found in the yolk. IgY and mammalian IgG differ in their molecular size, isoelectric point, susceptibility to proteolysis, and ability to bind to mammalian complement and protein A (Hassl *et al.*, 1987; Shimizu *et al.*, 1988; Hatta *et al.*, 1990; Otani *et al.*, 1991). It was found (Rose *et al.*, 1974) that the immunoglobulin content of the yolk is much higher than that of the hen's serum. A hen laying several eggs per week could provide antibodies equivalent to repeated bleedings, without any difficulty or harm to the animal. Up to 75 mg of IgY per egg yolk from immunised hens were extracted by the polyethylene glycol procedure (Carroll and Stollar, 1983).

Although it is recognised that using egg yolk as a source of antibodies offers many advantages as compared with conventional sources (Rose *et al.*, 1974; Polson *et al.*, 1980; Jensenius *et al.*, 1981; Gottstein and Hemmeler, 1985), IgY has not been used as frequently as one might expect. The reason for this may be the lack of satisfactory isolation procedures. Although several methods have been assessed for the isolation and purification of IgY from egg yolk (Polson *et al.*, 1980, 1985; Jensenius *et al.*, 1981; Bade and Stegemann, 1984; Hassl and Aspock, 1988; McCannel and Nakai, 1989, 1990; Hatta *et al.*, 1990; Akita and Nakai, 1992), most of them are of limited value for large-scale production of IgY since they are either tedious and expensive, lead to poor separation, or do not lend themselves to food applications. We therefore sought to develop a simple process for IgY isolation which may be automated and easily scaled up, using industrially separated yolk to take into account for contamination by egg white.

In this study, an automated process using cation exchange chromatography was developed for the separation of IgY from other egg proteins. The process is relatively efficient, simple, and inexpensive, and yet has demonstrated potential for use at a much larger scale.

Separation of Lipoproteins from Water-Soluble Egg Yolk Proteins

Egg yolk is a complex food which can be separated by high speed centrifugation into particles called granules and a clear fluid supernatant, the plasma (Stadelman and Cotterill, 1977). The plasma is about 78% of the total yolk and composed of low-density lipoproteins (LDL) and livetins which are lipid-free globular proteins. The livetins, which represent about 10.6% of the total yolk solids, were separated and identified as ß-livetin (α2-glycoprotein), α-livetin (serum albumin), and γ-livetin which is known as IgY (Martin and Cook, 1958; Williams, 1962). Therefore, the first step for IgY separation should be the removal of lipoproteins from the water-soluble proteins. Industrially separated egg yolk obtained from a local egg breaking plant (Vanderpols Eggs Ltd, Aldergrove, BC, Canada) was diluted with distilled water and the pH adjusted with 0.1 N HCl. After 24 h of sedimentation at 4°C, the supernatant was collected and analysed for total lipids (Hatta *et al.*, 1990) and the IgY concentration determined by radial immuno-diffusion (Kwan *et al.*, 1991) using chicken serum IgG (ICN, Cleveland, OH, USA) as a standard.

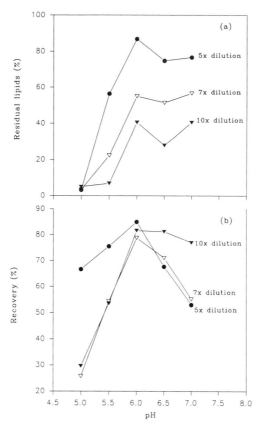

Fig. 18.1. Effect of pH and water dilution on residual lipids (a) and IgY recovery (b) in egg yolk supernatant after 24 h sedimentation.

Figure 18.1 shows the effect of pH and dilution on residual lipids (Fig. 18.1a) and IgY recovery (Fig. 18.1b). At pH 5.0, the effect of dilution on residual lipids is negligible (Fig. 18.1a); however, at higher pH values, the amount of residual lipids drops significantly when the dilution is increased. Both residual lipids (Fig. 18.1a) and IgY recovery (Fig. 18.1b) increase with increasing pH up to pH 6.0 at which maximum values occur for all dilutions. Although up to 80% IgY recovery may be obtained at pH 6.0 after 24 h sedimentation, the lipid content at this particular pH is very high, and only at pH 5.0 for all dilutions and pH 5.5 for 10× dilution is the residual lipid in the supernatant maintained below 10%. Under these conditions, where the residual lipids are below 10%, the best recovery is obtained for 10× dilution and pH 5.5 (54%). Therefore, the best compromise for minimising lipids and maximising IgY recovery obtained under the conditions studied is 10× dilution and pH 5.5.

IgY recovery may be increased to more than 90% using centrifugation (Akita and Nakai, 1992) or filtration. Nevertheless, sedimentation is considered as a simple and inexpensive technique for large-scale separation of proteins from a relatively inexpensive and readily available material such as egg yolk. It is also possible to improve recovery through an additional dilution/sedimentation step. Methods other than water dilution have been used for lipoprotein removal including ultracentrifugation (Burley and Vadehra, 1979), organic solvents (Bade and Stegemann, 1984; Polson et al., 1980), precipitation using sodium dextran sulphate (Jensenius et al., 1981), or using natural gums (Hatta et al., 1988, 1990). However, to produce food grade IgY, water dilution may be considered as the most appropriate technique and its efficiency for lipoprotein precipitation may be further improved by adding natural gums such as carrageenan (Hatta et al., 1990). In addition, water dilution is less expensive and allows the use of the remaining sedimented fraction in food applications or for the separation of other biologically active components (Kwan et al., 1991).

Automated Chromatography System

The use of microcomputers to control equipment and processing, acquire and analyse data, and rapidly produce reports is swiftly becoming commonplace in the research laboratory. In chromatography, fully automated analytical instruments such as gas chromatographs and HPLCs have been available for years. At the preparative and pilot scale, however, this high level of automation is often not as readily available due to the large number of methods for communicating between devices for the purposes of control and data acquisition: analogue and digital DC voltages of various ranges, variations in amperage of electrical currents, variation of frequencies of electrical waveforms, and serial and parallel port communication between computers.

In order to piece together non-standard automated chromatography systems, it is therefore often necessary to make adaptations and even construct custom interfaces. Figure 18.2 is a schematic diagram of the automated chromatography system used in this work. The system includes a low pressure chromatography column, feed tanks, two remote control Ismatec peristaltic pumps (Cole-Parmer, Chicago, USA), an on-line Pharmacia UV detector and monitor (Pharmacia-LKB,

Uppsala, Sweden), an ISCO Retriever II fraction collector (ISCO Inc., Lincoln, Nebraska), several solenoid valves (Burkert, Germany), and an IBM-compatible computer and monitor. Food grade Tygon tubing (B-44-4X, 6.4 mm o.d., 3.2 mm i.d.) and nylon fittings were used to connect the various components of the system. Control was carried out with ISCO ChemResearch data management/ system controller software specifically designed for HPLC, and adapted for use with our equipment. The software is menu driven, and allows calibration and control of devices (i.e. pumps and valves), data acquisition and analysis, and graphic display. An external distribution interface module, into which externally controlled instruments and accessories are connected, was included in the system. However, it was necessary to use a custom designed and built control interface unit to actuate the process-scale solenoid valves. The unit amplified the low/high (0/5 VDC) signal from the ISCO interface module by use of a Darlington transistor arrangement (Fig. 18.3) powered by a 24 V DC power supply.

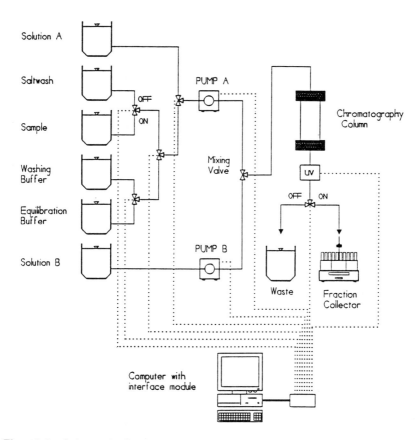

Fig. 18.2. Schematic diagram of the automated liquid chromatography system. Solution A (equivalent to equilibrium buffer) and solution B (equivalent to washing buffer) are used for gradient generation.

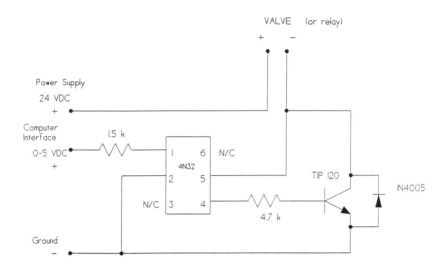

Fig. 18.3. Darlington transistor arrangement used in the interface unit for valve actuation.

Figure 18.4 is a photograph of the assembled system, which was utilised to carry out laboratory-scale experiments using a 50 ml column, and later bench-scale experiments using a 1.5 litre column shown here.

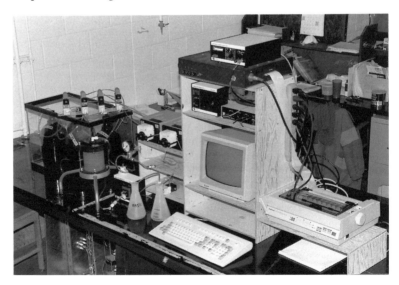

Fig. 18.4. The automated liquid chromatography system used for IgY separation having a 1.5 litre column.

Cation Exchange Chromatography

The automated system described above was first used with a 50 ml column (2.5 cm i.d.) packed with either CM-92 (Whatman Biosystems Inc., USA) or HC-2 (Phoenix Chemicals Ltd, Christchurch, New Zealand) both of which are cation exchange media. Preliminary tests using phosphate buffer showed that both cation exchangers had excellent flow properties, although the bed volume was affected by the pH and ionic strength of the buffer. For each run, the column was equilibrated with 0.2 M phosphate buffer (PB) at pH 5.0 and then washed with 0.01 M PB at pH 5.4. Filtered egg yolk supernatant was applied to the column, the column was washed again with 0.01 M PB, and the bound proteins were eluted using linear gradient with increasing PB molarity from 0.01 to 0.2 M. After elution, the column was washed with 0.01 M PB/0.5 M NaCl solution followed by 0.01 M PB. The flow rate was maintained at 1.95 ml/min with the exception of sample application where the flow rate was reduced to 0.93 ml/min. Fractions were collected for analysis every 7.5 min during elution and salt washing, while the total volume was collected and measured after both sample application and washing, and representative samples kept for analysis. The collected samples were analysed for total protein and IgY concentration, and the purity and recovery of each fraction were estimated.

Figures 18.5 and 18.6 show typical chromatograms obtained using HC-2 and CM-92, respectively. The first peak in each chromatogram occurs during sample application and washing, which represents the majority of impurities and about 30% of the IgY not initially bound to the column. The majority of the remaining IgY is eluted in a single peak using HC-2 (second peak in Fig. 18.5) while this peak is split into two using the CM-92 medium (Fig. 18.6).

Table 18.1 compares HC-2 and CM-92 in terms of IgY recoveries obtained in each chromatography step, and gives the mass balance estimations. The IgY recovery obtained with HC-2 during elution is somewhat higher than that obtained with CM-92 (64% versus 60%) as shown in Table 18.1. However, the purity of the eluted fraction using CM-92 is 69% while it is only 60% in the case of HC-2.

Figure 18.7 is a plot of purity as a function of recovery, if fractions are pooled starting with the highest purity fraction followed by fractions of decreasing purity. Better results were obtained using CM-92 where about 48% IgY at 78% purity or about 60% IgY at 70% purity could be recovered. Approximately similar results were observed using stepwise elution with 0.2 M PB with the exception of a slight improvement in recovery and sharper IgY-rich peaks obtained for both CM-92 and HC-2. Therefore, it would be easier to use step gradient elution for larger scale separation which might be applied to a batch process as well.

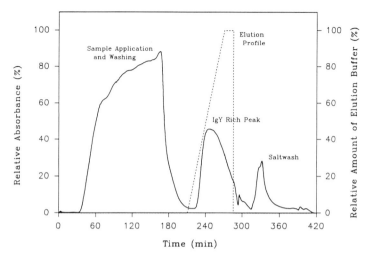

Fig. 18.5. Chromatogram of egg yolk supernatant applied to HC-2 cation exchanger column (50 ml) showing absorbance at 280 nm and elution profile. The column was equilibrated with 0.2 M PB and washed with 0.01 M PB, and IgY was eluted using linear gradient elution with 0.01-0.2 M PB.

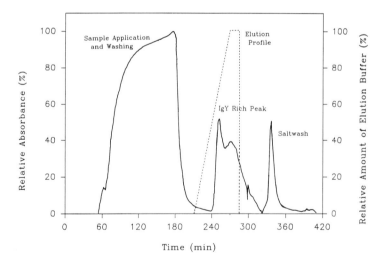

Fig. 18.6. Chromatogram of egg yolk supernatant applied to CM-92 cation exchanger column (50 ml) showing absorbance at 280 nm and elution profile. The column was equilibrated with 0.2 M PB and washed with 0.01 M PB, and IgY was eluted using linear gradient elution with 0.01-0.2 M PB.

Table 18.1. IgY recovery obtained in each chromatography step and mass balances for both HC-2 and CM-92 cation exchangers.

Step	IgY recovery (mg)		IgY recovery (%)[1]	
	HC-2	CM-92	HC-2	CM-92
Sample application and washing	31.3	36.2	31.0	34.6
Elution	64.8	62.6	59.9	63.9
Salt washing	6.7	8.1	6.6	7.8
Total	102.8	106.9	101.4	102.3
Total applied	101.4	104.5	100.0	100.0

[1] Percentages were estimated based on total IgY applied to the column.

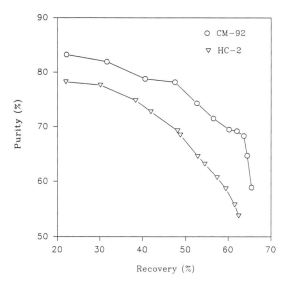

Fig. 18.7. IgY purity as a function of recovery obtained with HC-2 and CM-92 cation exchangers.

The fact that not all of the IgY was bound to the column may be due to its heterogeneity and tendency to aggregate (Martin and Cook, 1958). Different IgY subclasses may be present, and in addition, HPLC analysis using gel filtration columns shows the presence of IgY dimers and polymers with their ratio depending on pH and ionic strength of the buffer. Such phenomena, if present, will hinder the achievement of high recoveries with acceptable purities using ion exchange chromatography. Nevertheless, our results are a major improvement as compared with other chromatography studies. Very low recovery (16%) for a purity of about 60% was obtained on DEAE-Sephacel column using eggs from immunised chickens (McCannel and Nakai, 1990). Hatta and co-workers (1990) however, were able to achieve 76% recovery and 46% purity using the same chromatography medium, but with eggs from non-immunised hens. Affinity chromatography did not improve the results significantly as about 46% recovery for a purity of 65% was obtained (McCannel and Nakai, 1989). In addition, these results were obtained using manually separated egg yolk free from egg white contamination, while industrially separated yolk contains a significant amount of egg white.

Besides purity and recovery, the cation exchangers used (HC-2 and CM-92) possess additional advantages. Specifically, they have superior flow properties and a relatively low cost with the HC-2 being much cheaper (US$60/kg) than CM-92 (US$250/kg). These make HC-2 or CM-92 an excellent candidate for large-scale separation of food quality IgY.

Scale-up of Chromatographic Process

The first step in this process requires the separation of egg yolk from the other egg components. As we have chosen to work with industrially separated yolk, there is no need to consider scaling up this step; typical cracking machines are capable of processing 8 eggs per second, or about 2400 dozen per hour. This is equivalent to over 400 litres of yolk per hour assuming approximately 15 ml of yolk per egg. It is, however, important to consider the efficiency of the industrial separation of yolk. Based on information from a local cracking plant (Vanderpol's Eggs Ltd, Aldergrove, BC), the industrial yolk contains a significant amount of egg white (about 5.5% on a dry weight basis, 22% on a wet weight basis). This introduces egg white proteins as additional contaminants, especially ovalbumin, which constitutes approximately 50% of egg white proteins.

The next step in the purification of IgY, the separation of the water-soluble fraction from lipoproteins described earlier in this paper, is relatively easy to scale up as it involves dilution with distilled water and the use of a small amount of acid to adjust pH. This step could be easily carried out in a stainless steel batch tank.

The third separation step is somewhat more difficult to transfer to the large scale. At the process scale, the purity of the product must still be maintained, while the necessity of obtaining good recovery is crucial in order for the process to be economically feasible. Also, selecting equipment for large-scale chromatography is not a simple task since many 'package systems' do not provide sufficient flexibility for individual applications, and there are relatively few suppliers who offer a wide range of components (Johansson *et al.*, 1988).

The fabrication of a process-scale chromatography system represents a significant capital investment, and the major operational costs of such a system are generally associated with chromatographic media (Curling *et al.*, 1983). It is therefore necessary to determine the physical dimensions and running conditions that will minimise the overall cost of assembling and operating such a system.

As described earlier in this paper a number of experiments were carried out using HC-2 and CM-92 packed into 2.5 cm diameter columns where the exchanger formed a total bed volume of 50 ml. Having obtained an acceptable separation at the small scale, a 1.5 litre process-scale column (Pharmacia, Uppsala, Sweden) was packed with HC-2, and tested under similar conditions. The use of this particular column resulted in a bed height roughly 1.5 times that at the laboratory scale, while the cross-sectional area was increased by a factor of 20. In order to avoid overloading the column (and since a thorough study of the capacity of the exchangers to bind IgY has not yet been completed), the loading was reduced to 1 ml of egg yolk supernatant applied per ml of cation exchanger. The flow rate was increased to 15 ml/min for sample application and 30 ml/min for all other operations.

The pooled effluent from sample application and washing, IgY-rich peak, and salt wash peak were collected and analysed for total protein and IgY concentration. The results indicated that a similar separation to that of the laboratory scale was obtained with the major difference being a broadening of the IgY-rich peak. This was to be expected as the diameter of the column was increased significantly compared with that of the small-scale column, without a corresponding increase in column length.

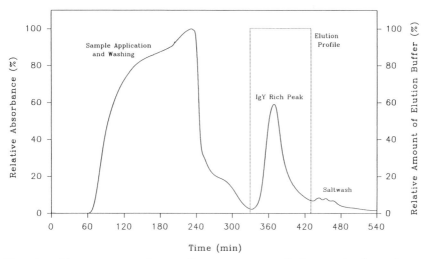

Fig. 18.8. Chromatogram of egg yolk supernatant applied to two 1.5 litre columns in series packed with HC-2 cation exchanger. The column was equilibrated with 0.2 M PB and washed with 0.01 M PB, and IgY was eluted using stepwise elution with 0.2 M PB.

Figure 18.8 is a representative chromatogram for an experiment involving two of these 1.5 litre columns connected in series. Analysis of the IgY-rich peak indicates a recovery of approximately 60% of IgY applied, with a purity of 70%. Thus using two 1.5 litre columns in series (i.e. increasing the length of the column) did improve the purity of the IgY-rich peak as compared with the results obtained with 50 ml or 1.5 litre column packed with HC-2 (60% purity) using step elution gradient, but at the expense of a slightly lower recovery.

A pressure gauge was connected just before the inlet to the first column in order to measure the gauge pressure at this point, but it indicated virtually zero pressure drop even during washing and elution when the flow rate was 30 ml/min. This prompted an investigation of the relationship between pressure drop and superficial velocity (the flow rate divided by the column cross-sectional area) for an individual 1.5 litre packed column. A single column was equilibrated with 0.2 M phosphate buffer, pH 5.0, and this same buffer was pumped through the column using a peristaltic pump at various flow rates, starting at 30 ml/min and increasing the flow rate stepwise to 500 ml/min. Once the pressure drop across the column had stabilised, a reading was taken and the flow rate was increased to the next level.

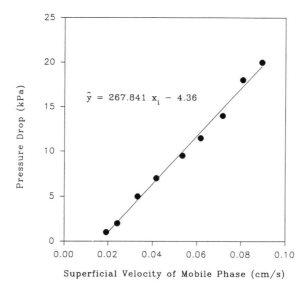

Fig. 18.9. Pressure drop versus superficial velocity using a 1.5 litre column packed with HC-2 cation exchanger.

The results are plotted in Fig. 18.9 which indicate that the pressure drop is linearly related to the superficial velocity of the mobile phase up to nearly 0.100 cm/s. Even at a flow rate of 500 ml/min (velocity of 0.083 cm/s), the pressure drop across the packed column was less than 20 kPa. The next logical step in this work should therefore be to study the effect of flow rate on the separation, since higher flow rates would lead to shorter cycle times and minimise the required column size.

Conclusion

This study demonstrated that the use of cation exchange chromatography for the isolation of IgY from industrially separated egg yolk is relatively simple, low in cost, and yields acceptable purities and recoveries. For instance, a purity of 70% was obtained for about 60% recovery. However, higher purities are possible when sacrificing recovery for purity is acceptable. The advantages of the system are enhanced through automation which allows the operator to monitor, control, and adjust the conditions, reducing separation time and labour expenses. Scaling up did not affect the separation power of the cation exchanger due mainly to its excellent flow properties. Considering these advantages in addition to the low cost of the exchange media, the method developed in this study may prove to be very useful for industrial separation of egg yolk antibodies.

Acknowledgements

The authors would like to thank the Canadian Egg Marketing Agency, Agriculture Canada, and the Natural Sciences and Engineering Research Council of Canada for their financial support, and Vanderpols Eggs Ltd for donating egg yolk.

References

Akita, E.M. and Nakai, S. (1992) Isolation and purification of immunoglobulins from egg yolk. *J. Food Sci.* 57:629-634.

Bade, H. and Stegemann, S. (1984) Rapid method of extraction of antibodies from hen yolk. *J. Immunol. Methods* 72:421-426.

Burley, R.W. and Vadehra, D.V. (1979) Chromatographic separation of the soluble proteins of hen's egg yolk: an analytical and preparative study. *Anal. Biochem.* 94:53-59.

Carroll, S.B. and Stollar, B.D. (1983) Antibodies to calf thymus RNA polymerase II from egg yolks of immunized hens. *J. Biol. Chem.* 258:24-26.

Curling, J.M., Low, D., and Cooney, J.M. (1983) Downstream processing of fermentation products by chromatography. In: Atkinson, A. (ed.), *BIOTECH '83*. Northwood Hills, UK, pp. 225-234.

Gottstein, B. and Hemmeler, E. (1985) Egg yolk immunoglobulin Y as an alternative antibody in the serology of echinococcosis. *Z. Parasitenkd.*

71:273

Hassl, A. and Aspock, H. (1988) Purification of egg yolk immunoglobulins. A two step procedure using hydrophobic interaction chromatography and gel filtration. *J. Immunol. Methods* 110:225-228.

Hassl, A., Aspock, H., and Flamn, H. (1987) Comparative studies on the purity and specificity of yolk immunoglobulin Y isolated from eggs laid by hens immunized with *Toxoplasma gondii* antigen. *Zbl. Bakt. Hyg. A* 267:247-253.

Hatta, H., Sim, J.S., and Nakai, S. (1988) Separation of phospholipids from egg yolk and recovery of water-soluble proteins. *J. Food Sci.* 53:425-427, 431.

Hatta, H., Kim, M., and Yamamoto, T. (1990) A novel isolation method for hen egg yolk antibody, "IgY". *Agric. Biol. Chem.* 54:2531-2535.

Jensenius, J.C., Andersen, I., Hau, J., Crove, M., and Koch, C. (1981) Eggs: conveniently packaged antibodies. Methods for purification of yolk IgG. *J. Immunol. Methods* 46:63-68.

Johansson, H., Ostling, M., Safer, G., Wahlstrom, H., and Low, D. (1988) Chromatographic equipment for large-scale protein and peptide purification. In: Mizrahi, A. (ed.), *Downstream Processes: Equipment and Techniques.* Alan R. Liss, Inc., New York, pp. 127-157.

Kwan, L., Li-Chan, E., Helbig, N., and Nakai, S. (1991) Fractionation of water-soluble and -insoluble components from egg yolk with minimum use of organic solvents. *J. Food Sci.* 56:1537-1541.

Losch, U., Schranner, I., Wanke, R., and Jurgens, L. (1986) The chicken egg, an antibody source. *J. Vet. Med., Series B* 33:609-619.

Martin, W.G. and Cook, W.H. (1958) Preparation and molecular weight of γ-livetin from egg yolk. *Can. J. Biochem. Physiol.* 36:153-160.

McCannel, A.A. and Nakai, S. (1989) Isolation of egg yolk immunoglobulin-rich fractions using copper-loaded metal chelate interaction chromatography. *Can. Inst. Food Sci. Technol. J.* 22:487-490.

McCannel, A.A. and Nakai, S. (1990) Separation of egg yolk immunoglobulins into subpopulations using DEAE-ion exchange chromatography. *Can. Inst. Food Sci. Technol. J.* 23:42-46.

Otani, H., Matsumoto, K., Saeki, A., and Hosono, A. (1991) Comparative studies on properties of hen egg yolk IgY and rabbit serum IgG antibodies. *Lebensm.-Wiss. u. Technol.* 24:152-158.

Polson, A., von Wechmar, M.B., and van Regenmortel, M.H.V. (1980) Isolation of viral IgY antibodies from egg yolks of immunized hens. *Immunol. Commun.* 9:475-493.

Polson, A., Coetzer, T., Krugar, J., von Maltzahn, E., and van der Merwe, K.J. (1985) Improvements in the isolation of IgY from the yolk of eggs laid by immunized hens. *Immunol. Invest.* 14:323-327.

Rose, M.E., Orlans, E., and Buttress, N. (1974) Immunoglobulin classes in the hen's egg: their segregation in yolk and white. *Eur. J. Immunol.* 4:521-523.

Shimizu, M., Fitzsimmons, R.C., and Nakai, S. (1988) Anti-*E. coli* immuno-globulin Y isolated from egg yolk of immunized chickens as a potential food ingredient. *J. Food Sci.* 53:1360-1366.

Stadelman, W.J. and Cotterill, O.J. (1977) *Egg Science and Technology.* 2nd edn. AVI Pub. Co., Westport, CT, pp. 65-91.

Williams, J. (1962) Serum proteins and the livetins of hen's-egg yolk. *Biochem. J.* 83:346-355.

Chapter Nineteen

Preparation and Purification of Fab' Immunoactive Fragments from Chicken Egg Immunoglobulin Using Pepsin and *Aspergillus saitoi* Protease

E.M. Akita and S. Nakai

Department of Food Science, University of British Columbia, Vancouver, British Columbia, Canada V6T 1Z4

Abstract Pepsin and a fungal protease from *Aspergillus saitoi* were used to produce immunoactive Fab' fragments from IgY preparations of different purities ranging from 15% to >99%. Prolonged hydrolysis using the enzymes led to complete digestion of the pFc' fragment leaving only the Fab' fragment as determined by SDS-PAGE. Subsequent purification was achieved by a combination of ultrafiltration and anion exchange chromatography. Fab' fragments produced by the fungal protease had the same molecular weight and showed complete identity with Fab' produced from peptic digestion by double immunodiffusion assay, hence these fragments are very similar. Use of low purity IgY and this protease may provide a more economical means for the preparation of Fab' fragments since it is much cheaper than pepsin.
(*Key words:* IgY, Fab', pepsin, *Aspergillus saitoi* protease, purification)

Introduction

The potential use of egg yolk immunoglobulin (IgY) and its immunoactive fragments in passive immunotherapy (Bartz *et al.*, 1980; Shimizu *et al.*, 1988; Ebina *et al.*, 1990) and immunodiagonistics (Larsson *et al.*, 1991; Larsson and Sjoquist, 1988) has been recognised. With respect to the immunoactive antibody binding fragments their use may be facilitated by developing efficient methods for their preparation and purification. These methods should not only be cost-efficient but also simple, rapid, and easily scaled up.

We have developed an efficient method for the large-scale production of Fab' fragments from pure IgY by peptic digestion (Akita and Nakai, 1993). The efficiency of the method could further be improved by using a cheaper source of enzyme and IgY preparations of lower purity.

Use of immobilised enzymes is one way to reduce the cost of enzymes. An alternative approach is the utilisation of cheaper microbial enzymes which are now available commercially. Yada (1984) studied the relationship between physicochemical properties and enzymatic activity of some aspartyl proteases. Comparing the milk clotting and proteolytic activity of these enzymes, the lowest milk clotting/proteolytic activity ratio was obtained with the aspartyl protease produced by the fungus *Aspergillus saitoi*. Yada attributed this to the higher proteolytic activity of the *Aspergillus saitoi* protease (ASP). ASP could therefore be used to produce immunoactive antibody fragments.

We have recently developed a method for the purification of IgY from egg yolk (Akita and Nakai, 1992). This involves simple water dilution of egg yolk under slightly acidic conditions, pH 5.0-5.2. The resulting supernatant or water-soluble fraction (WSF) after centrifugation or filtration contains the water-soluble proteins including IgY. We reasoned that use of this WSF without further purification as a source of IgY would provide a more economical way to produce Fab'. Although the purity of IgY in WSF is about 15%, the observation that the Fab' fragment was found to relatively stable against further pepsin digestion (unpublished work), could be exploited by allowing the enzyme to digest away the rest of the proteins in the WSF, therefore simplifying the subsequent purification. The objective of this study was to develop a more efficient way to produce and purify immunoactive antigen binding fragments by using IgY of lower purity and a cheaper microbial protease.

Materials and Methods

Immunisation

Enterotoxigenic *Escherichia coli* (ETEC) strain H10407 was cultured in brain heart infusion broth at 37°C for 18 h. The bacteria were killed by treating with formaldehyde (0.4%), washed with phosphate buffered saline (PBS, 0.01 M phosphate buffer containing 0.14 M NaCl, pH 7.0), and stored frozen in the same buffer until used. The antigen suspension (2×10^9 ETEC cells in PBS) was emulsified with an equal volume of incomplete Freund's adjuvant (Difco, Detroit, MI). Five laying hens (White Leghorn), 25 weeks old, were each immunised intramuscularly with 1 ml of 1×10^9 ETEC cells in adjuvant into four sites. A booster injection was given after 16 days. Eggs were stored at 4°C until used.

Preparation of IgY

IgY preparations of different purities as determined by SDS-PAGE, namely WSF (15%), ASP (30%), ultrafiltration (UF, 91%), and gel filtration chromatography (GF, >99%) were obtained by the method of Akita and Nakai (1992), as outlined in Fig. 19.1.

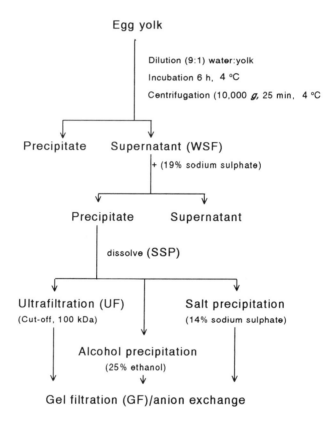

Fig. 19.1. Flow diagram for the isolation and purification of IgY from egg yolk.

Enzymatic Digestion of IgY

Digestion of IgY preparations with pepsin, P-7012, and *Aspergillus saitoi* protease, P-2143 (Sigma Chemical Co., St Louis, MO) were carried out under the following conditions.

Lyophilised proteins were dissolved in 50 mM sodium acetate buffer (SAB) pH 4.2, to give a final concentration of 10 mg/ml. The WSF (4 mg/ml) was adjusted to pH 4.2. The enzymes (10 mg/ml) in SAB pH 4.2 were added to give an enzyme:protein ratio of 1:50. Digestion was done at 37°C and stopped by adjusting the pH to 8.0. Digestion was monitored by SDS-PAGE.

Purification of Antibody Binding Fragments

The enzyme digest was ultrafiltered using a Diaflo ultrafiltration membrane (Amicon Corp., Lexington, MA) with a molecular weight cut-off of 10 kDa. The digest was initially concentrated to give a protein concentration of 8-10 mg/ml, followed by diafiltration with six concentrate volumes of 10 mM Tris-HCl buffer pH 8.0. Ultrafiltered digest was applied to a DEAE-Sephacel column, 2×8 cm, equilibrated with 10 mM Tris-HCl buffer pH 8.0. Anion exchange chromatography was done at ambient temperature and a flow rate of 1.0 ml/min.

SDS-Polyacrylamide Gel Electrophoresis

SDS-PAGE was done under non-reducing conditions on a Pharmacia Phast System (Pharmacia Biotech Inc., Uppsala, Sweden) using a 10-15% gradient or 12.5% homogeneous PhastGel, Phastgel SDS-Buffer Strips and Phastgel Blue R dye (Coomassie R 350 dye) according to the procedures described in the manufacturer's manual (PhastSystem Owner's Manual 1986, Pharmacia Biotech Inc., Uppsala, Sweden). The enzyme digests were added to a sample buffer to give a final protein concentration of 0.5-2 mg/ml, 2% SDS, and 0.005% bromophenol blue dye in 50 mM Tris-HCl pH 8.0 buffer. Samples were heated for 5 min at 100°C and 1 μl or 3 μl was applied. Molecular weight determination of proteins on PhastGels was done with a Pharmacia Phast Image Gel Analyser according to the manufacturer's recommendations (PhastImage Users' Manual, 1989, Pharmacia Biotech Inc., Uppsala, Sweden). Bio-Rad's broad range SDS-PAGE molecular weight standards, 6.5-200 kDa (Bio-Rad Lab., Richmond, CA), were used as molecular weight markers.

Immunodiffusion

Radial immunodiffusion was done using a plate containing 1% rabbit anti-chicken affinity purified IgG, F(ab')$_2$ specific (Jackson Immunoresearch Lab. Inc., Baltimore, MD), 1% agarose A (70 mg, Pharmacia-LKB, Uppsala, Sweden), and 0.02% sodium azide in PBS. Five microlitres of appropriately diluted samples and standards in the range of 0.1 mg to 1.0 mg Fab'/ml were added to 3 mm diameter wells. After incubation for 18 h, the gel was deproteinised by washing in 0.3 M NaCl for 18 h with several changes and visualised by staining with Coomasie Blue. A standard curve was obtained by plotting the squared diameter of the precipitation rings against $\log_{10}$ concentration. Concentration of unknown Fab' samples was determined by reference to this curve.

Double immunodiffusion was performed according to the method described by Ouchterlony (1962). Affinity purified rabbit IgG antibodies against F(ab')$_2$ and

Fc fragments of IgY used were purchased from Jackson Immunoresearch Lab. Inc., Baltimore, MD.

Enzyme-Linked Immunosorbent Assay (ELISA)

Formaldehyde-killed ETEC, strain H10407, whole cells were used as the antigen. Immulon I microtitre plate (96 wells, Dynatech Laboratories Inc., Chantilly, VA) was used as the solid support. Wells were coated with 100 µl of *E. coli* sonicated whole cell suspension (10^7 cells per well) in potassium phosphate buffered saline (PBS, 10 mM phosphate buffer, pH 7.0, 0.14 M NaCl), and incubated for 1 h at 37°C. Plates were washed three times with PBS-Tween (0.05% Tween 20), followed by a blocking step using 100 µl of 2% bovine serum albumin for 30 min at 37°C. Plates were incubated with the appropriate dilutions of Fab′ for 1 h at 37°C. The plates were then washed again three times with PBS-Tween and 100 µl of affinity purified rabbit anti-chicken IgG, $F(ab′)_2$ fragment specific IgG coupled to alkaline phosphatase (Jackson Immunoresearch Lab. Inc., Baltimore, MD, 1:1000 in PBS-Tween) was added to each well. After 1 h incubation at 37°C, the plates were washed again followed by addition of 50 µl freshly prepared substrate solution (0.1% *p*-nitrophenyl phosphate disodium in diethanolamine buffer, pH 9.8). The reaction was stopped by addition of 50 µl 2.5 N NaOH. For each plate, controls for non-specific binding of Fab′ fragments of IgY antibodies and enzyme-labelled antibodies were prepared. Absorbance was read at 405 nm using an ER-400 ELISA reader (SLT Labinstruments GmbH, Salzburg, Austria).

Protein Determination

Total protein content was determined by the Biuret method using Sigma's Total Protein Reagent (Sigma Chemical Co., St Louis, MO) as described by Akita and Nakai (1992).

Results

Pepsin and ASP were successfully used to digest pure IgY into the Fab′ fragments, as can be seen in Figs 19.2 and 19.3. The enzymes gave similar patterns of digestion of IgY with a slightly higher rate of digestion for pepsin. The pFc′ fragment was not stable against both enzymes in contrast to the Fab′ fragment. Consequently prolonged digestion of the IgY led to the complete digestion of the pFc′ fragment leaving only the Fab′ fragment. This observation suggests that the two aspartyl proteases have similar activities and may cleave the IgY at similar sites.

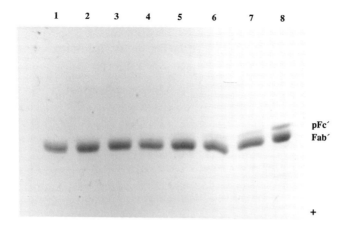

Fig. 19.2. SDS-PAGE (non-reducing) on 10-15% gradient Phastgel of peptic digest of IgY (>99%) at different times. Lane 1, 30 h; lane 2, 15 h; lane 3, 12 h; lane 4, 8 h; lane 5, 6.5 h; lane 6, 4 h; lane 7, 3 h; lane 8, 2 h.

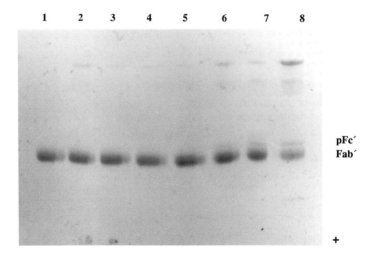

Fig. 19.3. SDS-PAGE (non-reducing) on 10-15% gradient Phastgel of ASP digest of IgY (>99%) at different times. Lane 1, 15 h; lane 2, 12 h; lane 3, 10 h; lane 4, 8 h; lane 5, 6.5 h; lane 6, 4 h; lane 7, 3 h; lane 8, 1 h.

The similarity in the activities of the two enzymes was further confirmed by their action on WSF, by giving similar digestion patterns as seen in Fig. 19.4. The other proteins in the WSF were found to be susceptible to digestion by pepsin and ASP. Consequently, Fab' could be produced upon prolonged digestion. The time it took to produce pure Fab' fragments from the various IgY preparation is summarised in Table 19.1. It took longer to produce pure Fab' from IgY preparations of lower purity, between 24 and 30 h for both enzymes, compared with IgY preparations of higher purity. Digestion was found to be faster by pepsin than ASP, 8-12 h and 12-15 h respectively, to give pure Fab' from the purer preparations of IgY.

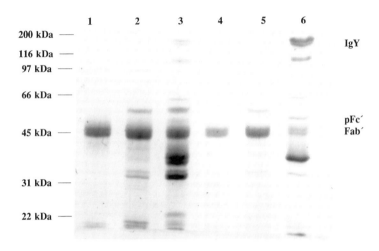

Fig. 19.4. SDS-PAGE (non-reducing) on 10-15% gradient Phastgel of peptic and fungal protease digests of WSF (15%) at different times. Lanes 1, 2, and 3 correspond to pepsin digestion after 24 h, 12 h, and 1 h, respectively. Lanes 4, 5, and 6 correspond to fungal protease digestion after 24 h, 12 h, and 1 h, respectively.

Table 19.1. Effect of enzyme and purity of IgY on the time to produce pure Fab' fragments.

Samples[1]	Time (hours)	
	Pepsin	Fungal protease
WSF (15)	24-30	24-30
SSP (30)	24-30	24-30
UF (91)	8-12	12-15
GF (>99)	8-12	12-15

[1] WSF, SSP, UF, and GF correspond to water-soluble fraction, sodium sulphate precipitation, ultrafiltration, and gel filtration respectively, with their % purities in parentheses.

Purification of Fab'

The objective of purification is to isolate the Fab' fragment from end products of digestion and enzymes. Separation of digested products, i.e. amino acids and possibly small peptides was done by ultrafiltration of the digest. Ultrafiltration also allowed concentration of the enzyme digest and buffer change. Consequently, the digest could be directly applied to the DEAE-Sephacel column. Application of the digest in 10 Tris-HCl pH 8.0 was found to cause binding of enzymes to the column, with the Fab' coming off in the eluent. A pure Fab' fragment could then be obtained (Fig. 19.5). Changing the buffer from Tris-HCl to phosphate buffer (10-100 mM pH 8.0) gave similar results.

However, equilibration of the anion exchange column took longer than with Tris-HCl buffer. Although, cationic buffers are recommended for anion exchange, phosphate buffer may be preferred from the safety view point. This becomes important for the *in vivo* uses of the purified Fab'.

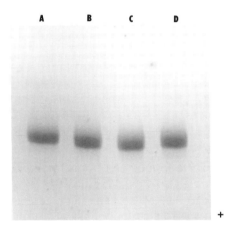

Fig. 19.5. SDS-PAGE (non-reducing) on 12.5% homogeneous Phastgel of purified Fab' fragments. Purification was done by ultrafiltration followed by anion exchange. a, b, c and d correspond to Fab' produced from WSF, SSF, UF, and GF respectively.

Immunochemical Assays

The purified IgY fragments were identified as Fab' fragments on the basis of being precipitated by rabbit anti-chicken IgG, F(ab')$_2$ fragment specific antibodies, and not precipitated by rabbit anti-chicken IgG, Fc fragment specific antibodies, in immunodiffusion assay.

The Fab' fragment produced by ASP showed complete identity with that produced by pepsin digestion. This observation, coupled with the fact that they have similar molecular weight, suggests that the Fab' fragments produced by the two enzymes may be similar. The fragments also demonstrated similar antigen binding activities (Fig. 19.6).

Yield of Fab′ fragments

The two enzymes were also compared on the basis of the yield of Fab′ fragment from purified IgY (Table 19.2).

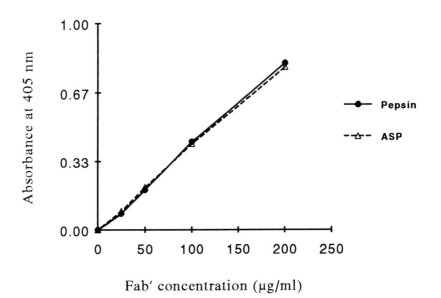

Fig. 19.6. Comparison of activities of Fab′ fragments produced by pepsin and ASP by direct ELISA. Antigen was incubated with doubling dilutions of Fab′ fragments, followed by incubation with rabbit anti-IgY, F(ab′)₂ fragment specific antibody conjugated to phosphatase and addition of substrate for colour development.

Although a slightly higher yield was obtained when Fab′ was produced from pure IgY, similar yields were obtained from the WSF. The yield of Fab′ was also found to be dependent on whether it was produced from specific or non-specific IgY and on the type of antigen used for immunisation. We observed that generally, higher yields were obtained from specific IgY. When the entero-toxin of *E. coli* was used as the antigen, a yield of over 90% was obtained (result not shown), compared with a yield of 50-60% when whole bacterial cells were used. It should be noted that in all these cases similar yields were obtained for ASP and pepsin.

Table 19.2. Effect of digestion time on the yield of Fab' fragments produced with ASP and pepsin from pure IgY (GF).

Time (hours)	Fungal protease		Pepsin	
	mg	% Yield	mg	% Yield
4	-	-	3.25 (0.02)	69.3
6.5	-	-	2.95 (0.05)	62.9
12	2.63 (0.02)	56.0	-	-
15	2.60 (0.02)	55.4	2.70 (0.04)	57.5
30	2.52 (0.06)	53.5	2.45 (0.03)	52.2

Quantification was done by radial immunodiffusion (RID) after complete digestion of IgY. Reported amount (mg) of Fab' was obtained from 9.6 mg of IgY and percentage yield was calculated based on molecular weights of 44,000 and 180,000 for Fab' and IgY respectively. Values are means of duplicates with their deviations in parentheses.

Discussion

The advantages of using antibody fragments have been widely recognised by many researchers. The Fab' fragment has a number of properties which make it suitable for a number of uses in a wide spectrum of research. It has a higher diffusion capacity than the original immunoglobulins in the extracellular space due to its lower molecular weight (Covell *et al.*, 1986). This makes it suitable in such areas as immuno-cytochemistry and membrane and whole cell binding assays where the ability of large antibodies to penetrate tissue has previously been a problem (Bidlack and Mabie, 1986).

Removal of the Fc portion of the antibody molecule allows its use, *in vitro*, in assay systems to minimise non-specific binding mostly associated with the Fc portion of the molecule. Examples include cell-binding immunoassays, generation of anti-idiotypic antibodies for use as vaccines (Nisonoff and Lamoyi, 1981), analysis of patient fluids and biopsy specimens, and the monitoring of patient response to administered monoclonal antibody (Milenic *et al.*, 1989).

The rapid clearing of the Fab' fragment and its smaller size which enable an extensive distribution into extracellular fluids make it suitable for the reversal of digoxin toxicity (Ochs and Smith, 1977).

Most of the uses indicated above involve monoclonal antibodies which are associated with high costs. Some of these uses could be replaced by affinity purified, antigen-specific polyclonal IgY and its Fab' fragment. It is known that IgY does not bind to mammalian Fc receptors (Jensenius *et al.*, 1981). The use of IgY and its fragments may therefore minimise the non-specific interactions observed between the Fc portion of mammalian antibodies and Fc receptors in a number of normal mammalian cells, associated with the use of mammalian antibodies. Furthermore, chicken antibodies do not activate mammalian complement, or react with human rheumatoid factor (RF). As pointed out by Larsson and Sjoquist (1988), this makes them useful in assays where RF could interfere and give false positive reactions e.g. nephelometry, latex agglutination, or ELISA. However, under conditions where rapid *in vivo* clearance, good distribution in the extracellular fluids, and less immunogenic product are desired, the Fab' fragment of IgY would offer a better replacement.

Since the preparation of the specific IgY or its Fab' fragment may require affinity chromatography which is relatively expensive, the other steps in the purification protocol should be quick and cheap in order to make the whole process of preparation of specific Fab' economically efficient. Consequently, we have sought to improve the economic efficiency of a method we developed for the preparation of antigen-specific polyclonal IgY.

Both pepsin and ASP were successfully used to produce Fab' from IgY preparations of varying purities. Our results show that there is no need to use pure IgY for the production of the Fab' fragment. In spite of the fact that more time is needed to produce pure Fab' fragment from the WSF compared with pure IgY, the energy and time savings gained by using the WSF make it a more economical way to produce the fragment.

It took similar amounts of time to generate pure Fab' with both enzymes from the WSF. Furthermore, ASP is much cheaper than pepsin and the Fab' fragment produced by the two enzymes demonstrated similar immunoactivities and yield. We therefore recommend the use of ASP for the preparation of Fab' from the WSF.

Conclusions

Pepsin and ASP were successfully used to produce immunoactive Fab' fragment from IgY preparations of purities ranging from 15 to >99%. Digestion of WSF with ASP is recommended for the preparation of Fab' fragment of IgY. Combination of UF and anion exchange provides an excellent purification procedure yielding very pure immunoactive Fab' fragments. Advantages include being easily scaled up, simple, rapid, and cheap.

References

Akita, E.M. and Nakai, S. (1992) Immunoglobulin from chicken egg yolk. Isolation and purification. *J. Food Sci.* 57:629-634.

Akita, E.M. and Nakai, S. (1993) Production and purification of Fab' fragments from chicken egg yolk immunoglobulin Y (IgY). *J. Immunol. Methods* 162: 155-164.

Bartz, C.R., Conklin, R.H., Tunstall, C.B., and Steel, J.H. (1980) Prevention of murine rotavirus infection with chicken egg immunoglobulin. *J. Infect. Dis.* 142:439-441.

Bidlack, J.M. and Mabie, P.C. (1986) Preparation of Fab' fragments from a mouse monoclonal IgM. *J. Immunol. Methods* 91:157-162.

Covell, D.G., Barbet, J., Holton, O.D., Black, C.D.V., Parker, R.J., and Weinstein, J.N. (1986) Pharmacokinetics of monoclonal immunoglobulin G_1, $F(ab')_2$ and Fab' in mice. *Cancer Res.* 46:3969-3978.

Ebina, T., Tsukuda, K., Umezu, K., Nose, M., Tsuda, K., Hatta, H., Kim, M., and Yamamoto, T. (1990) Gastroenteritis in suckling mice caused by human rotavirus can be prevented with egg yolk immunoglobulin (IgY) and treated with a protein-bound polysaccharide preparation (PSK). *Microbiol. Immunol.* 34:617-629.

Jensenius, J.C., Anderson, I., Hau, J., Crove, M., and Koch, C. (1981) Eggs: conveniently packaged antibodies. Methods of purification of yolk IgG. *J. Immunol. Methods* 46:63-68.

Larsson, A. and Sjoquist, J. (1988) Chicken antibodies: a tool to avoid false positive results by rheumatoid factor in latex fixation test. *J. Immunol. Methods* 108:205-208.

Larsson, A., Karlsson-Parra, A., and Sjoquist, J. (1991) Use of chicken antibodies in enzyme immunoassays to avoid interference by rheumatoid factors. *Clinical Chem.* 37:411-414.

Milenic, D.E., Esteban, J.M., and Colcher, D. (1989) Comparison of methods for the generation of immunoactive fragments of a monoclonal antibody (B72.3) reactive with human carcinomas. *J. Immunol. Methods* 120:71-83.

Nisonoff, A. and Lamoyi, E. (1981) Implications of the presence of an internal image of the antigen in anti-idiotypic antibodies: possible application to vaccine production. *Clin. Immunol. Immunopathol.* 21:397-406.

Ochs, H.R. and Smith, T.W. (1977) Reversal of advanced digitoxin toxicity and modification of pharmacokinetics by specific antibodies and Fab fragments. *J. Clin. Invest.* 60:1303-1313.

Ouchterlony, O. (1962) Diffusion in gel methods for immunological analysis. *Prog. Allergy* 6:30-154.

Shimizu, M., Fitzsimmons, R.C., and Nakai, S. (1988) Anti-*E. coli* immunoglobulin Y isolated from egg yolk of immunized chicken as a potential food ingredient. *J. Food Sci.* 53:1360-1366.

Yada, R.Y. (1984) A study of secondary structure predictive methods for proteins and the relationship between physical-chemical properties and enzymatic activity of some aspartyl proteinases. Ph.D. Thesis. University of British Columbia, Vancouver, Canada.

Chapter Twenty

Prevention of Fish Disease Using Egg Yolk Antibody

H. Hatta, K. Mabe, M. Kim, T. Yamamoto, M.A. Gutierrez,[1] and T. Miyazaki[1]

Central Research Laboratories, Taiyo Kagaku Co. Ltd, 1-3 Takaramachi, Yokkaichi and [1]Faculty of Bioresources, Mie University, Tsu, Mie 510, Japan

Abstract *Edwardsiella tarda* infection (Edwardsiellosis) of Japanese eel has been a serious problem to the eel farming industry. Egg yolk antibody (IgY) is being investigated to prevent this infectious disease. Anti-*E. tarda* IgY was prepared from egg yolk of hens upon immunisation with formalin-treated *E. tarda* as an antigen. In an inoculation experiment, *E. tarda* (10^5-10^6 c.f.u./eel) was orally administered to the eel after its intestinal mucosa had been damaged by introducing hydrogen peroxide through the anus. The infected eels died within 15 days, whereas eels orally administered with *E. tarda* and the active IgY survived throughout the experimental periods of 40 days, without any symptoms of *E. tarda* infection.

(*Key words:* immunoglobulins, egg yolk, immunisation, fish disease, Japanese eel, *Edwardsiella tarda*)

Introduction

Antigen-specific immunoglobulin G (IgG) isolated from the sera of hyper-immunised animals (rabbits, goats, etc.) has been widely used in the field of diagnosis as well as in biological research. Another important application of IgG is passive immunisation therapy in which pathogen-specific IgG is administered to individuals for prevention of infectious diseases. The practical application of this passive immunisation therapy requires an industrially effective method of preparation of antibody, because large amounts of antibodies may be required for administration.

Hens transfer serum IgG to their egg yolk to give immunity to their offspring (Jukes *et al.*, 1934). The antibody in egg yolk has been named IgY (Leslie and Clem, 1969). It is now possible to obtain antigen-specific IgY from egg yolk by hyperimmunising hens with particular antigens (Fig. 20.1). A hen lays 250-300 eggs per year and one egg yolk contains about 150 mg of IgY. We reported a large-scale isolation method of IgY using a natural gum (carrageenan) by which water-soluble protein (IgY fraction) was separated from egg yolk (Hatta *et al.*, 1990). This method does not require any chemicals or organic solvents. Therefore, the purified IgY is considered to be practically applicable for passive immunisation by oral administration.

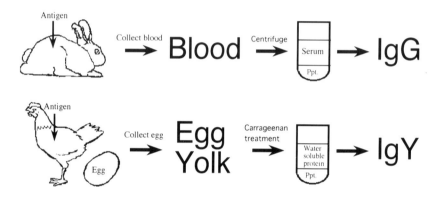

Fig. 20.1. Comparison of methods for preparing antigen-specific antibody.

The effectiveness of passive immunisation by oral administration of IgY to prevent infection has been reported for dental caries (Otake *et al.*, 1991; Hamada *et al.*, 1991) and rotavirus diarrhoea (Bartz *et al.*, 1980; Yolken *et al.*, 1988; Ebina *et al.*, 1990; Hatta *et al.*, 1993). Recently, we also were successful in exploring the possibility of passive immunisation to prevent infectious diseases in fish by using IgY (Gutierrez *et al.*, 1993).

The infection of Japanese eel with *Edwardsiella tarda* (Edwardsiellosis) is now a serious problem to the eel farming industry, because the disease often occurs in elvers, anguillets, and fish grown in farming ponds. The eels perish of Edwardsiellosis as shown in Fig. 20.2. This bacterial infection has been controlled by treatment with antibiotics, e.g. tetracyclines, and/or chemotherapy, e.g. with oxolinic acid. However, it has been found that the use of these antibiotics promotes the growth of drug-resistant bacterial strains (Aoki *et al.*, 1977). The widespread distribution of these strains adversely affects the effectiveness of chemotherapy. Recently, the annual loss in the eel industry in

Japan due to *E. tarda* infection was estimated to be several billion yens.

Edwardsiella tarda is found predominantly in water polluted with organic substances. Also it is known that the intestine of eels is a natural reservoir of *E. tarda*. Miyazaki and Egusa (1976) reported that the main pathological change of eels infected with *E. tarda* is abscess formation in either liver or kidney, as the causative bacterium thus becomes active to penetrate through the intestine. Miyazaki *et al.* (1992) have established an experimental inoculation method of Edwardsiellosis in eel by orally administering *E. tarda* after the eel intestine mucosa has been damaged by introducing a small quantity of hydrogen peroxide through the anus. These studies described above suggest that oral administration of anti-*E. tarda* antibody may provide an effective approach to prevent eel from *E. tarda* infection.

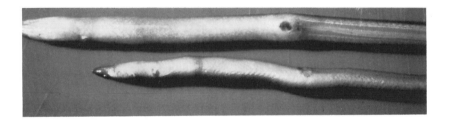

Fig. 20.2. Japanese eels that died as a result of Edwardsiellosis.

The present paper reports a simple method for preparation of anti-*E. tarda* IgY from eggs laid by hens which had been hyperimmunised with formalin-treated *E. tarda* and the effect of its oral administration on prevention of Edwardsiellosis of Japanese eels using an experimental infection model.

Materials and Methods

Immunisation Procedure

Edwardsiella tarda (SH-89198) isolated from naturally infected eel by Shizuoka Prefectural Experimental Fisheries Station in 1989 was used as the antigen. The bacterium was cultured on Heart Infusion Broth while shaking at 25°C for 48 h. The grown cells were harvested and allowed to stand overnight with 0.5% formalin. After washing three times with sterilised saline containing 0.05% sodium azide, the cell suspension was stored at 4°C before being used for immunisation.

Thirty White Leghorn hens (150 days old) kept in isolation were immunised by intramuscular injection with 1 ml of the bacterial cell suspension (ca.

10^8 cells) once a week over 5 weeks. The eggs laid were collected daily and stored in a cold room and the egg yolk was separated weekly and stored in a freezer before isolation of IgY.

Enzyme-Linked Immunosorbent Assay (ELISA)

The ELISA method reported by Cost *et al.* (1985) was applied with slight modifications to determine the antibody activity of egg yolk. The ELISA plate was coated with *E. tarda* whole cells and blocked with ovalbumin. Alkaline phosphate-conjugated rabbit IgG specific to chicken IgG (Zymed) and *p*-nitrophenyl phosphate (Sigma) were used as the second antibody and the substrate, respectively.

Preparation of Crude Anti-*E. tarda* IgY

The egg yolk from *E. tarda*-hyperimmunised hens (5-8 weeks after the initial immunisation) was separated into lipoprotein and water-soluble protein fractions by adding carrageenan (Fig. 20.3). The water-soluble protein fraction was filtered through filter paper and lyophilised. This lyophilised preparation was used as anti-*E. tarda* IgY in the subsequent experiments.

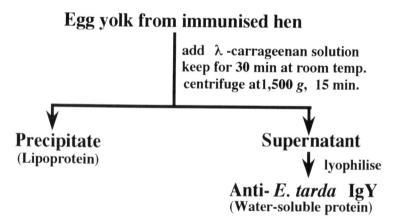

Egg yolk from immunised hen

add λ -carrageenan solution
keep for 30 min at room temp.
centrifuge at 1,500 *g*, 15 min.

Precipitate
(Lipoprotein)

Supernatant
lyophilise

Anti- *E. tarda* IgY
(Water-soluble protein)

Fig. 20.3. Preparation of anti-*E. tarda* IgY.

Stability of IgY Activity in the Intestinal Tract of Eel

The crude anti-*E. tarda* IgY (400 mg/eel) was mixed with a commercial eel diet powder (250 mg) and sterilised saline (1.75 g) to make a kneaded feed (2.4 g) for eels. This feed was administered orally to the eel stomach by cannulation while the eel was anaesthetised. The eels were then kept in 30 l tanks at 25°C until dissection. The entire content of the intestinal tract was sampled from the eels at 1, 2, 5, and 7 h after IgY administration, and stored at -20°C. The levels of antibody activity (anti-*E. tarda* IgY) in the intestinal tract were determined by the ELISA method.

Experimental Infection with *E. tarda*

The method of experimental infection developed by Miyazaki *et al.* (1992) is shown in Fig. 20.4. Japanese eels (ca. 200 g body weight) were anaesthetised in 1.5% urethane solution. A polyethylene tube (3-5 cm) with 1.0 mm diameter was inserted into the intestine through the anus and 0.1 ml hydrogen peroxide solution (30%) was infused to cause damage in the intestinal mucosa. After 18-24 h, 10^5-10^6 c.f.u./eel of viable *E. tarda* was mixed with a kneaded feed (2.4 g) and administered into the eel's stomach by cannulation while the eel was anaesthetised. In the experiment designed to examine the effect of anti-*E. tarda* IgY, the IgY (400 mg/eel) was mixed with the feed. These eels were kept in a tank containing 30 l water at 25°C for 40 days and the mortalities were assessed. Anatomy and histopathology of the dead or moribund eels were performed to look for accumulations of abscess formation due to Edwardsiellosis.

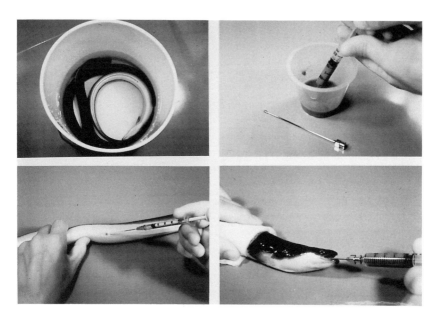

Fig. 20.4. Injection of *E. tarda* into Japanese eel.

Results

Changes in Antibody Levels of Egg Yolk

Changes in anti-*E. tarda* IgY levels of egg yolk were investigated by ELISA over 40 weeks after the initial immunisation (Fig. 20.5). By immunisation once a week, the level of anti-*E. tarda* IgY in the egg yolk markedly increased after the second injection, and reached its highest level in the fifth week, Thereafter, the IgY level gradually decreased. However, when an additional immunisation was performed, the active IgY again reached its maximum level. Thus, a shorter additional immunisation was found effective in maintaining the higher IgY level. It may be important to note that no reduction in egg-laying rate was observed during the immunisation period.

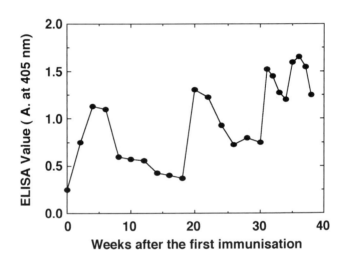

Fig. 20.5. Time course of change in IgY activity against *E. tarda*. Hens were immunised at 0, 1, 2, 3, 18, 30, and 34 weeks.

Anti-*E. tarda* IgY

The lyophilised preparation contained 0.93 mg IgY per 5 mg, based on the pure IgY when determined by a single radial immunodiffusion assay. The agglutination titre of the preparation (5 mg powder/ml) was 126 as determined by the microtitre agglutination method.

Stability of Anti-*E. tarda* IgY in the Eel Intestinal Tract

The anti-*E. tarda* IgY activity in the eel intestine after administration of IgY was detectable as long as the initial level was maintained at least for 7 h (Fig. 20.6). Administration of *E. tarda* (10^6-10^6 cells/eel) brought about severe mortalities to the eels within 15 days of the experiment (Fig. 20.7). All of the dead or surviving eels showed abscess formation in the liver or kidney indicating that they had Edwardsiellosis. In contrast, the eels administered with anti-*E. tarda* IgY (400 mg/eel) together with *E. tarda* (10^5-10^6 cells/eel) survived over the experimental period of 40 days, without showing any symptoms of Edwardsiel-losis (Fig. 20.7). Even the eels (three) previously treated with hydrogen peroxide infusion (surgical control) survived throughout the experimental period.

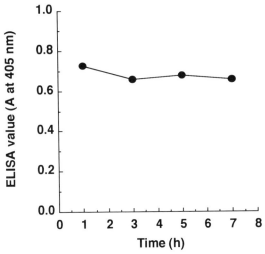

Fig. 20.6. Activity of IgY in the gastrointestinal tract of Japanese eel.

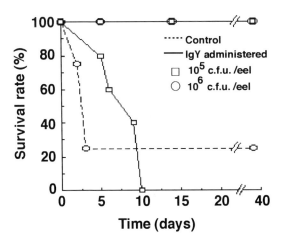

Fig. 20.7. Preventive effect of IgY against *E. tarda* infection in Japanese eel.

Discussion

It is now well known that an antigen (pathogen)-specific antibody can be obtained in large quantities from the eggs laid by hyperimmunised hens. In fact, a tremendous number of hens are immunised to protect them from several avian diseases and managed to lay eggs systematically. The eggs contain IgY and are consumed as food. Considering this fact, it is highly likely that it will be possible to safely apply IgY for passive immunisation therapy through oral administration.

In this study, hens were immunised with formalin-destroyed *E. tarda,* a pathogen known to cause Edwardsiellosis in eels. No decrease in egg-laying rate was observed in the hens during the immunisation period. The anti-*E. tarda* IgY was found to be produced in egg yolk and could be readily isolated from the yolk. Unlike the ordinary method of preparing antigen-specific IgG using mammals (Fig. 20.1), no bleeding is necessary for the production of antibody from egg yolk. An egg usually contains about 150 mg of IgY and thus the amount of IgY from 10-15 eggs may be comparable to the amount of IgG from the whole serum of a rabbit. Therefore, eggs are considered the most convenient source for large-scale production of pathogen-specific antibodies.

Of all the infectious diseases of eels, Edwardsiellosis causes the highest mortality rates. The prevalence of drug-resistant strains of *E. tarda* was recently reported to aggravate the ineffectiveness of chemotherapy. The method of vaccination (active immunisation) was also studied as an alternative control method for *E. tarda* infection. The method of intraperitoneal injection of formalin-treated *E. tarda* was found to be the most effective, thereby producing the antibody for inducing protective immunity of the eel. However, such methods may be difficult to practise in the eel culture industry.

In this paper, we demonstrated that Edwardsiellosis of Japanese eel can be prevented by oral administration of anti-*E. tarda* IgY. Although no immunological activity of IgY was detected in the serum of the eel by ELISA (unpublished data), a high level of anti-*E. tarda* IgY activity in the eel intestinal tract was found to be maintained for as long as 7 h after the oral administration.

Many pathogens of fish have been reported to spread by infection through intestinal mucosa. The oral supply of specific IgY against fish pathogens with feed will be an alternative to the method with antibiotics or chemotherapy for prevention of fish diseases. Moreover, the oral supply of active IgY would be a novel approach for preventing viral infectious diseases of fish because no medicine has been reported to be effective.

References

Aoki, T., Arai, T., and Egusa, S. (1977) Detection of R plasmids in naturally occurring fish pathogenic bacteria, *Edwardsiella tarda. Microbiol. Immunol.* 21:77-83.

Bartz, C.R., Conklin, R.H., Tunstall, C.B., and Steele, J.H. (1980) Prevention of murine rotavirus infection with chicken yolk immunoglobulins. *J. Infect. Dis.* 142:439-441.

Cost, K.M., West, C.S., Brinson, D., and Polk, H.C. (1985) Measurement of human antibody activity against *Escherichia coli* and *Pseudomonas aeruginosa* using formalin treated whole organisms in an ELISA technique. *J. Immunoassay* 6:23-28.

Ebina, T., Tsukada, K., Umezu, K., Nose, M., Tsuda, K., Hatta, H., Kim, M., and Yamamoto, T. (1990) Gastroenteritis in sucking mice caused by human rotavirus can be prevented with egg yolk immunoglobulin (IgY) and treated with a protein-bound polysaccharide preparation (PSK). *Microbiol. Immunol.* 34:617-629.

Gutierrez, M.A., Miyazaki, T., Hatta, H., and Kim, M. (1993) Protective properties of egg yolk IgY containing anti-*Edwardsiella tarda* against paracolo disease in the Japanese eel, *Anguilla japonica Temminck & Schlegel. J. Fish Dis.* 16:113-122.

Hamada, S., Horikoshi, T., Minami, T., Kawabata, S., Hiraoka, J., Fujiwara, T., and Ooshima, T. (1991) Oral passive immunization against dental caries in rats by use of hen egg yolk antibodies specific for cell-associated glucosyltransferase of *Streptococcus mutans. Infect. Immun.* 59:4161-4167.

Hatta, H., Kim, M., and Yamamoto, T. (1990) A novel isolation method for hen egg yolk antibody "IgY". *Agric. Biol. Chem.* 54:2531-2535.

Hatta, H., Tsuda, K., Akachi, S., Kim, M., Yamamoto, T., and Ebina, T. (1993) Oral passive immunization effect of anti-human rotavirus IgY and its behavior against proteolytic enzymes. *Biosci. Biotech. Biochem.* 57:1077-1081.

Jukes, T.H., Fraser, D.T., and Orr, M.D. (1934) The transmission of diphtheria antitoxin from hen to egg. *J. Immunol.* 26:353-360.

Leslie, G.A. and Clem, L.W. (1969) Phylogeny of immunoglobulin structure and function 3. Immunoglobulins of the chicken. *J. Exp. Med.* 130:1337-1352.

Otake, S., Nishihara, Y., Makihara, M., Hatta, H., Kim, M., Yamamoto, T., and Hirasawa, M. (1991) Protection of rats against dental caries by passive immunization with hen egg yolk antibody (IgY). *J. Dental Res.* 70:162-166.

Miyazaki, T. and Egusa, S. (1976) Histopathological studies of Edwardsiellosis of the Japanese eel (*Anguilla japonica*). *Fish Pathol.* 11:67-75.

Miyazaki, T., Gutierrez, M.A., and Tanaka, S. (1992) Experimental infection of Edwardsiellosis in the Japanese eel. *Gyobyo Kenkyu* 27:39-47.

Yolken, R.H., Leister, F., Wee, S., Miskuff, R., and Vonderfecht, S. (1988) Antibodies to rotaviruses in chickens' eggs: a potential source of antiviral immunoglobulins suitable for human consumption. *Pediatrics* 81:291-295.

Chapter Twenty-One

New Methods for Improving the Functionality of Egg White Proteins

A. Kato, H.R. Ibrahim, S. Nakamura, and K. Kobayashi

Department of Biochemistry, Faculty of Agriculture, Yamaguchi University, Yamaguchi 753, Japan

Abstract Novel physical approaches were used to improve the gelling and surface functional properties of egg white proteins. The first attempt was dry-heating of spray-dried egg white proteins. The gelling, foaming, and emulsifying properties increased in proportion to the heating time at 80°C in a dry state. The gel strength, foam stability, and emulsion stability of egg white proteins were increased about four times by dry-heating for 10 days. Thus heating in a dry state seems to be an effective method to improve the functional properties of egg white proteins. Another approach using dry-heating was developed for preparation of protein-polysaccharide conjugates which were readily formed in a controlled dry state (60°C, 65% relative humidity) through the Maillard reaction between the ε-amino groups in proteins and the reducing-end carbonyl residues in polysaccharides. Ovalbumin and lysozyme as egg white proteins and dextran and galactomannan as polysaccharides were employed for protein-polysaccharide conjugation. The resulting conjugates had excellent emulsifying properties superior to commercial emulsifiers, especially in acidic pH and high salt concentrations. Of the tested combinations of proteins with polysaccharides, a lysozyme-dextran (or lysozyme-galactomannan) conjugate yielded the best emulsifying properties. Another interesting bifunctional property of lysozyme-polysaccharide conjugates was antimicrobial activity for both Gram-positive and Gram-negative bacteria.
(*Key words:* dry-heat complexing, lysozyme-polysaccharide conjugate, antimicrobial activity, emulsification)

Introduction

Egg white proteins are extensively utilised as functional ingredients in food processing. Many attempts have been made to develop a rational molecular design using chemical and enzymatic modification of proteins to improve their gelling, foaming, and emulsifying properties. For food application, the use of chemicals should be avoided for the safety of modified proteins.

Proteins have unique surface properties due to their high molecular weight and their amphiphilic properties suitable as a potent surfactant. However, proteins are generally unstable to heating for pasteurisation, and shaking and homogenisation for emulsion formation. Therefore, if proteins could be partially denatured without losing their solubility or modified to a stable form for heat-induced coagulation, the use of proteins as emulsifying, foaming, or gelling agents might be further broadened for food applications. This could be an important goal of modern biotechnology. To achieve this goal, attempts were made to heat proteins in the absence of free water molecules. Although the mechanism of protein denaturation during dry-heating is unclear, it is possible that molecular unfolding is hindered and an intermediate conformation suitable for food functionality may be obtained.

In this study, physical approaches were employed to make new functional proteins without using chemicals. The first approach was the controlled dry-heating of spray-dried egg white proteins by which partial denaturation was expected to occur without loss of solubility. Secondly, the same process was applied to conjugates of proteins with polysaccharides through the Maillard reaction in a dry state. As expected, covalent cross-links were formed between ε-amino groups in proteins and the reducing-end carbonyl group in poly-saccharides. The resulting protein-polysaccharide conjugates had emulsifying properties better than those of commercial emulsifiers.

Dry-Heating of Egg White

Improvement of Gelling and Surface Properties of Egg White by Dry-Heating

Decarbohydrated egg white (DEW) was spray-dried at 60-70°C. DEW in a tightly sealed test tube was then incubated at 80°C for various time periods (days) (Kato *et al.*, 1989). The optimal heating conditions of DEW were preliminarily tested in the range 65-100°C. Heat treatment at 80°C in a dry state (7.5% moisture) yielded the best functional properties. As heating at 80°C is also an efficient way of reducing the microbial population in industrial applications, this condition was chosen for experiments in this study. Table 21.1 shows the solubility of DEW heat-treated at 80°C for different periods of time.

The solubility of DEW was not affected by heating in a dry state for 7 days, but it decreased slightly as a result of heating for 10 days. This result indicates that dry-heating has a relatively minor effect on the solubility of DEW proteins.

Table 21.1. Solubility of heat-treated DEW at 80°C in the dry state for different periods of time.

Heating time (days)	Solubility[a] (%)
0	100 ± 3
1	100 ± 4
3	100 ± 2
5	102 ± 4
7	99 ± 5
10	95 ± 3

[a] The values of solubility of heat-treated samples are represented as the ratio of heat-treated to untreated DEW samples. Reported values are means ± deviations, n = 4.

The effect on the gelling and surface properties of heating in dry state was also investigated. Figure 21.1 shows the effect of dry-heating time at 80°C on the gel strength of DEW. The gel strength of DEW greatly increased with the increase of heating time. The relative gel strength was about four times that of unheated DEW after 10 days of heating (Kato *et al.*, 1989).

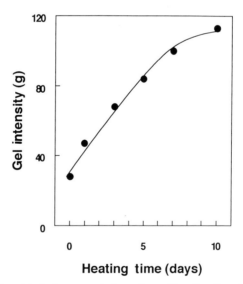

Fig. 21.1. Relationship between gel strength and heating time in dry state of DEW (Kato *et al.*, 1989).

As shown in Fig. 21.2, the foaming properties gradually increased with dry-heating time. The foaming properties were determined by the conductivity method (Kato *et al.*, 1983). The foaming power and foam stability were enhanced with increases of dry-heating time after DEW samples were heated for 10 days.

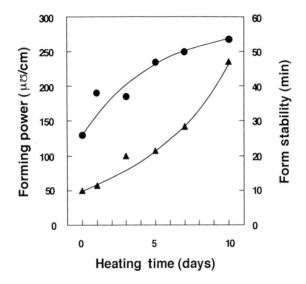

Fig. 21.2. Relationship between foaming properties and dry-heating time (Kato *et al.*, 1989). ●: foaming power, ▲: foam stability.

Similar relationships were observed between the emulsifying properties and dry-heating time of DEW, as shown in Fig. 21.3. These results suggest that controlled dry-heating is effective in improving the functional properties of DEW without loss of solubility. Good functional properties of DEW were obtained by heating at 80°C in 7.5% moisture content for 10 days. No detrimental effects on the emulsifying, foaming, and gelling properties were observed within this heating time range, although the solubility began to decrease upon heating for 7 days. Hence, dry-heating for 10 days is proposed as a way of improving the functional properties of DEW. The same effects were observed for the freeze-dried egg white without decarbohydrating treatment. However, the extent of improvement in the functional properties of freeze-dried egg white was less pronounced than that of DEW upon heating in a dry state. It is of particular interest to know what changes have occurred in the structure of DEW proteins during dry-heating, thereby causing great improvement in gelling and surface properties.

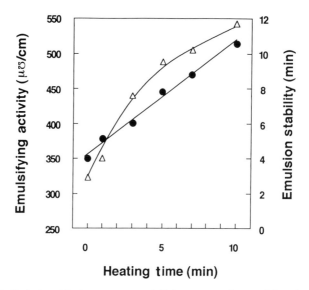

Fig. 21.3. Relationship between emulsifying properties and heating time in dry state of DEW (Kato *et al.*, 1989). ●: emulsifying activity, △: emulsion stability.

Structural Changes in DEW during Dry-Heating

Figure 21.4 shows the relationship between surface hydrophobicity of DEW and heating time in dry state. The surface hydrophobicity greatly increased to give a highly positive correlation with the increase of dry-heating time. The values of surface hydrophobicity (S_0) increased to 7 times the value of unheated DEW, when heated at 80°C for 10 days in a dry state. Taking into account that the S_0 value of unfolded egg white proteins was greater than 1000 (Kato and Nakai, 1980), the small changes observed here suggest the partial denaturation in DEW upon dry-heating. The structural changes of DEW were further investigated by differential scanning calorimetry (Kato *et al.*, 1990b). The DSC (differential scanning colorimeter) thermograms indicated apparent changes in the thermal properties between the untreated control and dry-heated DEW (Fig. 21.5). When the samples were heated in the dry state for 10 days, remarkable broadening of the endothermic peaks at T_d (denaturation temperature) values of 77.0, 66.7, and 60.2°C occurred, with a detectable shift of T_d to lower temperature. The sharpness of an endothermic peak is indicative of the cooperativity of the transition from the native to the denatured state of proteins. Therefore, the broadening of peaks indicates the existence of various intermediate, partially unfolded forms. Thus, the partially unfolded conformation formed by dry-heating may be attributed to the increase in the gelling and surface properties of DEW.

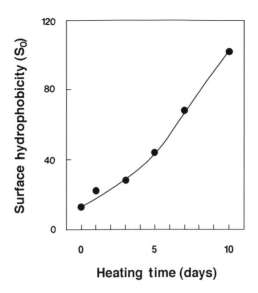

Fig. 21.4. Relationship between surface hydrophobicity and dry-heating time of DEW (Kato *et al.*, 1989).

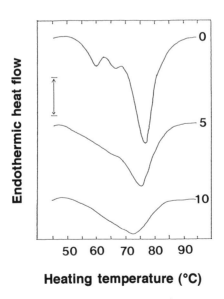

Fig. 21.5. DSC thermograms of DEW (Kato *et al.*, 1990b). 0: unheated, 5: heated for 5 days, 10: heated for 10 days. Vertical arrow represents 0.3 mW.

In addition, the polyacrylamide gel electrophoresis (PAGE) patterns of dry-heated DEW samples shown in Fig. 21.6A revealed the formation of solute aggregates which could migrate slowly into the gel when DEW samples were heated for 3 and 5 days, and the intensity of the bands of aggregates increased with dry-heating time. The result suggests that polymerisation of egg white proteins has occurred by heating in dry state for 5 days. SDS-PAGE was performed to assess the binding forces for aggregation of heat-treated DEW. In the SDS-PAGE patterns in the presence of mercaptoethanol (Fig. 21.6C), there are no changes in the bands at any heating time, indicating that the soluble aggregates were dissociated by SDS and mercaptoethanol. However, when SDS-PAGE was performed in the absence of mercaptoethanol (Fig. 21.6B), the bands with the highest staining intensity in the heated DEW considerably decreased and some aggregates remained undissociated. Accordingly, the binding forces are likely to be hydrophobic interactions in addition to the intermolecular disulphide bonds between egg white proteins. This result suggests that disulphide formation and/or sulphydryl-disulphide interchange reactions occurred during heating in a dry state.

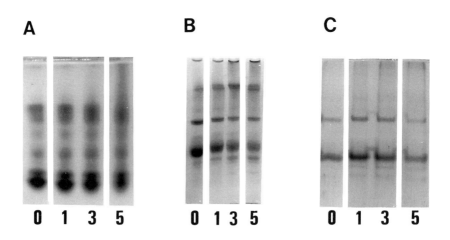

Fig. 21.6. Electrophoretic patterns of DEW dry-heated at 80°C for different periods of time (Kato *et al.*, 1989). Panel A: 7% PAGE patterns, panel B: SDS-PAGE patterns without 2-mercaptoethanol, panel C: SDS-PAGE patterns in the presence of 2-mercaptoethanol. 0: unheated sample, 1: sample heated for 1 day, 3: sample heated for 3 days, 5: sample heated for 5 days.

Table 21.3 shows the changes in free amino groups in DEW dry-heated for different periods of time. No significant changes in the free amino groups are observed at any heating time. This suggests that the controlled dry-heating used in this study apparently avoids browning reactions probably due to the decarbohydration treatment given prior to heating.

Table 21.2. Changes in free amino groups of DEW dry-heated at 80°C for various periods of time.

Heating time (days)	Absorbance at 420 nm[a]
0	0.39 ± 0.03
3	0.37 ± 0.04
5	0.39 ± 0.04
7	0.37 ± 0.03
10	0.37 ± 0.04

[a] The values of absorbance indicate the colour development with trinitrobenzenesulphonate (TNBS). Reported values are means ± standard deviation, n = 4.

The significant increase in surface hydrophobicity and broadening of endothermic peaks seen in DSC suggest that mild conformational changes are caused in DEW proteins by the dry-heating. The 'molten' intermediate structure that is partially unfolded and more flexible than the native form may be formed by controlled heating in dry state. In addition, protein-protein interactions due to disulphide formation and/or sulphydryl-disulphide interchange also occurred in DEW proteins upon dry-heating, as shown in electrophoresis patterns. Unexpectedly, the solubility of DEW was not affected by increasing heating time, although the surface hydrophobicity significantly increased. Thus, heating of DEW proteins in the absence of free water molecules seems to expose the hydrophobic residues to the surface of protein molecules. Since a positive correlation was observed between surface hydrophobicity and emulsifying properties (Kato and Nakai, 1980), the improvement of emulsifying properties observed in this study could be elicited by the increase in surface hydrophobicity. Structural factors reported to be important for foaming properties include protein flexibility (Kato *et al.*, 1985) and protein-protein interaction (Kato, 1991), in addition to surface hydrophobicity. Since these structural changes have been observed in DEW upon dry-heating, the foaming properties may be improved (Kato *et al.*, 1990c). However, the mechanism of the improvement of gelling properties is hard to elucidate. It is well known that the gelling properties may be affected by cross-linking of unfolded molecules through hydrogen bonding, and ionic and hydrophobic interactions. Therefore, it is possible that the 'molten' intermediate structure may strengthen these cross-links. In addition, the increase of cross-linking due to disulphide bond formation may be critical for gel formation. A possible mechanism is that the exposed and reactive residues of 'molten' intermediate molecules, in addition to disulphide bonds, are located in the appropriate positions to form a strong and stable gel matrix through intermolecular interactions during heating, while the reactive residues of the native molecules may remain buried in the interior of protein molecules.

In order to elucidate more precisely the mechanism by which the gelling properties of DEW proteins are improved by dry-heating, we investigated the molecular size of the soluble aggregates (progel) formed prior to gelation. The soluble aggregates are usually formed in an initial stage of gelation, although there is almost no information on this process. We were successful in determining changes in the molecular size and the distribution of soluble aggregates by using the low angle laser light scattering technique in conjunction with HPLC (Kato and Takagi, 1987; Kato *et al.*, 1990a). According to this method, the molecular weight distribution curves were drawn for the heat-induced aggregates of DEW and dry-heated DEW (Fig. 21.7). The molecular weight distribution of DEW aggregates shifted towards lower molecular weight and became sharper with increases in dry-heating time. These data suggest that the homogeneity of the molecular size in solute aggregates may be important in forming a strong gel. In addition, an expanded structure of the soluble aggregates was predicted for dry-heated DEW from the plots of molecular weight versus retention time of HPLC (Kato *et al.*, 1990a). Estimates of the molecular size and shape of progel derived from HPLC suggest that the intermediate unfolded forms induced by dry-heating of DEW may be suitable for the formation of a strong, swelling gel.

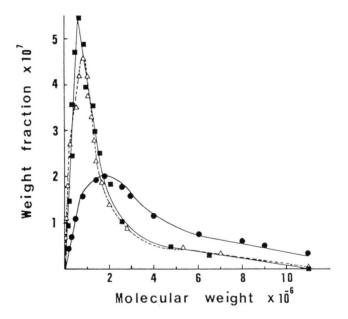

Fig. 21.7. Molecular distribution curves of heat-induced DEW aggregates determined by low angle laser light scattering technique (Kato *et al.*, 1990a). ●: unheated, △: preheated for 5 days in a dry state, ■: preheated for 10 days in a dry state.

Dry-Heating of Protein-Polysaccharide Mixtures

Preparation of Protein-Polysaccharide Conjugates through the Maillard Reaction in a Dry State

We found that protein-dextran conjugates were formed in controlled dry-heating through the Maillard reaction between the ε-amino groups in proteins and the reducing-end carbonyl group in dextran (Kato *et al.*, 1990d; Kato and Kobayashi, 1991). Of the many chemical and enzymatic modifications of proteins used to improve their functionality, this method could be one of the most promising approaches in food processing, because of the safety and the following advantages. Although proteins are generally unstable against heating, organic solvents, and proteolytic attack, they may be converted into stable forms by binding with polysaccharides. We, therefore, prepared a characteristic protein-polysaccharide conjugate through the Maillard reaction formed during storage at 60°C under 65-79% relative humidity. For this purpose, two types of egg white proteins known to have poor functional properties were used. First, an ovalbumin-dextran conjugate was prepared as a model sample. A 1:5 mixture of freeze-dried ovalbumin and dextran (mol. wt 60,000-90,000) was stored at 60°C under 65% relative humidity in a desiccator containing a cup of saturated KI solution for 2-3 weeks. Figure 21.8 shows the SDS-PAGE patterns of the ovalbumin-dextran conjugate. The electrophoresis patterns demonstrate single bands for protein and carbohydrate stains near the boundary between the stacking and separating gels, indicating the formation of a covalent conjugate between ovalbumin and dextran.

A lysozyme-dextran conjugate was also prepared in the same fashion except that the relative humidity at which the best conjugate was obtained was different (79%). Molecular weights of the ovalbumin-dextran conjugate (1:1) and the lysozyme-dextran conjugate (1:2) were 115,700 and 150,000, respectively. The molecular weight and the decrease in free amino groups of protein-dextran conjugates suggest that about two molecules of dextran are bound with ovalbumin and lysozyme, respectively. Since the binding is expressed as average values, it is probable that proteins bound with one to three molecules of dextran may also exist in the protein-dextran conjugates. The binding reaction (panel A) and mode of formation (panel B) of protein-dextran conjugates are proposed as shown in Fig. 21.9. Because there is only one active reducing-end group in dextran, protein and dextran would mutually react without the formation of a network structure. The number of bound polysaccharides thus limited may come from the steric hindrance of attached polysaccharide. This limitation is suitable for designing the functional properties of proteins. When glucose is attached to proteins in a similar manner, the function of proteins may be unfavourably lowered and the detrimental effects (brown colour development etc.) are brought about. However, when polysaccharides are bound to protein, the thermal stability is enhanced and good function is maintained without brown colour developing.

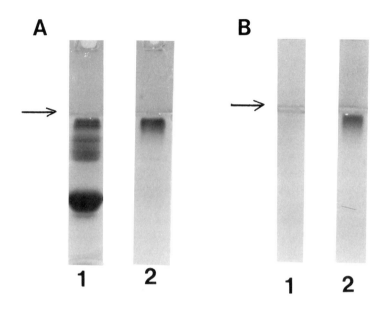

Fig. 21.8. SDS-PAGE patterns of an ovalbumin-dextran conjugate. Lane 1: ovalbumin obtained by dry-heating at 60°C for 3 weeks, lane 2: ovalbumin-dextran conjugate obtained by dry-heating at 80°C for 3 weeks. Panel A: protein stain, panel B: carbohydrate stain. Horizontal arrow shows the boundary between the stacking and separating gels.

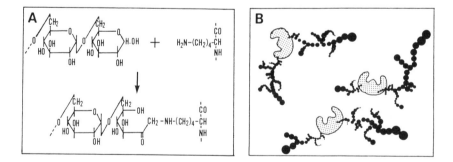

Fig. 21.9. Scheme for the binding of protein to polysaccharide through the Maillard reaction (panel A) and the binding mode (panel B). Dotted areas indicate protein molecules, whereas the branched solid circles represent polysaccharide molecules.

By screening various polysaccharides, galactomannan (mol. wt 15,000-20,000) obtained from mannase hydrolysate of guar gum was found to be a suitable polysaccharide in addition to dextran.

Emulsifying Properties of Protein-Polysaccharide Conjugates

The stability of protein structure can be enhanced by conjugation with polysaccharides, as has been confirmed by lysozyme-dextran conjugate formation (Kato and Kobayashi, 1991). Besides the stabilisation of protein structure, a dramatic enhancement of emulsifying properties for protein-polysaccharide conjugates was observed (Fig. 21.10). The turbidity of emulsion is plotted against standing time after emulsion formation according to the method of Pearce and Kinsella (1978). The value on the ordinate at time 0 is the relative emulsifying activity and the half-life of initial turbidity represents the stability of the emulsion. The conjugates of lysozyme with dextran and galactomannan revealed considerably superior emulsifying activity and emulsion stability to the conjugates of ovalbumin with each corresponding polysaccharide. The emulsifying properties of the conjugates of dried egg white protein with dextran or galactomannan were intermediate between those of lysozyme and ovalbumin as predicted from the composition of egg white protein (data not shown).

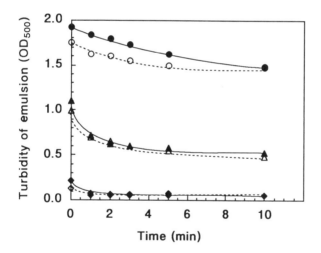

Fig. 21.10. Emulsifying properties of protein-polysaccharide conjugates obtained by dry-heating at 60°C for 2 weeks. ●: lysozyme-galactomannan, ○: lysozyme-dextran, ▲: ovalbumin-galactomannan, △: ovalbumin-dextran, ♦: ovalbumin, ◊: lysozyme. Emulsions were prepared by homogenisation (at 12,000 r.p.m. for 1 min at 20°C) of 1 ml of corn oil and 3 ml of a 0.1% sample solution. Emulsion (0.1 ml) was taken from the bottom of the test tube after standing for 0, 1, 2, 3, 5, and 10 min with 5 ml of 0.1% SDS solution. The turbidity at 500 nm is plotted in the figure.

In order to evaluate the potential industrial application, the emulsifying properties of lysozyme-polysaccharide conjugates were compared with those of commercial emulsifiers (Fig. 21.11a). The lysozyme-galactomannan conjugate prepared under 79% relative humidity revealed better emulsifying properties than those prepared under 65% relative humidity.

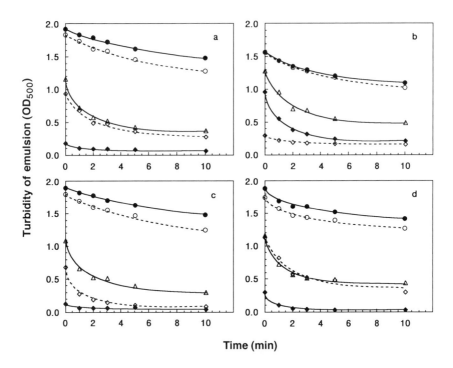

Fig. 21.11. Emulsifying properties of lysozyme-galactomannan conjugates and commercial emulsifiers (Nakamura *et al.*, 1991). In 0.067 M sodium phosphate buffer pH 7.4 (a), 0.067 M sodium citrate buffer pH 3.0 (b), 0.067 M phosphate buffer pH 7.4 containing 0.2 M NaCl (c), and heated samples at 90°C (d). ○: lysozyme-galactomannan conjugates obtained from 2 week incubation at 60°C under 65% relative humidity, ●: lysozyme-galactomannan conjugates obtained from 2 week incubation at 60°C under 78% relative humidity, ◊: commercial emulsifier (Sunsoft SE 11, sucrose-fatty acid ester), △ : commercial emulsifier (Sunsoft Q188, decaglycerol monostearate), ◆: lysozyme. The emulsifying properties were measured using 0.1% sample solution in a similar fashion.

The emulsifying properties of the lysozyme-galactomannan conjugate were much better than those of commercial emulsifiers (sucrose-fatty acid ester and glycerol-fatty acid ester). Furthermore, the emulsifying properties of the conjugates were not affected under acidic conditions (Fig. 21.11b), or by the presence of 0.2 M NaCl (Fig. 21.11c), or by heating of the conjugate (Fig. 21.11d). Since high salt conditions, acidic pH, and/or heating are commonly encountered in the industrial application, the lysozyme-galactomannan conjugate may be a suitable ingredient for uses in food processing.

These conjugates were nontoxic as judged by an animal feeding trial using rats and were also negative in reaction against the Ames test and Rec-assay. Since the commercial mannase hydrolysate of guar gum is contaminated with considerable amounts of low molecular weight carbohydrates, thereby resulting in deterioration of emulsifying properties, the low molecular weight galactomannan should be removed prior to the preparation of the lysozyme-polysaccharide conjugate.

Novel Antimicrobial Action of Lysozyme-Polysaccharide Conjugates

The antimicrobial action of lysozyme-polysaccharide conjugates was investigated (Nakamura *et al.*, 1991). The lytic activities of lysozyme-dextran conjugates and a lysozyme-galactomannan conjugate against typical Gram-positive bacteria were 32 and 78% respectively of that of a lysozyme control. Despite the steric hindrance due to the attachment of dextran or galactomannan, the lytic activity of lysozyme was fairly well preserved. Figure 21.12 shows the antimicrobial action of a lysozyme-dextran conjugate against typical Gram-positive bacteria. As predicted from the lytic activity, the antimicrobial effect on Gram-positive bacteria was about the same as that of control lysozyme. The antimicrobial effect of the lysozyme-dextran conjugate on five different Gram-negative bacteria was also investigated (Fig. 21.13). The number of live cells dramatically decreased with heating time at 50°C in the presence of the lysozyme-dextran conjugate and completely disappeared from the medium after 40 min. On the contrary, the bacterial effects were not observed in the presence of native lysozyme or in the control medium (buffer alone). Similar experiments were conducting using the lysozyme-galactomannan conjugate (Table 21.3); they demonstrated the antimicrobial effects of lysozyme-galactomannan conjugate on the Gram-negative bacterial strains measured after heat-treatment at 50°C for 30 min. Although all of the strains tested were slightly affected by heating in the absence of the conjugate, the antibacterial effects on all strains appeared in the medium supplemented with the lysozyme-galactomannan conjugate. Thus it was concluded that the lethal effect was effectively induced by exposing the cells to the lysozyme-galactomannan conjugate as well as the lysozyme-dextran conjugate (Nakamura *et al.*, 1991).

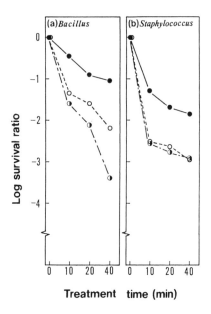

Fig. 21.12. Antimicrobial activity of a lysozyme-dextran conjugate for two Gram-positive bacteria (Nakamura *et al.*, 1991). a: *Bacillus cereus* IFO 13690, b: *Staphylococcus aureus* IFO 14462. ●: control medium without lysozyme or conjugate, ○: native lysozyme, ◐: lysozyme-dextran conjugate. Vertical lines represent standard deviation of the mean.

Lysozyme attacks only specific positions of glycosidic bonds between *N*-acetylhexamines of the peptidoglycan layer in bacterial cell walls. However, since the cell envelope of these bacteria contains a significant amount of hydrophobic materials such as lipopolysaccharide (LPS) associated with the thin peptidoglycan layer, lysozyme fails to lyse Gram-negative bacteria when it is simply added to a cell suspension in the native form. As discussed previously (Nakamura *et al.* 1991), synergistic factors such as detergents and heat treatment destabilise and consequently solubilise the outer membranes which are mainly composed of LPS. The excellent surfactant activity of the lysozyme-galactomannan conjugate seems to destroy the outer membrane synergistically along with thermal stresses in lysing Gram-negative bacterial cells.

Many attempts have been made to develop food preservatives having good antimicrobial effects without toxicities. For this purpose hen egg white lysozyme may be one of the most promising reagents. We have made an attempt to expand the antimicrobial spectrum of hen egg white lysozyme against Gram-negative bacteria and achieved the bifunctional properties of lysozyme-dextran conjugate (Nakamura *et al.*, 1991). In this paper, we also report the bacterial effects of a lysozyme-galactomannan conjugate. In addition to having antimicrobial activity, this lysozyme-galactomannan conjugate demonstrated emulsifying properties more stable than those of commercial emulsifiers under different conditions. The conjugates prepared without using chemicals can be applied for food formulation as a safe multifunctional food additive. Galactomannan may be expected to have some therapeutic effect, as Yamamoto *et al.* (1990) have reported that the oral administration of galactomannan decreases the total content of lipids in the liver of rats. Since galactomannan is not as expensive as dextran, a lysozyme-galactomannan conjugate could probably be used as a food preservative as well as an emulsifier.

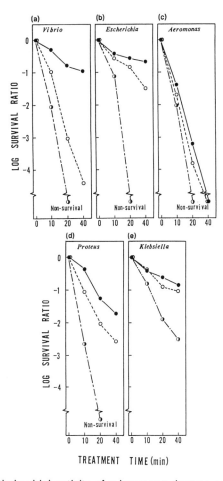

Fig. 21.13. Antimicrobial activity of a lysozyme-dextran conjugate for five Gram-negative bacteria (Nakamura *et al.*, 1991). a: *Vibrio parahaemolyticus* IFO 13286, b: *Escherichia coli* IFO 12713, c: *Aeromonas hydrophila* IFO 13286, d: *Proteus mirabilis* IFO 12668, e: *Klebsiella pneumoniae* IFO 14438. ●: control medium without addition of lysozyme or conjugate, ○: native lysozyme, ◐: lysozyme-dextran conjugate. Vertical lines represent standard deviation of the mean.

Table 21.3. Antimicrobial activity of lysozyme-galactomannan conjugate for five Gram-negative bacteria.

Test strains	Sample	Log survival ratio[a]
Aeromonas hydrophila IFO 13286	Control[b]	-3.95
	Conjugate[c]	Non-survival
	Lysozyme[d]	Non-survival
Vibrio parahaemolytucus IFO 12711	Control	-1.14
	Conjugate	Non-survival
	Lysozyme	-2.84
Escherichia coli IFO 12713	Control	-0.82
	Conjugate	-2.48
	Lysozyme	-0.79
Proteus miriabilis IFO 13300	Control	-2.05
	Conjugate	Non-survival
	Lysozyme	-2.15
Klebsiella pneumonia IFO 14438	Control	-0.59
	Conjugate	-2.89
	Lysozyme	-0.44

[a] Log survival ratio when tested strains were incubated at 50°C for 30 min. [b] In control medium (50 mM pottasium phosphate buffer). [c] In medium supplemented with 0.05% (for protein) lysozyme-galactomannan conjugate. [d] In medium supplemented with 0.05% native lysozyme.

Conclusion

Dry-heating was found to be useful in improving the functional properties of egg white proteins. A stable intermediate 'molten' conformation may have been formed upon heating of DEW in the absence of free water molecules, thereby exposing the functional groups such as hydrophobic domains of the protein molecules. The gelling and surface properties of DEW were greatly enhanced by dry-heating at 80°C for 10 days; this did not affect its solubility. Furthermore, the formation of conjugate was observed between protein and polysaccharide under controlled dry-heating conditions (60°C, relative humidity 65-79%). The protein-polysaccharide conjugate revealed excellent emulsifying properties superior to those of commercial emulsifiers. In addition, the lysozyme-polysaccharide conjugate exhibited antimicrobial activity against both Gram-positive and Gram-negative bacteria along with powerful emulsifying properties. Thus, these protein-polysaccharide conjugates have great potential to be useful food additives because of their safety, nutritional, or therapeutic effects, and their bifunctional properties as an emulsifier and antimicrobial agent.

Acknowledgements

The authors are grateful to Q.P. Company and Taiyo Kagaku (Japan) for supplying decarbohydrated egg white and mannase hydrolysate of guar gum (galactomannan, mol. wt 15,000-20,000), respectively. We gratefully acknowledge the technical assistance of Ms Yoko Sasaki and Mr Kazuaki Minagi. This work was supported by the Urakami Foundation, the Nakano Foundation, and Chemical Materials Research & Development Foundation, Japan.

References

Kato, A. (1991) Significance of macromolecular interaction and stability in functional properties for food proteins. In: Parris, N. and Barford, R. (eds), *Interactions of Food Proteins.* American Chemical Society, Washington, DC, pp. 13-24.

Kato, A. and Kobayashi, K. (1991) Excellent emulsifying properties of protein-dextran conjugate. In: El-Nokaly, M. and Cornell, D. (eds), *Microemulsions and Emulsions in Foods.* American Chemical Society, Washington, DC, pp. 213-229.

Kato, A. and Nakai, S. (1980) Hydrophobicity determined by a fluorescence probe method and its correlation with surface properties of proteins. *Biochim. Biophys. Acta* 624:13-20.

Kato, A. and Takagi, T. (1987) Estimation of the molecular weight distribution of heat-induced ovalbumin aggregates by the low-angle light scattering technique combined with high-performance gel chromatography. *J. Agric. Food Chem.* 35:633-637.

Kato, A., Takahashi, A., Matsudomi, N., and Kobayashi, K. (1983) Determination of foaming properties of proteins by conductivity measurements. *J. Food Sci.* 48:62-65.

Kato, A., Komatsu, K., Fujimoto, K., and Kobayashi, K. (1985) Relationship between surface functional properties and flexibility of protein detected the protease susceptibility. *J. Agric. Food Chem.* 33:931-934.

Kato, A., Ibrahim, H.R., Watanabe, H., Honma, K., and Kobayashi, K. (1989) New approach to improve the gelling and surface functional properties of dried egg white by heating in dry state. *J. Agric. Food Chem.* 37:433-437.

Kato, A., Ibrahim, H.R., Takagi, T., and Kobayashi, K. (1990a) Excellent gelation of egg white preheated in the dry state is due to the decreasing degree of aggregation. *J. Agric. Food Chem.* 38:1868-1872.

Kato, A., Ibrahim, H.R., Watanabe, H., Honma, K., and Kobayashi, K. (1990b) Structure and gelling properties of dry-heating egg white proteins. *J. Agric. Chem.* 38:32-37.

Kato, A., Ibrahim, H.R., Watanabe, H., Honma, K., and Kobayashi, K. (1990c) Enthalpy of denaturation and surface functional properties of heated egg white proteins in the dry state. *J. Food Sci.* 55:1280-1283.

Kato, A., Sasaki, Y., Furuta, R., and Kobayashi, K. (1990d) Functional protein-polysaccharide conjugate prepared by controlled dry-heating of ovalbumin-dextran mixture. *Agric. Biol. Chem.* 54:107-112.

Nakamura, S., Kato, A., and Kobayashi, K. (1991) New antimicrobial characteristics of lysozyme-dextran conjugate. *J. Agric. Food Chem.* 39:647-650.

Pearce, K.M. and Kinsella, J.E. (1978) Emulsifying properties of proteins: evaluation of a turbidimetric technique. *J. Agric. Food Chem.* 26:716-723.

Yamamoto, T., Yamamoto, S., Miyahara, I., Matsumura, Y., Hirata, A., and Kim, M. (1990) Isolation of β-mannan hydrolysing enzyme and hydrolysis of guar gum by isolated enzyme. *Denpun Kagaku* 37:99-105.

Chapter Twenty-Two

Preparation of Transparent Gels from Egg White and Egg Proteins

E. Doi and N. Kitabatake

Research Institute for Food Science, Kyoto University, Uji, Kyoto 611, Japan

Abstract The turbidity and hardness of heat-induced ovalbumin gels are controlled by the pH and ionic strength of the medium. A 5% ovalbumin solution containing 20 mM NaCl was heated for 1 h at 80°C (one-step heating method). Transparent gels were obtained at pH 3.0-3.5 and 7.0-7.5. At pH 7.5, 5% ovalbumin solution gave a transparent sol after heating in the absence of salt. Transparent gels were obtained within a relatively narrow range of NaCl concentrations (10-50 mM). The sol obtained by heating in the absence of salt was then mixed with NaCl solution and re-heated (two-step heating method). Using this method, transparent gels were obtained within a wide range of NaCl concentrations (10-500 mM). The one- and two-step heating methods were successfully applied to lysozyme, serum albumin, and egg white to produce transparent gels. Egg white was dialysed against water and the precipitate was removed by centrifugation. The supernatant gave transparent gels at acidic pH (2.5-3.5). The molecular mechanism which is believed to produce transparent gels is also described.
(*Key words:* egg white, ovalbumin, lysozyme, bovine serum albumin, transparent gel, gel)

Introduction

Egg white forms a white, turbid gel after heating: most globular protein solutions coagulate to form turbid gels after heating. For example, soybean proteins coagulate to form a bean curd which is called tofu in Japan. Transparent gels, like gelatin gels, had been thought to be difficult to obtain from globular protein solutions. However, we have obtained transparent gels from ovalbumin by controlling the medium pH, ionic strength, and protein concentration (Doi and Kitabatake, 1989; Doi, 1993). These transparent gels were obtained most readily by using a two-step heating method. Furthermore, the molecular mechanism which results in formation of the transparent gels has been elucidated. We have successfully applied our method to produce transparent gels with other proteins, including lysozyme, bovine serum albumin, and egg white.

Transparent Ovalbumin Gels

The structure and properties of heat-induced ovalbumin gel are affected by the temperature and duration of the applied heat. The temperature should be higher than the denaturation temperature of ovalbumin: i.e. about 70°C at pH 7.0 (Koseki *et al.*, 1989b). Three factors (pH, ionic strength, and protein concentration) play very important roles in determining gel structure and properties.

The pH profiles of the turbidity and hardness of the heat-induced gel of a 5% ovalbumin solution in 20 mM NaCl kept at 80°C for 1 h are shown in Fig. 22.1 (Hatta *et al.*, 1986) (one-step heating method). Either turbid gels with low hardness or turbid suspensions containing coagula were obtained near the isoelectric point (pI) of ovalbumin (pH 4.7). At both pH 3.0 and pH 7.0, transparent, hard gels were obtained. At more acidic or alkaline pH, transparent, very soft gels or transparent solutions were obtained. Gels were hardest at pH 3.5 and 6.5. These were also the pHs which were critical for determining gel turbidity.

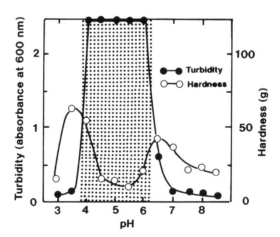

Fig. 22.1. Effects of pH on turbidity and hardness of ovalbumin gel. A 5% ovalbumin solution containing 20 mM NaCl was heated at 80°C for 1 h. Inside the stippled area, turbid gels were obtained (Hatta *et al.*, 1986).

The effect of NaCl concentration on the hardness and turbidity at pH 3.5, 5.5, and 7.5 is shown in Fig. 22.2a, b, and c, respectively. At pH 5.5, turbid soft gels were obtained at all salt concentrations. The absence of NaCl gave a turbid suspension containing a coagulum of ovalbumin. At pH 3.5 and 7.5, transparent solutions or very weak gels were obtained at low NaCl concentrations and in the absence of NaCl. At fairly low concentrations of NaCl, the gels were transparent and their hardness increased with an increase in NaCl concentration. When the concentration of NaCl was higher, an increase in NaCl concentration resulted in a rapid increase in gel turbidity and a gradual decrease in gel hardness. The results show that different pHs and NaCl concentrations of a 5% ovalbumin solution produced transparent solutions, transparent gels, turbid gels, and turbid suspensions. The transparent gels were obtained within a relatively narrow range of pH and salt concentration.

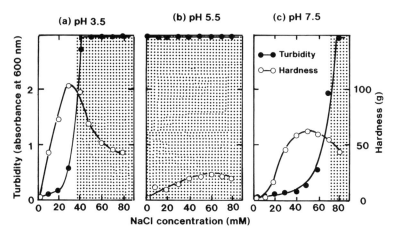

Fig. 22.2. Effects of NaCl concentration on turbidity and hardness of ovalbumin gel. (a) pH 3.5; (b) pH 5.5; (c) pH 7.5. Heating conditions were the same as in Fig. 22.1 except for salt concentrations. Inside the stippled area, turbid gels were obtained (Hatta *et al.*, 1986).

We used a two-step heating method to overcome this problem (Kitabatake *et al.*, 1988b). Figure 22.3a also shows the effect of salt concentration at pH 7.5. The range of salt concentration examined was broader than that represented in Fig. 22.2c. At the lowest ionic strength, i.e. without added salt, transparent solution was obtained after heating. However, the resulting solution was not the same with the original solution: it contained denatured proteins and had a high viscosity. After cooling, the solution to room temperature, NaCl was added to produce various salt concentrations and the mixture was re-heated as shown in Fig. 22.3b (two-step heating method). This treatment resulted in either a transparent gel or a slightly turbid gel with high hardness at wide range of salt concentrations.

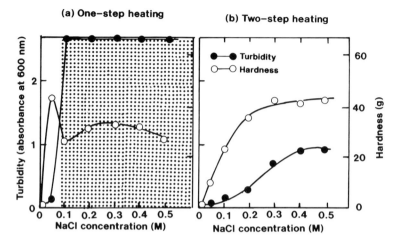

Fig. 22.3. Comparison of one-step heating method (a) and two-step heating method (b). (a) Ovalbumin solution (5%, pH 7.5) was heated at various NaCl concentrations for 1 h at 80°C. (b) Ovalbumin solution (5%, pH 7.5) was heated for 1 h at 80 C in the absence of NaCl. After being cooled, each sample was re-heated with various concentrations of NaCl. Inside the stippled area, turbid gels or sols were obtained (Kitabatake *et al.*, 1987).

The transparent solution obtained by the one-step heating method is believed to have already contained the basic structure necessary for creation of the gel network. This basic structure is thought to be formed irreversibly during the first heating. The conformation and molecular properties of ovalbumin in the transparent solution obtained after one-step heating (at pH 7.0, 20 mM phosphate buffer) was analysed by optical, hydrodynamic, and chromatographic procedures (Koseki *et al.*, 1989a,b).

The presence of large aggregated polymers in the heat-treated transparent solution was confirmed by sedimentation analysis and gel-permeation chromatography. Long linear polymers, without branches, were observed by transmission electron microscopy (Koseki *et al.*, 1989b). The weight average molecular weight of these polymers was estimated to be about 7,500,000 by light scattering. Experimental values obtained by gel permeation chromatography, light scattering, and viscometry were analysed with reference to dilute polymer solution theory. Ovalbumin linear aggregates can be described by the worm-like cylinder model with d = 120 Å and q = 230 Å, where d and q are the diameter and persistence length respectively. The linear polymers were not observed in solutions heated with 100 mM NaCl or at pH values near pI.

The formation of linear polymers or random aggregates is described in Fig. 22.4. The conformation of the heat-denatured ovalbumin is not much different from that of the native molecule. However, some of the hydrophobic areas that were buried in the molecule are now exposed on the surface. At pHs near the pI or at high ionic strength, hydrophobic interactions cause denatured proteins to aggregate randomly. At pHs far from the pI and at low ionic strengths, electrostatic repulsive forces hinder the formation of random aggregates, and linear polymers are formed.

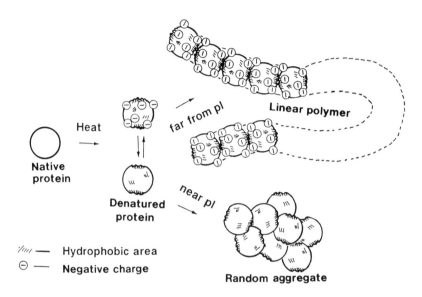

Fig. 22.4. Model for the heat denaturation and formation of linear polymer of ovalbumin (Doi and Kitabatake, 1989).

Figure 22.5 presents a model which explains the formation of either transparent gel or turbid gel and the role played by linear polymers. Heat-denatured proteins are present as monomers at very low protein concentrations, with pH far from pI and low ionic strengths. Linear polymers are formed at medium protein concentration, pH far from pI and low ionic strengths (Fig. 22.5a). With increasing ionic strengths, or decreasing electrostatic repulsive forces, a three-dimensional gel network is formed by interpolymer, perhaps hydrophobic, interactions (Fig. 22.5b). At high ionic strengths and pH near the pI, heat-denatured proteins aggregate to form turbid and soft gel or coagula (Fig. 22.5d). The presence of these coagula within the gel network results in the formation of opaque gels (Fig. 22.5c). The maximum gel hardness is obtained at the critical point of gel turbidity, where the interaction between linear polymers is balanced by electrostatic repulsive forces.

The model shown in Fig. 22.5 explains many of our experimental results. The universal applicability of this model was tested with other globular proteins.

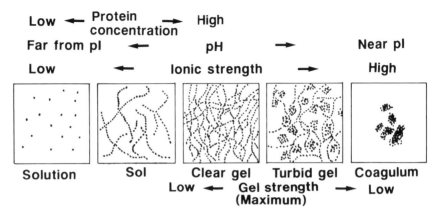

Fig. 22.5. Model for the formation of transparent gel of heated ovalbumin (Doi and Kitabatake, 1989).

Transparent Gels of Bovine Serum Albumin and Lysozyme

Bovine serum albumin (pI 5.2) gave turbid gels at pH 4.0-6.0, transparent gels at pH 3.0-3.5 and 6.5, and transparent solutions at other pH values (Murata *et al.*, 1993). Transparent gels were more readily obtained using the two-step heating method than by the one-step heating method. The presence of long linear polymers was detected by transmission electron microscopy of heated serum albumin solution at pH 7.5, 10 mM NaCl.

Lysozyme is an egg white protein containing four disulphide bonds per molecule. Lysozyme does not gel without reduction of disulphide bridges (Hayakawa and Nakamura, 1986). Heat-induced gelation of 5% lysozyme was examined at 0 to 20 mM dithiothreitol (DTT), pH 7.0 and various concentrations of NaCl (Tani *et al.*, 1993). Without addition of DTT, very soft turbid gels were obtained. However, in the presence of 7.5 mM DTT, transparent gels were obtained at NaCl concentrations of 30-48 mM. A translucent hard gel was obtained at 50 mM NaCl. Gel hardness was greater with 7.5 mM DTT than at higher concentrations of DTT. The two-step heating method was applied to the lysozyme gel. When the first heating was conducted without added salt, the second heating, with added NaCl, produced very hard gels. Transparent gels were obtained at NaCl concentrations between 20 mM and 70 mM. The presence of linear polymers in these gels were observed by transmission and scanning electron microscopy.

With serum albumin and lysozyme the two-step heating method produced transparent gels at a broader range of salt concentration than did the one-step heating method. While the range was not as broad as with ovalbumin gel, the general tendency proves the universal applicability of the model shown in Fig. 22.5 for globular protein gels.

Transparent Gel of Egg White

Heating egg white produces a white turbid gel which contains about 9-10% proteins at about pH 9. The turbidity of egg white before and after heating at various pH values is shown in Fig. 22.6. Egg white before heating was not very turbid at pH above 7. The turbidity was high at low pH. Heating egg white produced a turbid gel except at pH 11-12. In this alkaline pH, egg white became a brownish transparent solution (Kitabatake *et al.*, 1988a,c).

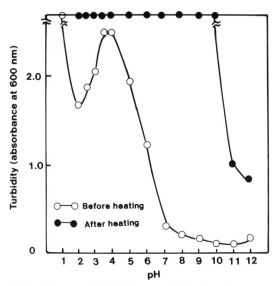

Fig. 22.6. Turbidity of egg white before and after heating at various pH values (Kitabatake *et al.*, 1988c).

As shown in Figs 22.2 and 22.5, the presence of salt in the medium inhibits the production of transparent gel. The original egg white contained many salts which were removed by dialysis against distilled water. Dialysis was continued until the electrical conductivity of a sample reached a constant value. The precipitate that formed during dialysis was removed by centrifugation at 15,000 r.p.m. (28,000 *g*). The turbidity of the supernatant before and after heating is shown in Fig. 22.7. Before heating, the solution was turbid at pH 5.0 and 6.0, and transparent at other pH values. After heating at 80°C for 1 h, transparent gels were formed at pH 2.0-3.5 and at 10.0. Thus, transparent heat-induced gels obtained from egg white by reducing the salt concentration at acidic pH.

Textural properties of heat-induced egg white gels produced after dialysis and centrifugation were examined at various pH values (Fig. 22.8). Hardness peaked sharply at about pH 4.0 at which the gel was translucent. The hardness of transparent gels obtained at pH 2.0, 2.5, 3.0, and 12.0 was low. Adhesiveness peaked at pH 3.5, which was slightly more acidic than where hardness peaked.

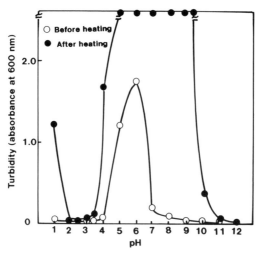

Fig. 22.7. Turbidity of egg white treated with dialysis and centrifugation. Egg white was dialysed against water and the precipitate that formed was removed by centrifugation at 15,000 r.p.m. (Kitabatake *et al.*, 1988c).

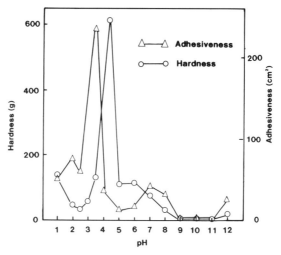

Fig. 22.8. Textural properties of egg white treated with dialysis, centrifugation, and heat. Conditions were the same as those in Fig. 22.7 (Kitabatake *et al.*, 1988c).

The effects on the turbidity and textural parameters of gels caused by adding NaCl to the dialysed egg white were examined at pH 2.5 (Fig. 22.9). Below 75 mM NaCl, transparent gels were obtained. Gel hardness was greatest at 75 mM NaCl. Turbid gels were obtained above 100 mM NaCl. The relationship between NaCl concentration and gel turbidity was very similar to that observed for ovalbumin (Fig. 22.2).

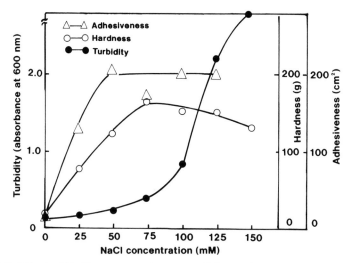

Fig. 22.9. Effect of NaCl concentration on turbidity and textural properties of egg white heated at pH 2.5. Egg white had been treated with dialysis and centrifugation as described in Fig. 22.7 (Kitabatake *et al.*, 1988c).

The two-step heating method was also used to prepare transparent gels from egg white. The method is schematically shown in Fig. 22.10 (Kitabatake *et al.*, 1988b).

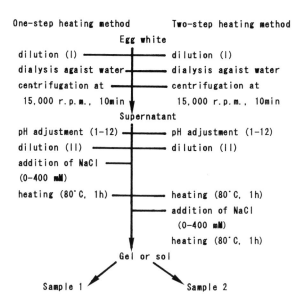

Fig. 22.10. Scheme of one-step and two-step heating methods used for experiments in Figs 22.11-13. The sample was diluted either before [dilution (I)] or after [dilution (II)] centrifugation (Kitabatake *et al.*, 1988b).

A typical example would begin with a dilution of egg white with three volumes of water (fourfold dilution). After dialysis with water and centrifugation, the pH of the supernatant was adjusted to pH 1-12 with addition of 1 N HCl or NaOH. The solution was then heated for 1 h at 80°C. After cooling for 30 min with tap water, NaCl was added (150 mM), and the mixture was reheated for 1 h at 80°C. As a comparison, the one-step heating method (where NaCl was added to the supernatant after dialysis and then heated) was also conducted (Fig. 22.10).

Figure 22.11 shows the turbidity of gels prepared by the one-and two-step heating methods. The one-step heating method (sample 1) gave translucent gels at pH 2.0 and 2.5. The two-step heating method (sample 2) gave transparent sols at 2.5 and transparent gels at pH 2.0, 2.5, 3.0, and 3.5. To prepare a firm, transparent gel, a high protein concentration was necessary.

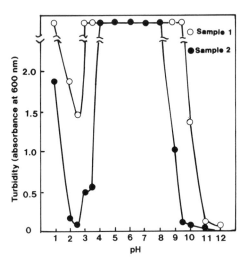

Fig. 22.11. Turbidity of samples 1 and 2 prepared with fourfold dilution at various pHs as shown in Fig. 22.10. The samples were diluted before centrifugation (Kitabatake et al., 1988b).

The effect of protein concentration on the turbidity and textural properties of gels prepared by the two-step heating method is shown in Fig. 22.12. Gels were prepared at pH 2.5. After the first heating, NaCl was added to produce a concentration of 150 mM. The protein concentration of egg white after dialysis and centrifugation at pH 2.5 was about 65 mg/ml (the original egg white contained about 95 mg/ml protein). Below a protein concentration of 15 mg/ml, gel was not formed. Above 45 mg/ml, it was difficult to mix NaCl with the sample after the first heating, because the sample had gelled. Between protein concentrations of 15 and 45 mg/ml transparent gel was obtained. The hardest gel was obtained at a protein concentration of 40 mg/ml.

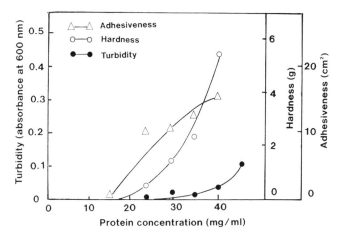

Fig. 22.12. Effects of protein concentration on turbidity and functional properties of sample 2 shown in Fig. 22.10. The sample was diluted after centrifugation. The sample contained 150 mM NaCl at pH 2.5 (Kitabatake *et al.*, 1988b).

The effect of NaCl was examined (pH 2.5, 40 mg/ml protein) (Figs 22.13 and 22.14) in gels prepared by the one-step (sample 1) and two-step (sample 2) heating methods. With sample 1, turbidity increased sharply at about 150 mM NaCl. Over 200 mM NaCl, the sample was completely turbid. Hardness peaked at about 125 mM NaCl, where the turbidity increased sharply. In sample 2, transparency was maintained up to 300 mM NaCl and hardness peaked at about 150 mM NaCl. Thus, transparent and hard gels were obtained by the two-step heating method within a broader range of salt concentrations.

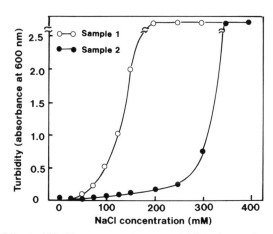

Fig. 22.13. Effect of NaCl concentration on turbidity of samples 1 and 2 shown in Fig. 22.10. The samples were diluted after centrifugation. The sample contained 40 mg/ml protein at pH 2.5 (Kitabatake *et al.*, 1988b).

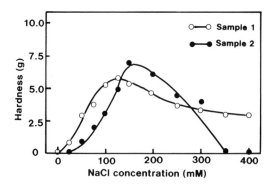

Fig. 22.14. Effect of NaCl concentration on hardness of sample 2 shown in Fig. 22.10. Conditions were the same as those described in Fig. 22.13 (Kitabatake *et al.*, 1988b).

After the first heating of egg white (supernatant which had been diluted after pH adjustment in Fig. 22.11), the clear solution was freeze-dried. The powder obtained by this process rehydrated easily and gave a transparent gel on a second heating with 150 mM NaCl. This procedure offers a convenient means of producing transparent ovalbumin gel for practical uses.

Conclusion

A transparent heat-induced gel was obtained from ovalbumin by adjustment of medium pH and salt concentration. We have given an explanation of the molecular mechanism for the formation of this transparent gel from ovalbumin which can be applied to other globular proteins. However, precise medium conditions are different with each protein.

Many factors affect the properties of gels, especially transparency as summarised in Table 22.1. In addition to pH and ionic strength, protein concentration, heating condition (duration and temperature), and heating method (one-step or two-step) affect the transparency and other textural properties of gels.

Different ionic species (chloride, sulphate, phosphate, and acetate) yield different results (data not shown). Additives such as sugars, fatty acid salts (Yuno-Ohta *et al.*, 1992), and other proteins would affect the transparency and properties. Although analysis of the effects of different factors is difficult, the production of transparent gels with a variety of textural properties is feasible under various conditions.

Table 22.1. Factors affecting transparency and other properties of egg white gel.

pH
Ionic strength, salt anions, and cations
Protein concentration, contaminating proteins
Heating conditions, temperature and duration of heating
Additives
   salts (sodium phosphate, sulphate, and chloride)
   organic acids (acetic, citric, oxalic, and lactic acids, etc.)
   fatty acids (oleic and linoleic acids, etc.)
Dialysis and methods for removing precipitates
   (electro-dialysis, membrane filtration, and centrifugation)
One- and two-step heating methods

References

Doi, E. (1993) Gels and gelling of globular proteins. *Trends Food Sci. Tech.* 4: 1-5.

Doi, E. and Kitabatake, N. (1989) Structure of glycinin and ovalbumin gels. *Food Hydrocol.* 3:327-337.

Hatta, H., Kitabatake, N., and Doi, E. (1986) Turbidity and hardness of a heat-induced gel of hen egg ovalbumin. *Agric. Biol. Chem.* 50:2083-2089.

Hayakawa, S. and Nakamura, R. (1986) Optimization approaches to thermally induced egg white lysozyme gel. *Agric. Biol. Chem.* 50:2039-2046.

Kitabatake, N., Hatta, H., and Doi. E. (1987) Heat-induced and transparent gel prepared from hen egg ovalbumin in the presence of salt. *Agric. Biol. Chem.* 51:771-778.

Kitabatake, N., Shimizu, A., and Doi, E. (1988a) Protein components precipitated from egg white by removal of salt. *J. Food Sci.* 53:292-293.

Kitabatake, N., Shimizu, A., and Doi, E. (1988b) Preparation of transparent egg white gel with salt by two-step heating method. *J. Food Sci.* 53:735-738.

Kitabatake, N., Shimizu, A., and Doi, E. (1988c) Preparation of heat-induced transparent gels from egg white by the control of pH and ionic strength of the medium. *J. Food Sci.* 53:1091-1095, 1106.

Koseki, T., Fukuda, T., Kitabatake, N., and Doi, E. (1989a) Characterization of linear polymers induced by thermal denaturation of ovalbumin. *Food Hydrocol.* 3:135-148.

Koseki, T., Kitabatake N., and Doi. E. (1989b) Irreversible thermal denaturation and formation of linear aggregates of ovalbumin. *Food Hydrocol.* 3: 123-134.

Murata, M., Tani, F., Higasa, T., Kitabatake, N., and Doi, E. (1993) Heat-induced transparent gel formation of bovine serum albumin. *Biosci. Biotech.*

Biochem. 57:43-46.

Tani, F., Murata, M., Higasa, T., Goto, M., Kitabatake, N., and Doi, E. (1993) Heat-induced transparent gel from hen egg lysozyme by a two-step heating method. *Biosci. Biotech. Biochem.* 57:209-214.

Yuno-Ohta, N., Maeda, H., Okada, M., and Hasegawa, K. (1992) Formation of transparent gels of sesame 13S globulin: effect of fatty acid salts. *J. Food Sci.* 57:86-90.

Chapter Twenty-Three

Gamma Irradiation and Physicochemical Properties of Eggs and Egg Products

C.-Y. Ma, M.R. Sahasrabudhe, L.M. Poste, V.R. Harwalkar, J.R. Chambers, and K.P.J. O'Hara

Centre for Food and Animal Research, Agriculture Canada, Ottawa, Ontario, Canada K1A 0C6

Abstract Fresh shell eggs were subjected to gamma irradiation at pasteurising dosages (1-3 kGy). There was significant loss in Haugh values and yolk colour, and a decrease in apparent viscosity of egg white. Polyacrylamide gel electrophoresis (PAGE) of irradiated egg white showed the appearance of minor bands. Differential scanning calorimetry (DSC) of egg white revealed no significant changes in denaturation temperature and enthalpy by irradiation, indicating that the protein was not extensively denatured. Irradiation of fresh eggs led to improvement of functional properties (emulsification activity, overrun, foam stability, gel rigidity, and angel cake volume) of the egg white protein. However, irradiation of frozen egg products (egg white and egg yolk) and spray-dried egg white at pasteurisation dosages (1-4 kGy for frozen products and 2-8 kGy for spray-dried egg white) caused no significant change or slight decrease in functionality of the proteins. The apparent viscosity of liquid egg white and yolk (from shell eggs) and frozen egg products was decreased by radiation treatment. PAGE patterns and DSC profiles of egg white proteins were not affected by irradiation. Angel cakes prepared with irradiated frozen egg white had increased cake volume. Although the foaming properties of dried egg white were improved by irradiation, the angel cake quality of both irradiated and non-irradiated egg white was inferior.
(*Key words:* irradiation, egg, physicochemical)

Introduction

Recent outbreaks of *Salmonella enteritidis* in the United Kingdom and north-eastern United States were traced to the consumption of raw or undercooked shell eggs (St Louis *et al.*, 1988; PHLS, 1989). There was evidence that eggs were contaminated internally with *S. enteritidis* through trans-ovarian infection (PHLS, 1989) which can only be controlled by pasteurisation. Extensive research has shown that ionising radiation at doses of up to 5 kGy suffices to reduce adequately the number of non-sporing pathogenic microorganisms in foods (Vernon, 1977). It therefore is an attractive alternative to heat pasteurisation which is of limited use in shell eggs. However, reports of irradiation on quality of shell eggs were conflicting (Parson and Stadelman, 1957; Morgan and Siu, 1957; Tung *et al.*, 1970; Rauch, 1971; Anonymous, 1989; T.A. Roberts, personal communication). Irradiation had also been used experimentally as an alternative process to pasteurise liquid, frozen, and dried egg products (Brogle *et al.*, 1957; Nickerson *et al.*, 1957; Brooks *et al.*, 1959; Mossel, 1960; Ijichi *et al.*, 1964; Grim and Goldblith, 1965; Bomar, 1970). The present study was carried out to investigate the effect of gamma irradiation on the physicochemical and functional properties of proteins from shell eggs and some egg products.

Materials and Methods

Fresh shell eggs were exposed to gamma radiation from a cobalt-60 source at Nordion International Inc., Kanata, Ontario. Eggs in cartons were packed in ice and irradiated at a minimum dose rate of 0.87-1.00 kGy/h. The ceric-cerous dosimetry system (ASTM, 1988) was used for absorbed dose measurements. The shell eggs received an average absorbed dose of 0.97, 2.37, and 2.98 kGy. Raw unpasteurised egg products, i.e. frozen and spray-dried egg white and frozen egg yolk, were irradiated at -30°C. Frozen egg products received an average absorbed dose of 1.0, 2.5, and 4 kGy, while spray-dried egg white received 2.0, 5.0, and 8.0 kGy. Heat pasteurised (non-irradiated) frozen egg products were also used as controls.

Internal Quality of Shell Eggs

Each shell egg was weighed and broken out onto a level glass plate and the albumen height was measured with a tripod-mounted dial micrometer. The egg weight and egg height determined were used to calculate the Haugh units. The yolk colour was compared against the Roche yolk colour fan (Marusich and Bauenfeind, 1970).

Proximate Analysis

The dry weights of egg white and yolk were determined by heating approximately 5 g samples in a forced air oven at 110°C for 16 h. The protein and free SH contents were determined by the method of Lowry *et al.* (1951) and Beveridge *et al.* (1974) respectively.

Flow Properties

The flow properties of liquid and reconstituted egg products were determined by a Carri-Med controlled stress rheometer (Mitech Corp., Twinsburg, Ohio). For egg white, a double concentric cylinder was used, while a cone and plate (2.0 cm radius, 2° angle) geometry was used for egg yolk. The flow data were analysed to determine the fluid models, using the Carri-Med flow analysis software.

Gel Electrophoresis

The egg albumen was analysed by native polyacrylamide gel electrophoresis (PAGE) and PAGE in the presence of sodium dodecyl sulphate (SDS-PAGE) using a Pharmacia Phast system (Pharmacia AB, Uppsala, Sweden). An 8-25% gradient gel was used and the gels were stained in Coomassie brilliant blue. Molecular size of the fractionated egg white proteins was determined in SDS-PAGE by comparing their electrophoretic mobilities with those of standard proteins of known molecular weights (14,400-94,000). Densitometric scanning of the stained gels was performed with an LKB 2002 Ultroscan laser densitometer (LKB-Produckter AB, Bromma, Sweden) equipped with a Hewlett-Packard 3390A reporting integrator.

Differential Scanning Calorimetry (DSC)

Thermal characteristics of albumen from non-irradiated and irradiated shell eggs were studied by DSC according to the method described by Ma and Harwalkar (1988), using a DuPont 1090 thermal analyser equipped with a high pressure DSC cell. Aliquots (10 µl) of liquid and reconstituted egg white containing approximately 1 mg protein were analysed. The samples were heated from 25 to 105°C at 10°C/min. The peak or denaturation temperature (T_d) and the heat of transition or enthalpy were computed from the thermograms by the 1090 analyser.

Functional Properties

The whippability and foam stability of egg white were determined by the method of Phillips *et al.* (1987). The emulsification activity index (EAI) of egg white and egg yolk was determined by a turbidimetric method (Pearce and Kinsella, 1978).

The gelling properties of egg white proteins were determined by methods described previously (Ma *et al.*, 1990). The angel cake-making quality of egg white was assessed and evaluated for specific gravity, cake volume, and shrinkage index according to AACC (1976) methods.

Results and Discussion

In this study, irradiation dose levels were selected based on previous reports which showed that dosages of 2-4 kGy and 5-7 kGy were effective in pasteurisation of liquid and dry egg products respectively (Proctor *et al.*, 1953; Brogle *et al.*, 1957; Morgan and Siu, 1957; Nickerson *et al.*, 1957; Brooks *et al.*, 1959; Bomar, 1970; Etienne *et al.*, 1973).

Internal Quality

The effect of gamma irradiation on the internal quality of shell eggs is shown in Table 23.1. The non-irradiated eggs had a high Haugh unit, typical of fresh grade A eggs. Irradiation caused a decrease in Haugh unit at all dose levels, and yolk colour at higher dosages. Both the shell and vitelline membrane were weakened by irradiation, and a definite off odour was observed in the treated egg white and yolk. The results indicate that irradiation caused deterioration in the internal quality of shell eggs.

Table 23.1. Internal quality of non-irradiated and gamma irradiated shell eggs.[a]

Dosage (kGy)	Haugh unit	Yolk colour
0	79.7[b]	4.5[b]
0.97	28.4[b]	n.d.
2.37	12.1[d]	3.5[c]
2.98	14.2[d]	3.3[c]
SEM	10.3	0.36

[a] Average of values from 15 (irradiated) and 45 (non-irradiated control) eggs.
[b,c,d] Means in a column with same superscript are not significantly different (*P* > 0.05). n.d., not determined. SEM, standard error of the mean.

Decreases in Haugh unit and weakening of egg shell and vitelline membrane by irradiation were observed by Tung *et al.* (1970). They found that the Haugh unit dropped rapidly with increase in dosage; at 15 kGy, the Haugh unit decreased to about 10, and the quality of the eggs was reduced to grade C (Tung *et al.*, 1970). Fading or discoloration of yolk had also been observed in irradiated hen and duck eggs (Anonymous, 1975)

Proximate and Chemical Analysis

Table 23.2 shows the effect of irradiation on the pH, dry weight, protein, and free SH contents of egg white and yolk in shell eggs. The pH and solid content were not affected by irradiation in either the egg white or egg yolk. The protein and SH contents of egg white were also unchanged by irradiation, but were progressively decreased in the egg yolk with increases in dose level. The data suggest some breakdown of proteins in egg yolk, and the changes in SH content may be attributed to SH oxidation in the yolk proteins. Mohamed (1969) showed that irradiation of ovalbumin solutions at 8.5 kGy caused a reduction in the number of reactive SH groups from four to three.

Table 23.2. Proximate and chemical analysis of non-irradiated and gamma irradiated shell eggs.[a]

Dosage (kGy)	pH	% solid	% protein	SH content (μmol/g)
Egg white				
0	8.52	11.3	10.1	42.6
0.97	8.63	11.4	10.4	42.9
2.37	8.45	11.0	10.0	43.6
2.98	8.50	11.3	10.2	43.5
SEM	0.012	0.25	0.38	0.20
Egg yolk				
0	5.98	53.2	16.1[b]	7.92[b]
0.97	5.98	52.5	15.0[bc]	7.61[bc]
2.37	5.93	53.2	15.6[bc]	7.69[bc]
2.98	5.86	52.9	14.3[c]	7.01[c]
SEM	0.011	0.40	0.53	0.032

[a] Averages of two or three determinations.
[b,c] Means in a column with same superscript are not significantly different ($P > 0.05$). SEM, standard error of the mean.

Table 23.3 shows the proximate and chemical analysis of non-irradiated and irradiated egg products. Neither pH nor solid content was significantly changed by irradiation. Heat pasteurisation caused a slight increase in protein content in the liquid egg products, while irradiation at 1 kGy led to a decrease in protein level. In contrast to shell eggs, the SH content of spray-dried egg white was increased by irradiation. Heat pasteurisation also caused an increase in SH content of frozen egg yolk. These may suggest a breakdown of the disulphide bonds by either irradiation or heat treatment.

Table 23.3. Proximate and chemical analysis of non-irradiated and irradiated egg products[a]

Dosage (kGy)	pH	% Solid	% Protein	SH content (µmol/g)
Frozen egg white				
0	9.42[b]	12.4[b]	12.2[c]	46.5[b]
1	9.40[b]	12.4[b]	12.5[bc]	45.2[b]
2.5	9.39[b]	12.5[b]	12.3[bc]	44.7[b]
4	9.36[b]	12.4[b]	12.4[bc]	45.5[b]
Heat-treated	9.35[b]	12.2[b]	12.7[b]	44.8[b]
SEM		0.050	0.095	0.66
Frozen egg yolk				
0	6.68[b]	48.7[b]	15.7[c]	11.9[c]
1	6.49[b]	48.7[b]	15.0[d]	11.8[c]
2.5	6.52[b]	48.7[b]	15.8[bc]	12.3[c]
4	6.63[b]	48.8[b]	15.8[bc]	11.7[c]
Heat-treated	6.54[b]	48.6[b]	16.0[b]	13.6[b]
SEM		0.052	0.075	0.14
Spray-dried egg white				
0	n.d.	93.2[b]	90.8[b]	48.8[c]
2	n.d.	93.3[b]	90.4[b]	51.5[b]
5	n.d.	93.2[b]	91.3[b]	50.4[b]
8	n.d.	93.3[b]	90.9[b]	50.8[b]
SEM	n.d.	0.060	0.136	0.72

[a] Averages of three determinations.
[b,c,d] Means in a column with the same superscript are not significantly different ($P > 0.05$). SEM, standard error of the mean. n.d., not determined.

Flow Properties

Table 23.4 summarises the flow properties of egg products prepared from non-irradiated and irradiated shell eggs. Both apparent viscosity and plastic viscosity (calculated from Bingham plastic model) of egg white were progressively decreased with increases in the level of irradiation. The apparent viscosity of egg yolk was decreased at 0.97 kGy but increased at 2.37 and 2.98 kGy. Table 23.5 shows the flow properties of non-irradiated and irradiated frozen egg products.

The apparent viscosity was not significantly changed in the frozen egg white and increased in the frozen egg yolk. Heat pasteurisation caused a significant increase in viscosity in both egg white and egg yolk.

Table 23.4. Flow characteristics of liquid egg white and liquid egg yolk from non-irradiated and gamma irradiated shell eggs.[a]

Dosage (kGy)	Apparent viscosity (cp)	Plastic viscosity (cp)	Consistency coefficient	Flow behaviour index
Liquid egg white				
0	12.4[b]	8.0[c]	0.22[b]	0.73[c]
0.97	6.5[c]	5.2[c]	0.08[c]	0.90[b]
2.37	5.0[d]	4.9[c]	0.08[c]	0.87[b]
2.98	4.0[d]	3.7[d]	0.08[c]	0.89[b]
SEM	0.45	0.21	0.0016	0.038
Liquid egg yolk				
0	1770[c]	-	25.5[d]	0.88[b]
0.97	1470[d]	-	18.3[e]	0.87[b]
2.37	2160[b]	-	40.1[c]	0.82[b]
2.98	2180[b]	-	66.4[b]	0.81[b]
SEM	93.5	-	2.06	0.043

Table 23.5. Flow properties of non-irradiated and irradiated egg products.[a]

Dosage (kGy)	Flow behaviour index	Consistency coefficient	Apparent viscosity (cp)
Frozen egg white			
0	0.43[b]	0.66[b]	5.02[c]
1	0.50[b]	0.60[b]	5.88[c]
2.5	0.60[b]	0.65[b]	4.88[c]
4	0.72[b]	0.59[b]	4.97[c]
Heat-treated	0.52[b]	0.69[b]	7.35[b]
SEM	0.035	0.027	0.325
Frozen egg yolk			
0	6.94[d]	0.55[b]	197[e]
1	8.04[cd]	0.48[c]	265[de]
2.5	10.69[b]	0.40[d]	430[b]
4	9.33[cd]	0.45[cd]	301[cd]
Heat-treated	10.38[bc]	0.43[cd]	385[bc]
SEM	0.422	0.021	15.5

[a] Averages of three determinations.

[b,c,d,e] Means in a column with the same superscript are not significantly different ($P > 0.05$). SEM, standard error of the mean.

The data on albumen from shell eggs were consistent with a previous report (Ball and Gardner, 1968) which showed a progressive drop in apparent viscosity of liquid egg white (from 5.26 to 2.81 centipoise) with increases in dosage (0 to 8.6 kGy). The lowering in egg white viscosity and the thinning of albumen (decrease in Haugh unit) may be due to breakdown of proteins, in particular ovomucin, a glycoprotein constituting 3.5% of total albumen proteins. The resistance of frozen egg white to changes in viscosity suggests that the low temperature protects the proteins from breakdown.

The increases in viscosity in the irradiated or heat pasteurised egg yolk may be attributed to aggregation or partial denaturation of the lipoproteins. The flow characteristics of both egg white and egg yolk follows the Power law for pseudoplastic (shear thinning) materials. The consistency coefficient (m), an index for viscosity, was significantly decreased in irradiated egg white from shell eggs and not changed in frozen egg white. The coefficient was increased in irradiated egg yolk from shell eggs and decreased in frozen egg yolk.

The flow behaviour index was increased in irradiated egg white from shell eggs and in frozen egg yolk indicating a lowering in pseudoplasticity. Irradiation of egg yolk from shell eggs and frozen egg white did not cause a significant change in flow behaviour index.

A decrease in consistency coefficient was also observed in irradiated egg white by Tung et al. (1970). They found that the flow behaviour index was not affected by irradiation.

Electrophoretic Patterns

Egg white proteins from shell eggs were fractionated into over ten bands by native PAGE, with the two major peaks corresponding to ovalbumin and conalbumin (Fig. 23.1), similar to the patterns reported by other workers (Chang et al., 1970; Galyean and Cotterill, 1979; Matsuda et al., 1981). Irradiation at 0.97 and 2.37 kGy did not cause significant changes in electrophoretic pattern. At 2.98 kGy, the protein bands became much more diffuse, particularly the minor bands.

When fractionated by SDS-PAGE, both non-irradiated and irradiated egg white showed the two major protein bands, ovalbumin and conalbumin, with estimated molecular weights of 43,000 and 76,000 respectively (Fig. 23.2). Irradiation, particularly at higher dosages, led to the appearance of some minor bands in the molecular weight range 15,000-30,000, and in the high molecular weight range. These suggest both aggregation and breakdown of the major egg white proteins. Irradiation did not cause significant changes in electrophoretic patterns in either frozen or spray-dried egg white (data not shown). The low temperature employed or low moisture content of the product may protect the proteins from aggregation and breakdown by irradiation treatments.

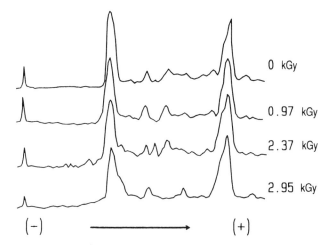

Fig. 23.1. Densitometric tracings of electrophoretic patterns of albumen from non-irradiated and gamma irradiated shell eggs.

Changes in electrophoretic patterns of egg white by gamma irradiation were reported by other workers. Starch gel electrophoresis of irradiated egg white showed a broadening of the protein bands with increases in dosage (Sato *et al.*, 1969). Paper electrophoresis of irradiated egg white showed a decrease in the relative concentration of ovomucin, conalbumin, and ovalbumin, and an increase in globulin fractions (Ball and Gardner, 1968).

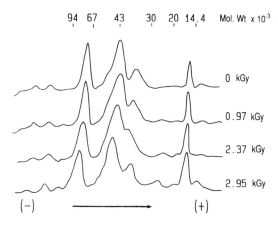

Fig. 23.2. Densitometric tracings of electrophoretic patterns of albumen from non-irradiated and gamma irradiated shell eggs. The molecular weights of standard proteins and their corresponding positions are shown.

Differential Scanning Calorimetry

The DSC characteristics of albumen from non-irradiated and irradiated shell eggs are presented in Table 23.6.

Table 23.6. DSC characteristics of albumen from non-irradiated and gamma irradiated shell eggs.[a]

Dosage (kGy)	pH	T_d (°C)			Enthalpy
		Peak 1	Peak 2	Peak 3	(J/g)
0	8.5	68.4	-	82.6	19.9
0.97	8.5	67.9	-	82.5	18.0
2.37	8.5	68.2	-	81.5	19.0
2.98	8.5	67.7	-	82.6	19.5
SEM		0.40	-	0.42	1.85
0	7.0	64.5	72.4	82.8	17.8
0.97	7.0	63.9	71.9	82.6	18.8
2.37	7.0	64.3	72.4	82.2	16.8
2.98	7.0	63.8	72.4	82.7	17.4
SEM		0.41	0.37	0.39	1.76

[a] Averages of three determinations. SEM, standard error of the mean.

At native pH (8.5), two peaks corresponding to ovalbumin (82.6°C) and conalbumin (68.4°C) (Donnovan et al., 1975) were observed. Irradiation did not significantly change the denaturation temperature, T_d, and total enthalpy of the two proteins. At pH 7, an additional minor peak (72.4°C), corresponding to lysozyme (Donnovan et al., 1975), was observed, and irradiation again did not affect the DSC characteristics of the proteins. Similar results were obtained in frozen and spray-dried egg white (Table 23.7). Apart from a slight decrease in the T_d of ovalbumin at 8 kGy (spray-dried egg white), the DSC characteristics were not affected by irradiation. The appearance of two peaks in frozen egg white and three peaks in spray-dried egg white might have been due to difference in the pH of the two products (Table 23.3). The data indicate that gamma irradiation did not cause significant changes in thermal stability (shift in T_d) of the albumen proteins.

Although the total enthalpy values of the albumen proteins were not affected by irradiation, decreases in enthalpy due to protein unfolding (an endothermic reaction) may be balanced by exothermic events such as protein aggregation and breakup of hydrophobic interactions (Arntfield and Murray, 1981).

Table 23.7. DSC characteristics of non-irradiated and irradiated egg products.[a]

Dosage (kGy) (J/g)	T_d (°C)			Enthalpy
	Peak 1	Peak 2	Peak 3	
Frozen egg white				
0	68.7[b]	-	83.4[b]	25.4[bc]
1	69.0[b]	-	84.8[b]	23.6[c]
2.5	69.1[b]	-	83.5[b]	27.2[b]
4	68.8[b]	-	83.4[b]	25.6[bc]
Heat-treated	69.0[b]	-	83.2[b]	24.9[bc]
SEM	0.14	-	0.48	0.79
Spray-dried egg white				
0	64.4[b]	74.5[b]	84.0[b]	12.1[b]
2	64.4[b]	73.8[b]	84.0[b]	13.2[b]
5	63.8[b]	74.2[b]	83.6[bc]	12.5[b]
8	63.9[b]	73.3[b]	83.4[c]	12.8[b]
SEM	0.20	0.67	0.15	0.48

[a] Averages of three determinations.
[b,c] Means in a column with the same superscript are not significantly different ($P > 0.05$). SEM, standard error of the mean.

Whipping and Emulsifying Properties

Table 23.8 shows the effect of irradiation of shell eggs on whipping and emulsifying properties of egg white. The overrun was not significantly changed by irradiation at 0.97 kGy but was increased at 2.37 and 2.98 kGy. The time for 50% drainage, an index of foam stability, was increased by irradiation at higher dosages indicating an improvement in foam stability. The emulsification activity index (EAI) of freeze-dried egg white was also significantly increased at 2.37 and 2.98 kGy.

Irradiation caused a decrease in overrun at 4 kGy in the frozen egg white but no change in foam stability (Table 23.9). EAI was also lowered by irradiation. Heat pasteurisation did not change the whipping or emulsifying properties of frozen egg white. Both overrun and foam stability of spray-dried egg white were significantly increased by irradiation while EAI was not changed (Table 23.9).

Table 23.8. Whipping and emulsifying properties of albumen from non-irradiated and gamma irradiated shell eggs.[a]

Dosage (kGy) (m²/g)	% Overrun	Time for 50% drainage (min)	EAI
0	1146[b]	30[d]	6.74[c]
0.97	981[b]	35[cd]	7.07[c]
2.37	1354[c]	42[c]	
10.94[b]			
2.98	1446[c]	52[b]	11.05[b]
SEM	91.3	3.4	0.485

[a] Averages of two or three determinations.
[b,c,d] Means in a column with same superscript are not significantly different ($P > 0.05$). EAI, emulsification activity index. SEM, standard error of the mean.

Table 23.9. Whipping and emulsifying properties of non-irradiated and irradiated egg products.[a]

Dosage (kGy)	% Overrun	Time for 50% drainage (min)	EAI (m²/g)
Frozen egg white			
0	815[bc]	40[bc]	10.6[b]
1	870[b]	35[c]	8.1[c]
2.5	779[c]	42[bc]	8.0[b]
4	666[d]	42[bc]	8.5[c]
Heat-treated	857[b]	46[b]	10.3[b]
SEM	19.3	1.8	0.50
Spray-dried egg white			
0	627[e]	27[d]	13.6[b]
2	848[d]	29[cd]	13.4[b]
5	953[c]	30[c]	13.4[b]
8	1105[b]	34[b]	13.2[b]
SEM	22.0	0.70	0.62

[a] Averages of three determinations.
[b,c,d,e] Means in a column with the same superscript are not significantly different ($P > 0.05$). SEM, standard error of the mean.

Gelling Properties

The effect of irradiation on the rheological properties of heat-induced egg white gels is presented in Tables 23.10 and 23.11. For egg white from irradiated shell eggs, the gel hardness was significantly increased at 0.97 and 2.37 kGy, and the rigidity (G') and loss modulus (G") were increased at all dose levels (Table 23.10). All the gels had low tan δ values (G"/G') <0.1 indicating a viscoelastic gel matrix. For frozen egg white, irradiation did not cause significant changes in rheological properties (Table 23.11).

Table 23.10. Rheological properties of heat-induced protein gels prepared with albumen from non-irradiated and gamma irradiated shell eggs.[a]

Dosage (kGy)	Gel hardness (N)	G' (dynes/cm²)	G" (dynes/cm²)	Loss tangent
0	7.02[b]	52,600[d]	3740[d]	0.071[d]
0.97	7.59[c]	80,200[b]	7800[b]	0.098[b]
2.37	8.60[d]	82,600[b]	7140[b]	0.086[c]
2.98	7.27[bc]	71,200[c]	6440[c]	0.090[bc]
SEM	0.202	1050	380	0.0036

[a] Gels were prepared by heating 10% (w/v) reconstituted freeze-dried albumen at 90°C for 20 min. Averages of three to six determinations.
[b,c,d] Means in a column with same superscript are not significantly different (*P* > 0.05). SEM, standard error of the mean.

Table 23.11. Rheological properties of heat-induced protein gels prepared with albumen from non-irradiated and gamma irradiated egg products.[a]

Dosage (kGy)	Gel hardness (N)	G' (dynes/cm²)	G" (dynes/cm²)	Loss tangent
Frozen egg white				
0	15.8	149,300	14,350	0.096
1	15.7	158,000	13,350	0.084
2.5	15.9	153,750	13,540	0.088
4	15.6	149,600	15,060	0.101
Heat-treated	15.5	142,100	15,010	0.106
SEM	0.12	8100	480	0.025
Spray-dried egg white				
0	2.55[b]	44,900	5200	0.115
2	2.35[c]	43,600	4400	0.124
5	2.47[bc]	43,400	5100	0.116
8	2.60[b]	47,400	5500	0.115
SEM	0.015	2750	210	0.024

[a] Averages of three determinations.
[b,c] Means in a column with the same superscript are not significantly different (*P* > 0.05). SEM, standard error of the mean.

For spray-dried egg white, gel hardness was decreased at low dosage but not affected at higher dosages. The data show that the enhancement in thermal gelation property of egg white proteins by gamma irradiation was only observed in the liquid albumen.

Angel Cake Quality

Table 23.12 shows the effect of gamma irradiation of shell eggs on the angel cake-making properties of egg white.

There was no significant difference in meringue specific gravity which was within the range (0.13-0.14) specified in the AACC method. Irradiation led to progressive decreases in batter density and increases in cake volume. These may be related to the enhancement in foaming properties of egg white proteins in the irradiated eggs, since the major functions of egg white proteins in angel cake are the incorporation of air into the batter during whipping and the stabilisation of the expanding foam in baking (Ma *et al.*, 1986). The shrinkage index was not significantly changed. Angel cakes prepared from albumen of irradiated shell eggs exhibited large and non-uniform air cells, and an off-odour was detected when the cakes were removed from the oven.

Table 23.12. Properties of angel cakes prepared with albumen from non-irradiated and irradiated shell eggs.[a]

Dosage (kGy)	Meringue specific gravity (g/ml)	Batter density (g/ml)	Cake volume (ml)	Shrinkage index (%)
0	0.136[b]	0.365[bc]	854[d]	2.02[b]
0.97	0.135[b]	0.374[b]	936[c]	1.40[b]
2.37	0.138[b]	0.335[cd]	1005[b]	1.99[b]
2.98	0.135[b]	0.317[d]	1009[b]	2.87[b]
SEM	0.0151	0.0113	14.4	0.636

[a] Averages of four (irradiated) and 12 (control) determinations.
[b,c,d] Means in a column with same superscript are not significantly different ($P >$ 0.05). SEM, standard error of the mean.

Table 23.13 shows the properties of angel cakes prepared with frozen and spray-dried egg white. For frozen egg white, irradiation or heat pasteurisation caused an increase in meringue specific gravity and cake volume, while the batter density and shrinkage index were not altered. For spray-dried egg white, significant decreases in meringue specific gravity were observed, but both batter density and cake volume were decreased by irradiation.

Nickerson *et al.* (1957) reported that angel cakes prepared from irradiated egg white products (liquid or powder) were not different from the control in volume, flavour, or texture. Ball and Gardner (1968), however, showed that cakes made from gamma irradiated (8.64 kGy) egg white had larger and fewer air cells with thicker cell wall structure, but the finished cakes had no detectable off-odour.

Table 23.13. Properties of angel cakes prepared with non-irradiated and irradiated egg products.[a]

Dosage (kGy)	Meringue specific gravity (g/ml)	Batter density (g/ml)	Cake volume (ml)	Shrinkage index (%)
Frozen egg white				
0	0.197[d]	0.379[bc]	902[c]	1.65[b]
1	0.201[cd]	0.362[c]	926[c]	2.68[b]
2.5	0.212[c]	0.375[bc]	961[b]	1.05[b]
4	0.233[b]	0.400[b]	985[b]	2.03[b]
Heat-treated	0.210[c]	0.390[b]	970[b]	1.03[b]
SEM	0.0039	0.0081	10.2	0.76
Spray-dried egg white				
0	0.172[b]	0.317[b]	1028.8[b]	n.d.
2	0.157[c]	0.281[c]	965.7[b]	n.d.
5	0.151[d]	0.277[cd]	868.2[c]	n.d.
8	0.148[d]	0.260[d]	859.5[c]	n.d.
SEM	0.0009	0.0057	24.54	

[a] Averages of three determinations.
[b,c,d] Means in a column with the same superscript are not significantly different ($P > 0.05$). SEM, standard error of the mean. n.d., not determined.

Conclusion

The present study confirms previous reports (Parson and Stadelman, 1957; Morgan and Siu, 1957; Tung *et al.*, 1970) which showed that gamma irradiation of shell eggs caused a substantial deterioration in quality. Some physicochemical properties of the shell egg components were also affected, leading to general improvement in functional properties. However, the effect of gamma irradiation on frozen and spray-dried egg products was less pronounced, indicating that temperature and moisture content influence the conformational changes, aggregation, and breakdown of proteins. In general, the data are in agreement with previous reports (Brogle *et al.*, 1957; Nickerson *et al.*, 1957; Brooks *et al.*, 1959), and indicate that irradiation of frozen and dried egg products at pasteurising dosages did not cause marked deterioration in whipping or cake baking performance. The present results also show that heat pasteurised liquid egg products have physicochemical and functional properties similar to those of irradiated products. These suggest that irradiation can be used as an alternative procedure for pasteurising frozen and dried egg products.

Acknowledgement

The skilful technical assistance of V. Agar, D. Raymond, and G. Khanzada is acknowledged.

References

AACC (1976) *Approved Method 10-15.* American Association of Cereal Chemists, St Paul, MN.

Anonymous (1975) Information relating to the wholesomeness of irradiated foods: effect of gamma-irradiation on local acceptability and microbial quality of fresh chicken and duck shell eggs when cooked and salt-cured. *Food Irradiation Information* No. 4, 69. International Atomic Energy Agency, Viena Austria.

Anonymous (1989) Ionizing energy in food processing and pest control. II. Applications. *Task Force Report #115*, Council of Agricultural Science and Technology, p. 36.

Arntfield, S.D. and Murray, E.D. (1981) The influence of processing parameters on food protein functionality. I. Differential scanning calorimetry as an indicator of protein denaturation. *Can. Inst. Food Sci. Technol. J.* 14:289-294.

ASTM (1988) *Standard Test Method (E 1205-88) for Using the Ceric-cerous Sulfate Dosimeter to Measure Absorbed Dose in Water.* American Society for Testing and Materials, Philadelphia, PA.

Ball, H.R. and Gardner, F.A. (1968) Physical and functional properties of gamma irradiated liquid egg white. *Poultry Sci.* 47:1481-1487.

Beveridge, T., Toma, S.J., and Nakai, S. (1974) Determination of SH- and SS-groups in some food proteins using Ellman's reagent. *J. Food Sci.* 39:49-51.

Bomar, M.R. (1970) *Salmonella* eradication in egg powder by irradiation. *Arch. Lebensmittelhygiene* 21(5):97-103.

Brogle, R.C., Nikerson, J.T.R., Proctor, B.E., Pyne, A., Campbell, C., Charm, S., and Lineweaver, H. (1957) Use of high-voltage cathode rays to destroy bacteria of the *Salmonella* group in whole egg solids, egg yolk solids, and frozen egg yolk. *Food Res.* 22:572-589.

Brooks, J., Hannan, R.S., and Hobbs, B.C. (1959) Irradiation of eggs and egg products. *Int. J. Appl. Radiat. Isotopes* 6:149-154.

Chang, P., Powrie, W.D., and Fennema, O. (1970) Disc gel electrophoresis of proteins in native and heat-treated albumin, yolk, and centrifuged whole egg.

J. Food Sci. 35:774-778.

Donnovan, J.W., Mapes, C.J., Davis, J.G., and Garibaldi, J.A. (1975) A differential scanning calorimetric study of the stability of egg white to heat denaturation. *J. Sci. Food Agric.* 26:73-83.

Etienne, J.C., Biltiau, J., and Rombaux, J.P. (1973) *Irradiation of Egg and Egg Products.* CEC Information Booklet, Monograph #29, Eurisotop Office. Commision of European Communities, p. 106.

Galyean, R.D. and Cotterill, O.J. (1979) Chromatography and electrophoresis of native and spray-dried egg white. *J. Food Sci.* 44:1345-1349.

Grim, A.C. and Goldblith, S.A. (1965) The effect of ionizing radiation on the flavor of whole egg magma. *Food Technol.* 18:1594-1596.

Ijichi, K., Hammerle, O.A., Lineweaver, H., and Kline, L. (1964) Effects of ultraviolet irradiation of egg liquids on *Salmonella* destruction and performance quality with emphasis on egg white. *Food Technol.* 17:1628-1632.

Lowry, O.H., Rosebrough, N.J., Farr, A.L., and Randle, R.J. (1951) Protein measurement with the Folin phenol reagent. *J. Biol. Chem.* 193:265-275.

Ma, C.-Y. and Harwalkar, V.R. (1988) Studies of thermal denaturation of oat globulin by differential scanning calorimetry. *J. Food Sci.* 53:531-534.

Ma, C.-Y., Poste, L.M., and Holme, J. (1986) Effects of chemical modifications on the physicochemical and cake-baking properties of egg white. *Can. Inst. Food Sci. Technol. J.* 19:17-22.

Ma, C.-Y., Sahasrabudhe, M.R., Poste, L.M., Harwalkar, V.R., Chambers, J.R., and O'Hara, K.P.J. (1990) Gamma irradiation of shell eggs. Internal and sensory quality, physicochemical characteristics, and functional properties. *Can. Inst. Food Sci. Technol. J.* 23:226-232.

Marusich, W.L. and Bauenfeind, J.C. (1970) Oxycarotenoids in poultry pigmentation. 1. Yolk studies. *Poultry Sci.* 49:1555-1566.

Matsuda, T., Watanabe, K., and Sato, Y. (1981) Heat-induced aggregation of egg white proteins as studied by vertical flat-sheet polyacrylamide gel electrophoresis. *J. Food Sci.* 46:1829-1834.

Mohamed, S.Y. (1969) The effect of ionizing radiation on selected chemical, physical, and microbiological characteristics of egg white proteins. Ph.D. Dissertation, Texas A & M University, Texas.

Morgan, B.H. and Siu, R.G.H. (1957) Action of ionizing radiation in individual foods. In: Bailey, S.D., Davies, J.M., Morgan, B.H., Pomerantz, R., Siu, R.G.H. and Tischer, R.G. (eds), *Radiation Preservation of Food.* U.S. Government Printing Office, Washington, DC, pp. 268-294.

Mossel, D.A.A. (1960) The destruction of *Salmonella* bacteria in refrigerated liquid whole egg with gamma irradiation. *Int. J. Appl. Radiat. Isotopes* 9:109-112.

Nickerson, J.T.R., Charm, S.E., Brogle, R.C., Lockart, E.E., Proctor, B.E., and Lineweaver, H. (1957) Use of high-voltage cathode rays to destroy bacteria of the *Salmonella* group in liquid and frozen egg white and egg white solids. *Food Technol.* 11:159-166.

Parson, R.W. and Stadelman, W.J. (1957) Ionizing radiation of fresh shell eggs. *Poultry Sci.* 36:319-322.

Pearce, K.N. and Kinsella, J.E. (1978) Emulsifying properties of proteins: evaluation of a turbidimetric technique. *J. Agric. Food Chem.* 26:716-723.

Phillips, L., Haque, Z., and Kinsella, J. (1987) A method for the measurement of foam formation and stability. *J. Food Sci.* 52:1047-1077.

PHLS (1989) Memorandum of evidence to the Agricultural Committee Inquiry on *Salmonella* in eggs. *Public Health Laboratory Service Microbiol. Digest* 6:1.

Proctor, P.E., Joslin, R.P., Nickerson, J.T.R., and Lockhart, E.E. (1953) Elimination of *Salmonella* in whole egg powder by cathode ray irradiation of egg magma prior to drying. *Food Technol.* 7:291-296.

Rauch, W. (1971) Influence of ionizing radiation on egg quality. *Archiv für Gefluegelkunde,* 35:112-115.

Sato, Y., Umenoto, Y., and Kume, T. (1969) The effect of gamma irradiation on egg white proteins. Preliminary report. *Food Irradiation (Shokuhin-Shosha)* 4:42-46.

St Louis, M.E., Morse, D.L., Potter, M.E., DeMelfi, T.M., Guzewich, J.J., Tauxe, R.V., and Blake, P.A. (1988) The emergence of grade A eggs as a major source of *Salmonella enteritidis* infections. *J. Amer. Med. Assoc.* 259: 2103-2107.

Tung, M.A., Richards, J.F., Morrison, B.C., and Watson, E.L. (1970) Rheology of fresh, aged and gamma-irradiated egg white. *J. Food Sci.* 35:872-874.

Vernon, E. (1977) Food poisoning and *Salmonella* infection in England and Wales, 1973-1975. *Publ. Health* (London) 91:225-232.

Chapter Twenty-Four

^{31}P-NMR Study on the Interfacial Adsorptivity of Ovalbumin Promoted by Lysophosphatidylcholine and Free Fatty Acids

Y. Mine, K. Chiba, and M. Tada[1]

Research Institute of Q.P. Corporation, 5-13-1 Sumiyoshi-cho and [1]Tokyo University of Agriculture and Technology, Laboratory of Bioorganic Chemistry, 3-5-8 Saiwai-cho, Fuchu-shi, Tokyo 183, Japan.

Abstract Dynamic structures and interfacial adsorptivity of ovalbumin and ovalbumin-phospholipid complexes were investigated by observing their phosphate residues with ^{31}P-NMR. The complex of ovalbumin and lyso-phosphatidylcholine (LPC) showed separated phosphorus signals of SerP-68 SerP-344 of ovalbumin and phosphate residues of LPC in the ^{31}P-NMR. The line widths of the phosphorus signals of ovalbumin or the complex in emulsion were closely correlated with their interfacial adsorptivity and differed markedly between the highly and less adsorptive protein-phospholipid complexes. The interfacial adsorptivity of ovalbumin increased with the interaction of phosphatidylcholine (PC) or LPC to form smaller droplets, and the formation of the microemulsion was further promoted by the addition of free fatty acids (FFAs) to a greater extent as the degree of unsaturation and temperature increased. The results suggested that LPC and FFA mutually changed the structure of protein to increase the interfacial adsorption.
(Key words: ovalbumin, lysophosphatidylcholine, free fatty acid, emulsion, NMR)

Introduction

Proteins and phospholipids (PLs) are the principal surface-active compounds in nature and many proteins have good emulsifying properties which are beneficial for manufacturing various emulsified foods. They have a very important physical property in common, an inherent amphiphilic property. This dual character is important for the formation of composite biological structures, such as cell membranes and lipoproteins. Interactions between proteins and PLs have received intensive interest for many years.

In the field of food chemistry, physical and chemical properties of protein-PL complexes in emulsions are interesting topics of study. Phosphatidylcholine (PC) is one of the most common PLs for constructing biomembranes and exists in egg yolk and soybean at high concentrations. Phosphatidylcholine plays an important role in the manufacture of foods as an emulsifier. Lysophosphatidylcholine (LPC), although it is a minor component in PLs, is also an interesting emulsifier which possesses high water solubility and emulsifying properties. Phospholipase A_2 specifically hydrolyses the fatty acids at glycerol-*sn* 2 of the PC molecules and LPC is widely utilised industrially as an emulsifier with high surface activity (von Nieuwenhuyzer, 1981).

A number of reports have appeared on the interactions between PC and food proteins (Ohtsuru *et al.*, 1976, 1979; Kanamoto *et al.*, 1977; Schenkman *et al.*, 1981; Brown *et al.*, 1983; Beckwith, 1984; Hanssens *et al.*, 1985; Cornell and Patterson, 1989; Ericsson, 1990). The emulsifying properties of several proteins were greatly enhanced by sonicating the proteins along with egg yolk PC (Nakamura *et al.*, 1988). In contrast, only few investigations have been reported on the interaction between LPC and food proteins and on the interfacial adsorptivity of the complex of LPC and proteins.

The emulsifying ability and heat stability of egg yolk lipoprotein are improved by phospholipase A_2 (Hell *et al.*, 1970; Dutilh and Groger, 1981); consequently, the higher emulsifying property of the modified egg yolk lipoprotein may be closely correlated with the formation of the complex composed of LPC, free fatty acids (FFAs), and proteins. Phospholipase A_2 is also secreted from the pancreas when we eat dietary oil. As phospholipase A_2 produces an equimolar mixture of LPC and FFAs, there is a possibility that LPC enhances the interfacial adsorptivity of proteins by interacting with FFAs to promote the absorption of dietary oil or protein digestion. Thus, it is an interesting problem to elucidate the dynamic state of protein or protein-lipid complexes on the interface, but it is difficult to evaluate the interfacial adsorptivity of protein or protein-lipid complex using conventional physico-chemical techniques. Phosphorus NMR of phospholipids has given new insight into lipid-protein interaction in membranes and provided useful information about lipid-protein interactions (Seelig, 1978; Yeagle, 1982). We have previously investigated the interfacial adsorptivity of PC and LPC by ^{31}P-NMR, and found that the headgroup motion of PC and LPC was changed with the changes in interfacial adsorptivities and emulsion stability (Chiba and Tada, 1989, 1990a,b). The line width of ^{31}P-NMR spectra of LPC was closely correlated with the interfacial adsorptivity of the LPC. Moreover, stable emulsions were effectively formed by LPC in the presence of FFAs (Chiba and Tada, 1990b).

Egg white proteins are extensively utilised as functional food products in food processing, and ovalbumin, a major protein of egg white, is a critical factor for the functional properties of egg white such as gelling, foaming, and emulsifying capacities. Ovalbumin contains two phosphoserine residues at serines 68 and 344 that give rise to two well-resolved signals in a ^{31}P-NMR spectrum (Vogel and Bridger, 1982). Thus the NMR technique can be applied to evaluate the effects of interaction between ovalbumin and PLs on the interfacial adsorptivity. In this paper, the interfacial adsorptivity of complexes of ovalbumin and PLs was investigated by ^{31}P-NMR to elucidate the role of LPC and FFAs on the promotion of interfacial adsorption of proteins.

Materials and Methods

Materials

Ovalbumin was prepared from fresh egg white by crystallising it out of aqueous sodium sulphate and recrystallising from aqueous ammonium sulphate five times (Kekwich and Cannan, 1936). p-Ovalbumin and plakalbumin were prepared by pepsin digestion (Kitabatake et al., 1988) and subtilisin digestion (Ottesen, 1958), respectively. PC and LPC were obtained from egg PC (Q.P. Corporation, PC-98N) and egg LPC (Q.P., LPC-I) which was obtained by phospholipase A$_2$ treatment of egg PC. The purity of PC and LPC was higher than 98%. Triolein (Wako Pure Chemicals, Tokyo) was purified by silica-gel column chromatography (hexane-diethylether 97:3), and was evaporated to dryness under N$_2$ gas. FFAs were purchased from Wako Pure Chemicals.

Preparation of Vesicles and Micelles

Unilamellar PC vesicles (SUV-PC) were prepared as follows. PC (37 mg/ml) in 20 mM HCl or 20 mM NaOH was subjected to a Microfluidizer (Model M11 OF, Microfluidics Corporation, Newton, MA). The pressure was raised to 8000 p.s.i. at 4°C. After centrifugation (5000 g × 30 min, 4°C), a clear or slightly translucent dispersion was obtained. The sample was flushed with nitrogen, put in vials, capped tightly, and placed in a refrigerator at 4°C. The size distribution of SUV-PC was 20-50 nm, as measured using a laser light scattering photometer (submicron particle sizer Nicomp Model 370-HPL, Pacific Scientific Corporation, Silver Spring, MD). The LPC globular micelles (25 mg/ml) in 20 mM HCl or 20 mM NaOH were prepared by sonicating (Branson Sonifier Model 250, output 40 W) for 5 min at 60°C. The LPC/FFA vesicles were prepared as follows; an aqueous dispersion of an equimolar mixture of LPC (25 mg/ml) and linoleic acid in 20 mM HCl or 20 mM NaOH was sonicated for 5 min at 60°C. The size distribution of LPC/linoleic acid vesicles was 50-200 nm. The phosphorus signal of NMR in SUV-PC and LPC/linoleic acid vesicles was split by the addition of $PrCl_3$ to make 5 mM in the aqueous phase, indicating the bilayer vesicle.

Preparation of Complexes

The SUV-PC, LPC micelles, and LPC/FFA vesicles were mixed with ovalbumin solution (45 mg/ml in 20 mM HCl or 20 mM NaOH) to yield various molar ratios. The total volume of this mixture was adjusted to 2 ml with 20 mM HCl or 20 mM NaOH. Next, the mixture was sonicated for 3 min at 25°C. The ratio of PC or LPC bound to protein was measured as follows: the mixture was subjected to chromatography on a Sephacryl S-200HR column (2.8 × 42 cm) and eluted with 20 mM HCl or 20 mM NaOH. The fractions eluted from the column were combined and the PC or LPC bound to protein was extracted with chloroform-methanol (2:1, v/v). The amount of protein was measured by the Lowry method (Lowry *et al.*, 1951) and PC or LPC was determined using an Iatroscan-TLC/FID analyser (Iatron model TH-10, Tokyo; $CHCl_3$:CH_3OH:H_2O = 70:30:3).

Emulsification and Evaluation of Emulsion

The emulsifying properties of ovalbumin were determined by the turbidimetric technique (Pearce and Kinsella, 1978). Three millilitres of 1 or 5% (w/v) protein solution in distilled water, adjusted to pH 3.0-9.0 using HCl or NaOH, was mixed with various amounts of soybean oil to prepare the emulsion. The mixture was homogenised with a Physcotron instrument (Nition Rikaki Corp., Tokyo) equipped with generator shaft NS-10 at 12,000 r.p.m. for 3 min at 20°C. A 25 µl emulsion sample was taken from the bottom of the container and diluted with 5 ml of 0.1% SDS solution. The emulsifying activity was determined by the absorbance measured immediately after emulsification. Emulsification of the ovalbumin/FFA/LPC complex was achieved by addition of triolein (135 µl) dropwise to the complex solution (2 ml) in 20 mM HCl with agitation by a disperser at 12,000 r.p.m. for 2 min. The size distribution of emulsion droplets was measured by a laser light scattering photometer. The data were expressed as the mean volume droplet size.

Measurements of ^{31}P-NMR Spectra

^{31}P-NMR measurements were performed at 20°C on a Varian VXR-4000s spectrometer at 161.0 MHz, fitted with the probe (10 mm 45-165 MHz frequency), using a 45° angle pulse (25 µs), with 32K data points, a 40,000 Hz spectral window, 20 rotations/s spinning rate, and 2.0 s pulse delay. Protons were fully decoupled by a 9900 Hz decoupler modulation frequency. NMR samples were 3.1 ml in 10 mm precision tubes. The line widths were measured from the resonances at half-height.

Results and Discussion

Figure 24.1 shows the emulsifying activities of ovalbumin, *p*-ovalbumin, and plakalbumin.

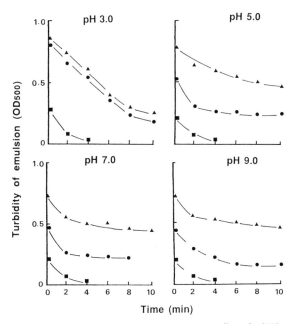

Fig. 24.1. Emulsifying activity of ovalbumin (●), *p*-ovalbumin (■), and plakalbumin (▲) at different pH. 1% (w/v) protein was emulsified with 25% (w/v) soybean oil.

The emulsifying activity of ovalbumin was high at pH 3.0 and decreased with an increase of pH. The emulsifying activity of plakalbumin was about 1.5 times higher than that of ovalbumin at pH 5.0-9.0. In contrast, the emulsifying activity of *p*-ovalbumin was markedly decreased at all pHs used herein. The emulsion stability of *p*-ovalbumin was very low, and oil was separated immediately after emulsification. These results suggest that the N-terminal residues are important for the emulsifying property of ovalbumin. The interfacial adsorption of protein and the mean droplet size of emulsion were dependent on the pH at which the protein was dispersed. The amount of the adsorbed protein decreased, and the droplet size of emulsion became larger with increasing pH. The amount of the plakalbumin adsorbed on the interface was 2-3 times larger and formed smaller droplets than that of ovalbumin. On the other hand, the amount of *p*-ovalbumin adsorbed on the interface was lower than that of ovalbumin (Mine *et al.*, 1992). From these observations, the difference in emulsifying properties of ovalbumin and its partially hydrolysed forms can be predicted to correlate with their dynamic states on the interface.

We expected that differences in phosphorus signals would be seen by ^{31}P-NMR between highly adsorptive and less adsorptive proteins. Thus, the aqueous dispersion and emulsion composed of ovalbumin, *p*-ovalbumin, and plakalbumin prepared at pH 3.0 or 9.0 were subjected to ^{31}P-NMR analysis, and the line widths of ^{31}P-NMR spectra were measured (Fig. 24.2).

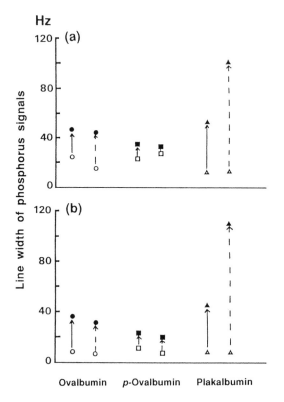

Fig. 24.2. Phosphorus line widths of ovalbumin (●), *p*-ovalbumin (■), and plakalbumin (▲) in aqueous dispersion and emulsion at pH 9.0 (a) and 3.0 (b). (—), SerP-68; (...), SerP-344. 1% (w/v) protein was emulsified with 25% (w/v) soybean oil.

The line widths of ^{31}P-NMR spectra of ovalbumin and plakalbumin obviously broadened following emulsification and the broadening ratios of the line widths of ^{31}P-NMR spectra were larger at pH 3.0 than those at pH 9.0. The phosphorus line widths of plakalbumin in emulsion were broader than those of ovalbumin. In contrast, the line widths of ^{31}P-NMR of *p*-ovalbumin were scarcely changed by emulsification relative to those of ovalbumin and plakalbumin. These data are well correlated with the results of the amount of adsorbed protein on the interface and the mean droplet size of emulsion of ovalbumin and its partially hydrolysed forms (Mine *et al.*, 1992).

The phosphorus signals of ^{31}P-NMR spectra are influenced by the motional properties of the phosphate moiety in the molecules and peak broadening of phosphorus signals of ^{31}P-NMR spectra is observed by increasing the magnitude of the negative phosphorus chemical anisotropy (Wu *et al.*, 1984; Gorenstein, 1982). From these results, the peak broadening of ^{31}P-NMR spectra was interpreted as indicating that the motions of the phosphoserine moiety in ovalbumin and plakalbumin were restricted by adsorbing on the interface. It was also suggested that the extensive peak broadening of ^{31}P-NMR spectra of plakalbumin in emulsion was correlated with its higher interfacial adsorptivity and the amount of protein adsorbed on the interface.

Consequently, we attempted to evaluate the relationship between the line widths of ^{31}P-NMR spectra of ovalbumin and its interfacial adsorptivity at two oil:protein ratios. Figure 24.3 shows typical ^{31}P-NMR spectra of ovalbumin in an aqueous dispersion and in emulsion at the two oil:protein ratios.

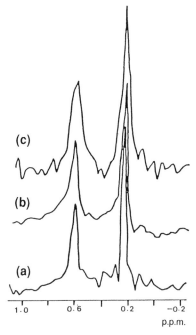

Fig. 24.3. ^{31}P-NMR spectra of ovalbumin in an aqueous dispersion and emulsions. (a) Aqueous dispersion of 5% ovalbumin at pH 3.0, (b, c) emulsion at pH 3.0. The oil:protein ratio was 4 and 8, respectively. Chemical shifts were referenced to 85% H_3PO_4 at 0 p.p.m.

In the emulsion of ovalbumin, the phosphorus line widths became broader with the increasing oil:protein ratio. The relationship between the line widths of ^{31}P-NMR spectra and the mean droplet size in the emulsion of ovalbumin and *p*-ovalbumin at various oil:protein ratios is shown in Fig. 24.4.

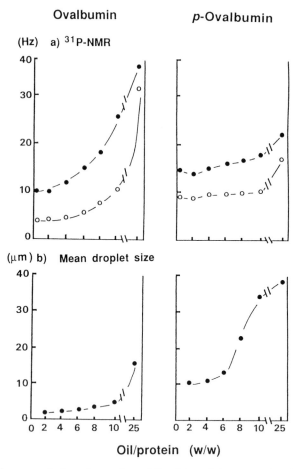

Fig. 24.4. Changes of phosphorus line widths (a) and the mean droplet size of the emulsion (b) of ovalbumin or *p*-ovalbumin at different oil:protein ratios. (a) (●) SerP-68, (○) SerP-344. The emulsions were prepared with a 5% ovalbumin dispersion in distilled water at pH 3.0 and with soybean oil. The droplet size was measured using a submicron particle sizer or particle size distribution analyser.

When the oil:ovalbumin ratio was under 8, the line width of ^{31}P-NMR spectra gradually increased by emulsification and the mean droplet size was less than 3.0 μm. The mean droplet size became larger as the oil:protein ratios increased to >10, and the emulsion became unstable. It was suggested that the amount of ovalbumin was excess when the oil:protein ratio was under 6, and excess ovalbumin existing in an aqueous phase could be adsorbed on the interface with increasing oil content. The peak broadening of ³¹P-NMR spectra was interpreted as indicating that the motions of the phosphoserine moiety in ovalbumin were restricted by adsorption onto the oil:water interface. On the other hand, the amount of ovalbumin became deficient as the oil:protein ratios increased >10 and larger droplets were formed. Moreover, the phosphorus line widths of these emulsions were much broadened. The reason for these phenomena was assumed to be that the motions of the phosphoserine moiety in ovalbumin became more restricted due to the changes of curvature of the droplet surface by increasing the droplet size.

Interestingly, the line widths of ³¹P-NMR spectra of p-ovalbumin scarcely changed despite the increasing amount of oil. The mean droplet size was very large (>10 µm) and the emulsion was unstable at all stages employed here. These results suggested that the interfacial adsorptivity of p-ovalbumin was fairly low and that most of the p-ovalbumin was not adsorbed on the interface by emulsification despite the increasing amount of oil. The line width of ³¹P-NMR spectra of ovalbumin and partially hydrolysed ovalbumin correlated well with their interfacial adsorptivities. ³¹P-NMR spectra in emulsion of ovalbumin would provide an effective measure of its interfacial adsorptivity.

Figure 24.5 shows the changes in mean droplet size of emulsion composed of triolein and an aqueous dispersion of PL-protein complexes. Protein concentration was maintained at 45 mg/2 ml in 20 mM HCl. The mean droplet size of an emulsion made simply from aqueous dispersion of ovalbumin and triolein was about 5.8 µm and greater than 95% of the total volume of oil was turned into an emulsion with droplet sizes between 3.3 µm and 8.6 µm. As the molar ratio of PLs:protein increased, the mean droplet diameter of emulsion became smaller. The LPC-ovalbumin complex formed smaller droplets than did the PC-ovalbumin complex. Moreover, the LPC-ovalbumin complex in the presence of linoleic acid formed smaller droplets than without linoleic acid. These results indicated that interfacial absorptivity of ovalbumin was promoted by the interaction with PC or LPC, and that the formation of microemulsion was further promoted by linoleic acid.

The aqueous dispersion and emulsion composed of PC-ovalbumin, LPC-ovalbumin, and LPC-ovalbumin complexes in the presence of linoleic acid in 20 mM HCl were subjected to ³¹P-NMR analysis (Fig. 24.6). The molar ratio of PC and LPC to protein was 10. The complex of protein and PLs gave three separated phosphorus signals and these signals were as signed to SerP-68, SerP-344 of ovalbumin (Vogel and Bridger, 1982) and to the phosphate residues of PLs, by addition of PLs to ovalbumin solution, as shown in Fig. 24.6. In the case of the PC-ovalbumin complex, the line width of the phosphorus signal of PC in aqueous dispersion became broader (130 Hz wide) with the formation of the complex with ovalbumin. The line width of PC again became narrower (74 Hz wide) on emulsification. The increasing ratio of line widths of the two phosphoserines of ovalbumin in the PC-ovalbumin complex were only 1.2- to

1.4-fold wider in emulsion than in aqueous dispersions. Moreover, the phosphorus signal of LPC in an aqueous dispersion of the LPC-ovalbumin complex and the LPC-ovalbumin complex with linoleic acid were scarcely changed with the interaction of ovalbumin.

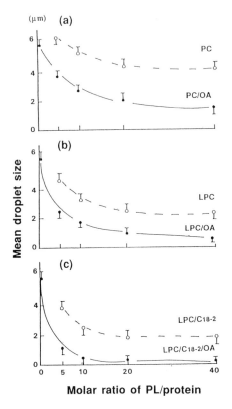

Fig. 24.5. Changes of the mean droplet size of emulsion composed of PLs and protein (—) or just PLs (...) as the emulsifier. 20 mM HCl and triolein were used as aqueous and oil phase, respectively. The ratio of oil:protein was 3. (a) PC (O), PC-ovalbumin (●); (b) LPC (O), LPC-ovalbumin (●); (c) LPC-linoleic acid (O), LPC-linoleic acid-ovalbumin (●).

The results indicate that the headgroup motion of LPC was scarcely restricted in spite of the interaction with protein. On the other hand, ^{31}P-NMR spectra of the two phosphoserines of ovalbumin at the interface differed markedly between the LPC-ovalbumin complex and the LPC/linoleic acid/ovalbumin complex. In the presence of linoleic acid, the mean droplet size of the emulsion became smaller (< 0.5 μm) and >95% of the total volume of oil formed a fine emulsion with droplet size between 0.09 μm and 0.8 μm. The line width of SerP-68 of ovalbumin was much broadened, becoming about 3-fold wider than that in aqueous dispersion.

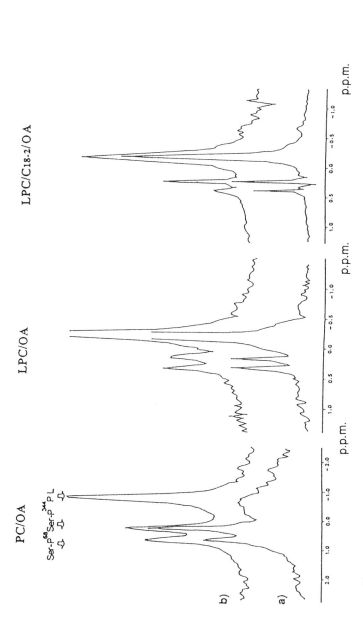

Fig. 24.6. ^{31}P-NMR spectra of complexes in an aqueous dispersion (a) and emulsion (b). The molar ratio of PL: protein was 10. The emulsions were prepared with triolein and the complex of PL and protein dispersed in 20 mM HCl. The ratio of oil: protein (w/w) was 3. A total of 12, 000 acquisitions was made, and additional line broadening 0.1 Hz. Chemical shifts were referenced to 85% H_3PO_4 at 0 p.p.m.

The results suggested that the specific restriction of the SerP-68 moiety of ovalbumin was important for formation of a fine emulsion with the interaction of LPC and linoleic acid. LPC and linoleic acid both changed the structure of ovalbumin to increase the interfacial adsorption of ovalbumin. As mentioned above, the linoleic acid promoted the interfacial adsorptivity of the LPC-protein complex. The surface activity of lysophospholipid was influenced by the degree of unsaturation and chain length of FFAs which coexisted with lysophospholipid (Fujita and Suzuki, 1990). We investigated the effect of the kind of FFA on the interfacial adsorptivity of the LPC-protein complex.

Figure 24.7 shows the changes of mean droplet size under the influence of emulsifying temperature and the kind of, or degree of unsaturation of FFAs which coexisted with LPC. The mean droplet size of the emulsion was influenced little by saturated fatty acids, while it decreased as the degree of unsaturation of FFAs increased. Smaller droplets were obtained by heating during emulsification. The size distribution of LPC-linoleic acid bilayer vesicles was 50-200 nm. We also observed that the bilayer vesicles were destroyed by the interaction with protein and that the size distribution of the LPC/linoleic acid/ovalbumin complex was 10-20 nm, as measured with a laser light scattering photometer.

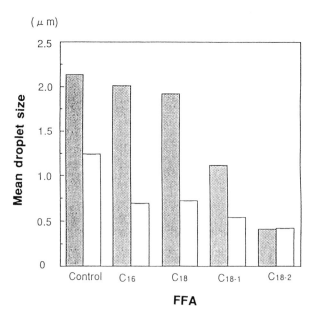

Fig. 24.7. Mean droplet size of emulsions composed of a complex and triolein. The complex was composed of LPC/free fatty acid/ovalbumin (10:10:1 mol/mol) in 20 mM HCl. The ratio of oil:protein was 3. The emulsions were prepared at 25°C (dotted bars) or 80°C (open bars). Control: LPC-ovalbumin (10:1 mol/mol), C16: palmitic acid, C18: stearic acid, C18-1: oleic acid, C18-2: linoleic acid.

In conclusion, the interfacial adsorptivity of ovalbumin was promoted by interacting with PLs, and the formation of the microemulsion was further promoted by the interaction with LPC and FFAs. It was found that the SerP-68 moiety of ovalbumin was specifically restricted at the interface when the complex formed a microemulsion. Moreover, bilayer vesicles composed of LPC and FFA were destroyed by the formation of a complex with ovalbumin, i.e. higher interfacial adsorptivity resulted from a complex of LPC, FFA, and the protein (Fig. 24.8). It is suggested that the LPC/FFA/protein complexes show highly interfacial adsorptivity.

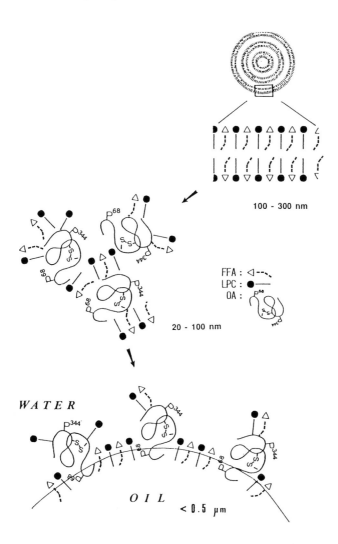

Fig. 24.8. A proposed model of the interfacial adsorption of ovalbumin promoted by LPC and FFAs.

References

Beckwith, A. (1984) Interaction of phosphatidylcholine vesicles with soybean 7S and 11S globulin proteins. *J. Agric. Food Chem.* 32:1397-1402.

Brown, E.M., Carroll, R.J., Pfeffer, P.E., and Sampugna, J. (1983) Complex formation in sonicated mixtures of α-lactoglobulin and phosphatidylcholine. *Lipids* 18:111-118.

Cornell, D.G. and Patterson, D.L. (1989) Interaction of phospholipids in monolayers with α-lactoglobulin adsorbed from solution. *J. Agric. Food Chem.* 37:1455-1459.

Chiba, K. and Tada, M. (1989) Study of the emulsion stability and headgroup motion of phosphatidylcholine and lysophosphatidylcholine by ^{13}C- and ^{31}P-NMR. *Agric. Biol. Chem.* 53:995-1001.

Chiba, K. and Tada, M. (1990a) Effect of the acyl chain length of phosphatidylcholines on their dynamic states and emulsion stability. *J. Agric. Food Chem.* 38:1177-1180.

Chiba, K. and Tada, M. (1990b) Interfacial adsorptivity of lysophosphatidylcholine measuring its interaction with triacylglycerols and free fatty acids. *Agric. Biol. Chem.* 54:2913-2918.

Dutilh, C.E. and Groger, W. (1981) Improvement of product attributes of mayonnaise by enzymic hydrolysis of egg yolk with phospholipase A 2. *J. Sci. Food Agric.* 32:451-458.

Ericsson, B. (1990) Lipid protein interaction. In: Larsson, K. and Friberg, S. (eds), Food Emulsion. 2nd edn. Marcel Dekker, Inc., New York, pp.181-201.

Fujita, S. and Suzuki, K. (1990) Surface activity of the lipid products hydrolysed with lipase and phospholipase A-2. *J. Am. Oil Chem. Soc.* 67:1008-1014.

Gorenstein D.C. (1982) *Phosphorus-31 NMR; Principles and Application.* Academic Press, London, pp. 1-51.

Hanssens, I., Vanebroech, J.C., Pottel, H., Preaux, G., and Vanwelaert, F. (1985) Influence of the protein conformation on the interaction between α-lactalbumin and dimyristoylphosphatidylcholine vesicles. *Biochim. Biophys. Acta* 817:154-164.

Hell, R., Menz, H., and Wieske, T. (1970) Food Emulsion. UK Patent No. 1215868.

Kanamoto, R., Ohtsuru, M., and Kito, M. (1977) Diversity of the soybean protein-phosphatidylcholine complex. *Agric. Biol. Chem.* 41:2021-2026.

Kekwich, R.A. and Cannan, R.K. (1936) The hydrogen ion dissociation curve of the crystalline albumin of the hen's egg. *Biochem. J.* 30:227-234.

Kitabatake, N., Indo, K., and Doi, E. (1988) Limited proteolysis of ovalbumin by pepsin. *J. Agric. Food Chem.* 36:417-420.

Lowry, O.H., Rosebrough, N.J., Farr, A.L., and Randall, R.J. (1951) Protein measurement with the folin phenol reagent. *J. Biol. Chem.* 193:265-275.

Mine, Y., Chiba, K., and Tada, M. (1992) Effects of a limited proteolysis of ovalbumin on the interfacial adsorptivity studied by ³¹P-NMR. *J. Agric. Food Chem.* 40:22-26.

Nakamura, R., Mizutani, R., Yano, M., and Hayakawa, S. (1988) Enhancement of emulsifying properties of protein by sonicating with egg yolk lecithin. *J. Agric. Food Chem.* 36:729-732.

Ohtsuru, M., Kito, M., Takeuchi, Y., and Ohnishi, S. (1976) Association of phosphatidylcholine with soybean protein. *Agric. Biol. Chem.* 40:2261-2266.

Ohtsuru, M., Yamashita, Y., Nakamoto, R., and Kito, M. (1979) Association of phosphatidylcholine with soybean 7S globulin and its effect on the protein conformation. *Agric. Biol. Chem.* 43:765-770.

Ottesen, M. (1958) Transformation of ovalbumin into plakalbumin. *C.R. Trav. Lab. Carlsberg, Ser. Chim.* 30:211-270

Pearce, K.N. and Kinsella, J.E. (1978) Emulsifying properties of proteins; evaluation of a turbidimetric technique. *J. Agric. Food Chem.* 26:716-723.

Schenkman, S., Arauje, P.S., Dijkman, R., Quina, F.H., and Chaimorich, H. (1981) Effects of temperature and lipid composition on the serum albumin-induced aggregation and fusion of small unilamellar vesicles. *Biochim. Biophys. Acta* 649:633-641.

Seelig, J. (1978) Phosphorus-31 nuclear magnetic resonance and the headgroup structure of phospholipids in membranes. *Biochim. Biophys. Acta* 515: 105-140.

Vogel, H.J. and Bridger, W.A. (1982) Phosphorus-31 nuclear magnetic resonance studies of the two phosphoserine residues of hen egg white ovalbumin. *Biochemistry* 21:5825-5831.

von Nieuwenhuyzer, W. (1981) The industrial uses of special lecithin. *J. Am. Oil Chem. Soc.* 58:886-888.

Wu, W., Stephenson, F.A., Mason, J.T., and Huang, C. (1984) A nuclear magnetic resonance spectroscopic investigation of the headgroup motions of lysophospholipids in bilayers. *Lipids* 19:68-71.

Yeagle, P.L. (1982) Phosphorus-31 nuclear magnetic resonance studied of the phospholipid-protein interface in cell membranes. *Biophys. J.* 37:227-239.

Chapter Twenty-Five

Low-Strain Rheology of Egg White and Egg White Proteins

H.R. Ball, Jr., J.M. Michaels, and S.E. Winn

Department of Food Science, North Carolina State University, Raleigh, NC 27695-7624, USA

Abstract The transition of egg white or egg white protein sols to elastic gels during heating or chemical denaturation was followed with low-strain rheology, differential scanning calorimetry (DSC), circular dichroism analysis, and UV-absorption. There was good agreement between observed rheological transitions and transitions observed with other methods. Increases in the elastic modulus and decreases in the viscous component followed transition temperatures for denaturation determined by DSC. Chemically gelled egg white exhibited significant reductions in the viscous component prior to development of rigidity. The more elastic chemically set gels were highly translucent. Melt curves indicated structural changes coincide with denaturation. Dispersing egg white in 0.85% NaCl did not affect denaturation temperatures but affected the ultimate rigidity of the gels formed.
(*Key words:* egg white, rheology, gelation, denaturation)

Introduction

Beveridge *et al.* (1984) and Koseki *et al.* (1989) presented models for thermal gelation of egg white and ovalbumin. The outlined multiple-step models began with denaturation of protein. Hydrophobic interactions moderated by change were suggested as being the predominant mechanism directing gelation during the first step. Sulphydryl-disulphide interchanges were believed to predominate in the second step, building gel strength and rigidity. On cooling, hydrogen bonding predominated and increased the gel's rigidity and elasticity.

   Advances in biochemically oriented research on protein gelation have been concomitant with rheologically oriented research to determine physical properties of protein gels (Hamann, 1983; Montejanno *et al.*, 1984a,b; Beveridge *et al.*, 1984; Vigdorth and Ball, 1988). Applications of the testing procedures described in the above citations allow determination of fundamental physical properties of protein gels. It follows that the proper choices of biochemical and physical examinations may allow further elucidation of the relative importance of protein interaction mechanisms in the formation of gels and of their contribution to the rheological properties of the gels.

Materials and Methods

Egg white (EW) was prepared from 1 day old eggs. Guanidine hydrochloride (GHCl), urea, and sodium dodecyl sulphate (SDS) were added to EW with stirring.

Chemical gelation was followed using the Instron Universal Testing Machine as described by Hamann (1987) for thermal scanning rigidity modulus (TSRM). Rheological transitions of ovalbumin and conalbumin were obtained with a Bholin rheometer. Samples were 10% protein in 0.85% saline. The heating rate was 1°C/min.

Differential scanning calorimetry (DSC) was performed using a Perkin-Elmer DSC II at a heating rate of 10°C/min. Circular dichroism was carried out for solutions containing 0.025 mg protein/ml. Samples were scanned from 195 to 300 nm while heating at 1°C/min up to 90°C. Molar ellipticities at selected wavelength were plotted as a function of temperature to construct melt curves. UV-absorption spectra for ovalbumin and conalbumin were obtained using a Hewlett Packard 8452 Diode Array spectrophotometer. Protein concentration and heating rates were as described above.

Results and Discussion

Figures 25.1 and 25.2 present rheological and spectral evaluations of ovalbumin and conalbumin heated from 25 to 90°C. Theta from circular dichroism melt curve, UV-absorption spectra, modulus of elasticity (G′), and phase angle (tan δ) are presented as percent changes from their respective initial unheated value at 25°C and are shown as a function of temperature. In Fig. 25.1, it is clear that there was early denaturation of ovalbumin as indicated by changes in theta and UV-absorbance prior to changes in rheological properties. As expected, changes in tan δ preceded significant development of G′. The denaturation onset temperature for ovalbumin as determined by calorimetry was 83°C which is about the temperature at which a rapid increase in G′ occurred.

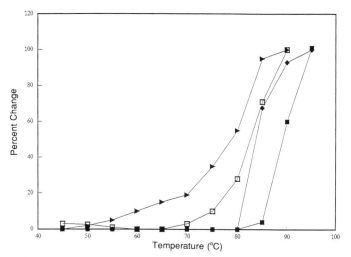

Fig. 25.1. Percent change in rheological and spectral properties of ovalbumin during thermal gelation. ■: G′, ◆: tan δ, ▶: theta, □: absorbance.

There was a similar pattern for the early phases of heating conalbumin (Fig. 25.2). The denaturation onset temperature was 65°C. Spectral changes occurred at lower temperatures than rheological changes indicating protein rearrangement leading to structures maintained by lower bond energy forces (hydrophobic etc.). The major development of G′ occurred at temperatures above 65°C where stronger forces (covalent bonds) would be more stable.

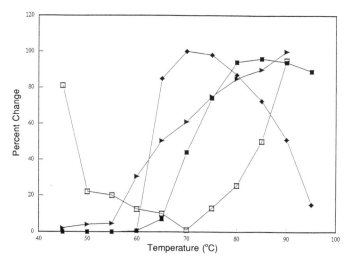

Fig. 25.2. Percent change in rheological and spectral properties of conalbumin during thermal gelation. ■: G′, ◆: tan δ, ▶: theta, □: absorbance.

Theta and UV-absorbance continued to change indicating considerable structural rearrangement at temperatures above the onset of denaturation determined by calorimetry and large increase in rigidity (Figs 25.1 and 25.2). There was also a major transition from an elastic solid to more viscous behaviour (tan δ) (Fig. 25.2). The shift in the rheological characteristics, i.e. the loss of rigidity and the increase in viscosity, occurred at temperatures from 65 to 90°C where ovalbumin denaturation and formation of rigidity were maximum (Fig. 25.1). The apparent melting of conalbumin gels at temperatures where ovalbumin is undergoing denaturation allowed the two proteins to cooperate in forming the basic structure for egg white gels (Michaels and Ball, 1990). Heating ovalbumin and conalbumin together, in the natural ratio in which they are found in egg white, resulted in gels with rheological properties mimicking those made with egg white (Michaels and Ball, 1990). The data (Figs 25.1 and 25.2) and Michaels and Ball (1990) suggest that those two proteins interact during gelation of egg white and determine the properties of the gel.

Figure 25.3 shows the time course for chemical gelation of egg white at room temperature. The ultimate rigidity values after 60 min were low relative to thermally-set gels observed in earlier studies.

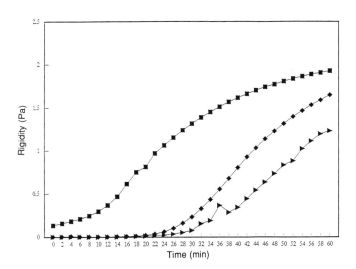

Fig. 25.3. Time course for chemical gelation of egg white at room temperature. ■: 2.8 mol GHCl/kg EW, ♦: 9 mol urea/kg EW, ▶: 0.08 mol SDS/kg EW.

The energy loss data in Table 25.1 clearly show significant transition from a viscous sol to an elastic solid. Because of the gelation method, it can be assumed that only the weaker forces were involved in forming the networks. In the case of the chemically-set gels, those formed with urea and SDS were translucent (Table 25.1).

Table 25.1. Properties of chemically gelled egg white.

Treatment[a]	Rigidity (Pa)[b]	% Energy loss	OD[c]	pH
Control	0.003	70	0.157	8.8
GHCl 2.8 mol	1.915	9	3.104	7.9
Urea 9.0 mol	1.544	13	0.147	9.3
SDS 0.08 mol	1.268	29	0.375	9.0

[a] Moles of denaturant per kg of egg white.
[b] Average of three observations after 60 min in TSRM.
[c] OD, optical density at 550 nm.

Conclusion

Beveridge *et al.* (1984), Koseki *et al.* (1989), and others have proposed that the initial phases of protein gelation initiate with relatively mild denaturation. The denaturation results in changes in protein conformation exposing hydrophobic groups. The results of this study agree with the above models. The data clearly show that there is significant denaturation early in heating and that small forces are most likely involved in initiation of gel networks formed by heating or by chemical denaturation. The results also indicate that there is substantial opportunity for conalbumin and ovalbumin to interact at temperatures above those generally recognised as denaturation temperatures for conalbumin.

References

Beveridge, T., Jones, L., and Tung, M.A. (1984) Progel and gel formation and reversibility of gelation of whey, soybean, and albumin protein gels. *J. Agric. Food Chem.* 32:307-313.
Hamann, D.D. (1983) Structural failure in solid foods. In: Peleg, M. and Bagley, E.B. (eds), *Physical Properties of Foods*. AVI Publishing, Westport, CT, pp. 351-383.
Koseki, T., Fukuda, T., Kitabatake, N. and Doi, E. (1989) Characterization of linear polymers induced by thermal denaturation of ovalbumin. *Food Hydrocol.* 3:135-148.
Michaels, J.M. and Ball, H.R., Jr (1990) Contribution of ovalbumin, conalbumin and lysozyme to gelation of egg white. Annual Meeting of the Institute of Food Technologists, Anaheim, CA.
Montejano, J.G., Hamann, D.D., and Ball, H.R., Jr (1984a) Mechanical failure characteristics of native and modified egg white gels. *Poultry Sci.* 63:1969-1974.

Montejano, J.G., Hamann, D.D., Ball, H.R., Jr, and Lanier, T.C. (1984b) Thermally induced gelation of native and modified egg white-rheological changes during processing; final strengths and microstructure. *J. Food Sci.* 49:1249-1257.

Vigdorth, V.L. and Ball, H.R., Jr (1988) Relationship of disulfide bond formation to altered rheological properties of oleic acid modified egg white. *J. Food Sci.* 53:603-608, 640.

Chapter Twenty-Six

A New Process for Preparing Transparent Alkalised Duck Eggs and Assessment of Their Quality

H.-P. Su and C.-W. Lin

Department of Animal Science, National Taiwan University, Taipei, Taiwan, Republic of China

Abstract When fresh duck eggs (pH 8.0-8.5) are heated, their albumens develop into a turbid gel. Through appropriate alkalisation (pH 11.5-12.8), the gel's transparency can be increased. The transparency of the heated duck egg white is affected by pH value, heating temperature, heating rate, and salt concentration. This research deals with the process for preparing transparent alkalised duck eggs and the change in their quality when stored. If fresh duck eggs are pickled in a solution of 4.2% NaOH + 5.0% NaCl (25 ± 3°C), taken out 8 days later, put in a water bath, and heated to 70°C for 10 min, transparent alkalised duck eggs result whose hardness increases while penetrability decreases with storage. Total bacterial count and volatile basic nitrogen also increase with storage. The total bacterial count and the volatile basic nitrogen were 4.6×10^6 c.f.u./g and 32 mg% respectively when stored at a temperature of 25°C for 4 weeks.
(*Key words:* transparent, alkalised, duck egg, hardness, penetration, total bacterial count, volatile basic nitrogen)

Introduction

Only a few alkalised foods are available. Pidan ('one-thousand-year egg') is a popular traditional alkalised duck egg in China (Lin, 1979). Its egg white generally appears brown and turbid. To increase the variety of alkalised egg products, to satisfy consumers' demand for novelty, and also to enhance the development of the fowl production business, we conducted research into how to produce a transparent alkalised duck egg.

When a fresh duck egg is heated, its egg white becomes a turbid gel. With proper alkalisation, its transparency can be increased. The transparency of heated duck egg white gel is influenced by pH, heating temperature, heating rate, and salt concentration (Hegg, 1982). Based on previous results (Su and Lin, 1991), this research deals with the process for preparing transparent alkalised duck eggs and the resulting change in the quality of the eggs during storage. Our aim was to produce transparent alkalised duck eggs of consistent quality.

Materials and Methods

Egg Samples

Fresh duck eggs were taken directly from the poultry farm. The shells were cleaned and those without cracks were selected. Average weights were about 62 ± 3 g.

Formulation of Alkaline Solution

The alkaline solution contained the following, where the quantity of each chemical is expressed as a percentage of the total weight of the eggs (10 kg): H_2O 10 kg (100%), NaOH 420 g (4.2%), and NaCl 500 g (5.0%).

Coating Material

A 10% aqueous solution of polyvinyl alcohol (PVA) was prepared by heating until the PVA had dissolved, and was then set aside prior to coating.

Preparation of Transparent Alkalised Duck Eggs

Fresh duck eggs were pickled in the alkaline solution described above (25 ± 3°C), removed 8 days later, and heated in a water bath to 70°C for 10 min. After allowing the eggs to cool and dry, they were coated with the 10% PVA solution.

Microstructure of Transparent Alkalised Duck Albumen Gel

The microstructure of transparent alkalised duck albumen gel was examined with a scanning electron microscope (SEM) (Hitachi, S-550, Japan) as described previously (Su and Lin, 1991).

Hardness and Penetrability

The hardness and penetrability of transparent alkalised duck albumen gel were determined using a rheometer (Fudoh, Japan) as described previously (Su *et al.*,

1985). Probes with broad and narrow tips were used for measuring hardness and penetrability, respectively. The peak force to depress or penetrate 2 cm in depth into the gel was expressed in grams.

Determination of Total Bacterial Count and Volatile Basic Nitrogen

The procedure used here for determining the total bacterial count and volatile basic nitrogen of transparent alkalised duck albumen gel was as described previously (Department of Food Science and Technology, Kyoto University, 1970).

Results and Discussion

Preparation of Transparent Alkalised Duck Egg

The pH value of fresh duck albumen was about 8.0-8.5 (Lin, 1979). After heating, it became a turbid gel. The gel's transparency increased with pH. When duck eggs were pickled in a solution of 4.2% NaOH + 5.0% NaCl (25 ± 3°C) for 8 days, the pH value of albumen was about 12.0-12.2. When these pickled eggs were heated, transparent alkalised duck eggs (Fig. 26.1) could be obtained.

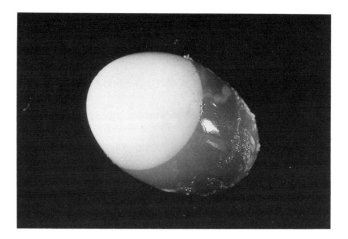

Fig. 26.1. Appearance of transparent alkalised duck egg.

The best transparency was attained at pH values between 12.0 and 12.8. The disulphide bond of the gel was destroyed at higher pH values (Su and Lin, 1985). The gel released sufficient H_2S to combine with the iron ions in the albumen gel to become brown and turbid, thereby reducing the transparency of the gel. Heating temperature, heating rate, and other factors also affected the transparency of alkalised duck egg (Su and Lin, 1991). Lower heating temperatures produced more transparent alkalised duck albumen gel (Su and Lin, 1991) because the protein denaturation rate was lower than its aggregation rate. This can be corroborated by scanning electron microscopy. Considering the fact that the denaturation temperature for duck albumen is about 60°C (Lin, 1979), better transparency could be attained by heating to 70°C for 10 min and the albumen was more transparent than that of eggs heated at 90°C for 10 min. Figure 26.2 shows the processing flow chart for transparent alkalised duck egg.

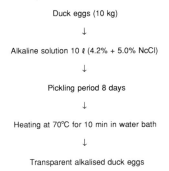

Duck eggs (10 kg)

↓

Alkaline solution 10 ℓ (4.2% + 5.0% NcCl)

↓

Pickling period 8 days

↓

Heating at 70°C for 10 min in water bath

↓

Transparent alkalised duck eggs

Fig. 26.2. Procedure for making transparent alkalised duck eggs.

Microstructure of Transparent Alkalised Duck Albumen Gel

The transparency of protein gel is closely regulated by protein denaturation rate and aggregation rate (Ferry, 1948). The gel is coarser and more opaque when the protein denaturation rate is greater than the protein aggregation rate. Conversely, when the protein denaturation rate is lower than the protein aggregation rate, the gel is smoother and more transparent.

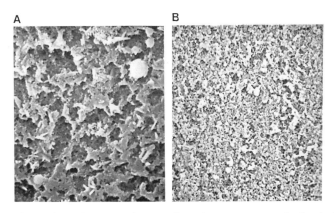

Fig. 26.3. Scanning electron micrographs of transparent alkalised duck egg albumen. A: heated at 70°C for 10 min, B: heated at 95°C for 10 min.

Figure 26.3 shows the influence of different heating temperatures on the microstructure of transparent alkalised duck albumen gel. The microstructure of transparent alkalised duck albumen gel heated at 70°C for 10 min is coarser than that of gel heated at 90°C for 10 min.

Changes in Hardness and Penetrability during Storage

The hardness and penetrability of the transparent alkalised duck albumen gel were 55 ± 6 g and 47 ± 8 g, respectively. The hardness and penetrability both increased during storage (Fig. 26.4). When stored at a temperature of 25°C for 4 weeks, the hardness and penetrability became 82 ± 4 g and 78 ± 6 g, respectively. This might be due to water vaporisation during storage (Su and Lin, 1985).

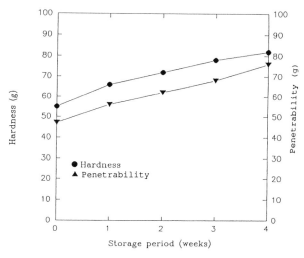

Fig. 26.4. Changes in hardness and penetrability of transparent alkalised duck egg albumen during storage period.

Changes in Total Bacterial Count and Volatile Basic Nitrogen during Storage

The total bacterial count of the transparent alkalised duck albumen gel gradually increased during storage (Fig. 26.5). When stored at a temperature of 25°C for 4 weeks, the total bacterial count increased to 4.6×10^6 c.f.u./g. Volatile basic nitrogen also increased to 32 mg% (Fig. 26.5), when stored at a temperature of 25°C for 4 weeks. As the critical value for judging meat freshness (Ishida and Kadota, 1976) is 30 mg% volatile basic nitrogen, it may be important to note that the volatile basic nitrogen of alkalised duck eggs becomes higher than that of common fresh food. Accordingly, when stored for 5 or 6 weeks, putrefaction begins.

In summary, when a duck egg is pickled in a solution of 4.2% NaOH + 5.0% NaCl (25 ± 3°C) for 8 days, optimal transparency of the alkalised duck egg albumen gel can be attained by heating to 70°C for 10 min. The shelf-life of transparent alkalised duck eggs can be increased by keeping them refrigerated.

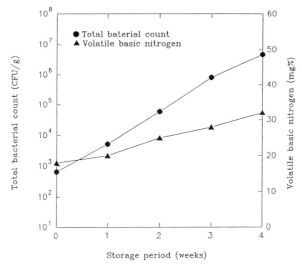

Fig. 26.5. Changes in total bacterial count and volatile basic nitrogen of transparent alkalised duck egg during storage.

Acknowledgement

Our sincere thanks go to the Agriculture Council of the Republic of China for financial support.

References

Department of Food Science and Technology, Kyoto University (1970) *Experiments in Food Science and Technology*, Volume 1. Yokendo Publishing Co., Japan, p. 50.

Ferry, J.D. (1948) Protein gels. *Adv. Protein Chem.* 4:2-78.

Hegg, P.-O. (1982) Conditions for the formation of heat-induced gel of some globular food proteins. *J. Food Sci.* 47:1241-1249.

Ishida, Y. and Kadota, H. (1976) Microbiological studies on salted fish stored at low temperature. I. Chemical and microbial changes of salted fish. *Bull. Jap. Soc. Fish* 42:351-359.

Lin, C.W. (1979) *Chemistry and Utilization of Egg*. Hwa Shiang Yuarn Publishing Co., Taipei, Republic of China, pp. 226-229.

Su, H.P. and Lin, C.W. (1991) Studies on the factors affecting the transparency of alkalized duck egg white gel. *J. Chin. Agric. Chem. Soc.* 29(4):523-532.

Su, H.P., Huang, C.C., and Lin, C.W. (1985) Studies on the shell appearance and physical qualities of pidan. *J. Chin. Agric. Chem. Soc.* 23(3-4):227-235.

Chapter Twenty-Seven

Improvement of Solubility and Foaming Properties of Egg Yolk Protein and Its Separation from Yolk Lipid by Liquid-Liquid Extraction

A.C. Germs

DLO Spelderholt Centre for Poultry Research and Information Services, Agricultural Research Department (DLO-NL), 7361 DA Beekbergen, The Netherlands

Abstract The water-solubility of egg yolk protein was improved by partial proteolysis of egg yolk prior to its separation into protein and lipid. Susceptibility of hydrophobic egg yolk vitellenins to proteolysis had to be enhanced with urea (7 M) as a denaturant. Foaming capacity as well as foam stability were considerably improved by partial proteolysis and scored even better than egg albumen. Partial proteolysis of vitellenins was achieved with the industrial enzyme preparations Alcalase and Auxillase but not with Neutrase which might not be stable to 7 M urea. Partial proteolysis of egg yolk containing 15% egg albumen ('industrially prepared egg yolk') also seemed possible, but not proteolysis of frozen egg yolk after thawing, due to its gelation. Separation of the protease-treated egg yolk into protein and lipid was found to be possible by liquid-liquid extraction with chloroform. The use of one organic solvent, e.g. chloroform, is preferable to a mixture of solvents for solvent reuse in industrial application. Unfortunately, the polarity of *n*-hexane, a less objectionable solvent than chloroform and diethylether with respect to food safety, is too low to extract lipids by liquid-liquid extraction.
(*Key words:* egg yolk protein, solubility, foaming, partial proteolysis, separation, liquid-liquid extraction, egg yolk lipid)

Introduction

Egg yolk solids are composed of about 31% proteins and 69% lipids (Table 27.1). Of these proteins two-thirds are associated with lipids in the form of lipoproteins. Most of the hydrophobic apoproteins (vitellenins) from the major egg lipoprotein, the low density lipoprotein (LDL), are insoluble in water (Burley and Sleigh, 1983).

Table 1. Components of egg yolk solids in % (Parkinson 1966)

Proteins		Lipids	
Livetins	4-10	Triglycerides	46
Vitellin (HDL apoprotein)	4-15	Phospholipids	20
Vitellenin (LDL apoprotein)	8-9	Sterols	3
Phosvitin	5-6		

To isolate them from egg yolk by delipidation, these proteins were dissolved in aqueous solutions of urea and guanidine HCl (Burley and Sleigh, 1983; Scheumack and Burley, 1988). The proteins dissolve because the alteration in the chemical environment causes them to take up a randomly coiled conformation. However, after removal of the denaturants by either dialysis or ultrafiltration, the apoproteins of the LDL will become insoluble again (Greene and Pace, 1974; Germs, 1991). Their solubility can therefore be improved by modifying the primary structure.

Chemical modification of the primary structure of proteins can improve their solubility and functional properties. Preliminary results (Germs, 1991) showed that the isolation method by Scheumack and Burley (1988) yielded much more soluble protein from succinylated than from native egg yolk. Chemical modification of proteins, however, may have serious drawbacks, including damage to nutritional value, formation of toxic amino acid derivatives, and the incorporation of toxic reagent residues.

Improvement of food proteins has also been attempted by enzymic modification using proteases. A well-known example is the tenderisation of meat with papain. Fish, vegetable, and denatured proteins can be solubilised by partial proteolysis with enzymes (Cheftel et al., 1985). Partial proteolysis of egg yolk protein might also facilitate its separation by delipidation. Due to loss of hydrophobicity, the protein will then no longer be bound to (phospho)lipid by hydrophobic interactions and thus be separated (Evans et al., 1968a,b).

Whether modification and isolation of egg yolk protein will be economically feasible depends on production costs and the functional properties of the proteins, which should be superior to those of other less expensive food proteins, e.g. whey protein concentrates. Economical feasibility will also depend on the value of the remaining lipid.

Modification of the Egg Yolk Protein by Partial Proteolysis

'The susceptibility of a membrane protein to proteolysis is a function of the membrane structure and the orientation of the protein in the membrane, in addition to the structure of the protein and the presence of susceptible peptide bonds in the protein sequence' (Butler, 1989). Consequently, susceptibility will also depend on the hydrophobicity of the protein. According to Butler (1989), proteolytic solubilisation is merely a quick, clean, and simple way of removing proteins from a membrane if they possess only a small proportion of hydrophobic peptides.

The susceptibility of the egg yolk vitellenins to proteolysis will also be reduced by the structure of the LDL and their hydrophobicity. Therefore, it might be necessary to change their structure to a random coil prior to proteolysis. This can be done with urea or guanidine HCl. Preferably, urea should be used, because of the toxicity of guanidine HCl.

Egg yolk was treated with proteolytic enzymes to investigate improvement in solubility and foaming properties of the yolk protein. Three commercial enzyme preparations, Alcalase Food Grade, Neutrase, and Auxillase, were used for proteolysis.

Specifications of Enzyme Preparations Used for Proteolysis

The major enzyme component in Alcalase Food Grade is subtilisin A (= subtilisin Carlsberg), an endopeptidase of the serine type. No other enzymic activities are present in appreciable amounts. Standard conditions for application are a temperature of 50°C and pH 8.30 (specified by Novo Nordisk, Denmark). Subtilisin A is stable in 4 M urea and retains activity in 8 M urea for at least 3 h (Herbert and Dunnill, 1988).

Neutrase contains only the neutral part of proteases produced by a selected strain of *Bacillus subtilus*, whereas most other commercial preparations contain the alkaline protease. The optimal working conditions are at 45-55°C and pH 5.5-7.5 (specified by Novo Nordisk). Stability to 7 M urea was not specified.

Auxillase activity corresponds to that of papain. Optimal working conditions are 40-65°C and pH 3-9 (specified by Merck). Papain is an endopeptidase of the cysteine type (Bond, 1989) and is stable in urea solutions; no inactivation occurs after 24 h in 9 M urea at 30°C and pH 6.5 (Long, 1961).

Separation of the Protease-Treated Protein by Lipid Extraction

Apart from centrifugation, the use of three different solvents (chloroform, methanol, and petroleum ether) is a disadvantage of the delipidation method used by Scheumack and Burley (1988) to isolate egg yolk proteins. Reuse of the extraction solvent (a prerequisite for industrial application) will only be possible if complete delipidation can be obtained with one solvent.

Extraction of all lipids (phospholipids and cholesterol included) from egg yolk, without denaturation of proteins is impossible with organic solvents (Hatta *et al.*, 1988; Kwan *et al.*, 1991). After partial proteolysis of the yolk protein, a sufficient extraction of the lipid might be achieved without the use of an alcohol as a denaturant (Evans *et al.*, 1968a,b). If so, one non-polar, water-immiscible solvent might be sufficient for the delipidation of egg yolk by liquid-liquid extraction.

Solvents generally used for the extraction of lipids from dried egg yolk are diethyl-ether and chloroform. Because of their immiscibility with water, both might be used for liquid-liquid extraction. However, they would probably not be safe for food products (Larsen and Froning, 1981; Tokarska and Clandinin, 1985). For that reason, in the solvent mixtures used in these two papers hexane was used with an alcohol as a denaturant to extract lipid from liquid egg yolk. Suitability of chloroform and *n*-hexane to separate proteolytically modified egg yolk into protein and lipid fractions by a continuous liquid-liquid extraction were investigated.

Materials and Methods

Pure egg yolk was prepared by separating the egg and then removing adhering albumen from the yolk by rolling the unbroken yolk over filter paper. Alcalase Food Grade 0.6 L and 2.4 L, Neutrase 0.5 L (Novo Nordisk A/S, Denmark), and Auxillase (Merck cat. no. 7138) were used as proteolytic enzyme preparations.

Partial Proteolysis of Egg Yolk

Egg yolk solutions containing 7 M urea and aqueous solutions of the same yolk content (6.6 g egg yolk/24 ml) were used. When a buffer was used during Alcalase treatment, the yolk solutions contained also tris(hydroxy-methyl)aminomethane. pH was then adjusted to 8.5 with diluted HCl. When no buffer was used, the pH was adjusted with diluted NaOH. No buffer and no pH adjustment were necessary for the treatments with Neutrase and Auxillase. Proteolysis was carried out in a waterbath shaker at 50°C for 1 h, when Alcalase and Auxillase were used. For proteolysis with Neutrase the temperature was 45°C.

Isolation of the Protease-Treated Egg Yolk Protein by Delipidation

After proteolysis, the egg yolk solution was cooled in water to ambient temperature and poured into a continuous liquid-liquid extraction apparatus (Fig. 27.1). When chloroform was used for extraction, the apparatus was previously filled with solvent to the lowest overflow level.

Delipidation of the yolk solution was achieved by recirculation of the extraction solvent by heating and refluxing. 75 ml solvent was used during the first delipidation step and 60 ml during the following steps.

For the extraction with chloroform, a short inner tube (with a sintered glass disc) suitable for solvents with higher specific gravity than water was used. The inner tube and position of the level regulator (= tap) were changed when extraction was carried out with *n*-hexane, which is lighter than water. Extraction was stopped when for a second time a colourless extract was obtained in a 1 h delipidation step.

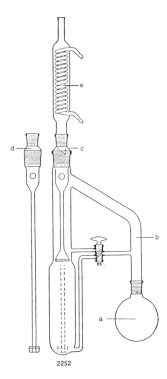

2252

Fig. 27.1. Liquid-liquid extraction apparatus (Witeg, Wertheim, Germany). a, heating flask containing extraction solvent; b, extractor containing aqueous solution to be extracted; c, inner tube for extraction with solvent heavier than water; d, inner tube for extraction with solvent lighter than water; e. condenser.

The isolated yolk protein solution was subsequently dialysed in Visking tubing (∅ 14.3 mm and pore ∅ 2.4 nm) against 3 1 demineralised water at 4°C for 8 h. The water was continuously mixed with a magnetic stirrer and replaced every 2 h.

Determination of Delipidation

After cooling, solid material was removed from the extracts by filtration over a folded filter (Schleicher and Schuell no. 595½) into a 250 ml extraction flask. The solvent was evaporated from the extracts and quantities of lipid were determined by drying the residues to constant weight in an oven at 102 ± 2°C. Delipidation was calculated as percentage of lipid based on egg yolk.

Estimation of Solubility Improvement of Yolk Protein

Improvement in solubility was approximated by relating the height decrease of the protein precipitate in the tubing after dialysis to the height of the precipitate obtained without previous proteolysis. The quantity of isolated protein was not sufficient for both the measurement of foaming properties and the standardised solubility test by Morr et al. (1985). The improvement in solubility was calculated with the formula:

Solubility improvement (%) = (h.w.p. - h.a.p.)/(h.w.p.) × 100

where h.w.p. = height of protein precipitate without proteolysis and h.a.p. = height of protein precipitate after proteolysis. The h.w.p. of solution of 6.60 g egg yolk = 10.4 cm (10.0-10.8 cm; n = 5)

Measurement of Foaming Properties of Protein

Foaming capacity and foam stability were measured simply by shaking the urea-free protein solution in a 100 ml stoppered graduated cylinder. Not enough material was available for the measurement of these properties with mechanical equipment. Maximum foam volume was achieved by shaking the solution twice, for 30 s each time. Foam volume was measured 30 s after shaking the second time. Drainage was measured after standing for 1 h to determine foam stability. The following formulae (Janssen, 1971) were used:

Foam capacity (%) = (volume of foam)/(volume of initial liquid phase)
× 100

Foam stability (%) = (vol. initial liq. phase - vol. drainage)
/(vol. initial liq. phase) × 100

Drainage = (volume of liquid phase after 60 min) - (volume of liquid phase after 30 s.)

Results and Discussion

Improvement of Solubility and Foaming Properties by Partial Proteolysis

Results of solubility improvement and foaming properties of total protein separated from egg yolk by liquid-liquid extraction of the lipids with chloroform are shown in Table 27.2.

Partial proteolysis of the yolk protein without use of 7 M urea yielded less improvement of solubility and no improvement in foaming properties. So, susceptibility of yolk vitellenins to proteolysis is indeed enhanced by the use of urea as a denaturant. Improvements in solubility of >33% by proteolysis in 7 M urea did not yield a higher foaming capacity or foam stability. Further investigations are necessary to reveal optimum proteolysis conditions in 7 M urea with respect to foaming properties.

Both Alcalase and Auxillase can be used to modify yolk protein enzymatically to improve its solubility and foaming properties. Neutrase, however, did not improve these properties. Probably, this enzyme is not stable to 7 M urea.

Comparison between the foaming properties of the protein solutions (theoretical maximum 4%) obtained from pure egg yolk and those of egg albumen (10.5% protein) as a standard (Cheftel *et al.*, 1985) shows that these properties are better for the modified yolk protein solutions. Foaming capacity and foam stability obtained were on average 363% (n = 10, coefficient of variation (c.v.) = 8.4%) and 55% (n = 10, c.v. = 8.0%) respectively. Albumen, measured with the same method yielded 173% (n = 12, c.v. = 6.6%) and 39% (n = 12, c.v. = 15.0%), respectively.

Proteolysis with Auxillase can be achieved without a buffer, as the pH value did not shift much during treatments with this enzyme preparation. Another advantage of Auxillase is the broad pH working range. Treatment with Alcalase required a high buffer capacity to maintain pH within the predetermined range. However, satisfactory results were obtained for foaming properties even when no buffer was used.

Besides pure egg yolk, yolk containing 15% egg albumen was also treated proteolytically, because industrially separated egg yolk is always contaminated with albumen. The presence of egg albumen seems to have no negative effect on the proteolysis of the yolk protein.

Table 27.2. Improvement of solubility and foaming properties of egg yolk protein at different proteolytic treatments and pH shift.

Solution and treatment	pH shift	Solubility improve- ment	Foaming capacity	Foam stability
Egg yolk in water		-----------%-----------		
390 mg Alcalase 0.6 L	8.53-8.42	29	<5	0
390 mg Auxillase	6.62-6.42	53	<5	0
Egg yolk in 7 M urea with buffer				
80 mg Alcalase 2.4 L	8.50-8.18	97	357	57
80 mg Alcalase 2.4 L	8.46-8.17	86	335	60
30 mg Alcalase 2.4 L	8.48-8.31	75	318	61
15 mg Alcalase 2.4 L	8.53-8.25	33	357	56
Egg yolk in 7 M urea without buffer				
390 mg Alcalase 0.6 L	8.61-6.83	93	342	54
80 mg Alcalase 2.4 L	8.44-6.86	98	404	47
390 mg Auxillase	-6.13	100	396	50
150 mg Auxillase	6.17-6.23	91	352	54
80 mg Auxillase	6.25-6.33	87	409	51
80 mg Auxillase	6.23-6.35	79	355	56
240 mg Neutrase	-6.72	-[†]	0	0
240 mg Neutrase	6.44-6.81	-[†]	55	82
85% Egg yolk + 15% egg albumen in 7 M urea without buffer				
30 mg Alcalase 2.4 L	8.48-7.32	-[§]	355	51
60 mg Alcalase 2.4 L	8.52-7.35	-[§]	375	51
30 mg Auxillase	6.68-6.63	-[§]	346	38

[†] Not measured, because gelated solution could not be removed quantitatively from the extraction apparatus.
[§] Could not be measured with method used, because in presence of egg albumen no precipitate had been formed from turbid solution.

Frozen egg yolk after thawing was not suitable for the isolation of protein by this method due to its gelation. The gelated yolk did not dissolve in 7 M urea and thus could not be solubilised by proteolysis and could not be delipidated by chloroform extraction.

Separation of Egg Yolk into Protein and Lipid by Liquid-Liquid Extraction

Delipidation by liquid-liquid extraction was investigated as a method of separating egg yolk into lipid and protein, both with and without prior proteolysis of the protein.

Simple dissolution of egg yolk in 7 M urea (Table 27.3), thus without proteolysis of the yolk protein, diminished the extractability of lipids - both in quantity and rate - compared with aqueous egg yolk solutions. This must be due to the gelation of egg yolk by the denaturant urea. The observed gelation in 7 M urea explains also the slower and incomplete extraction after Neutrase treatment, which, obviously, did not succeed in performing a sufficient proteolysis (see Table 27.2).

Table 27.3. Delipidation of egg yolk solutions by liquid-liquid extraction with chloroform, both with and without previous proteolysis of egg yolk.

Egg yolk solutions	% lipids extracted in successive steps				% Total lipid
	1° step	2° step	3° step	4° step	
Without proteolysis					
in 7 M urea	13.93 (2)	10.25 (2)	0.99 (2)		25.17
in water	30.12 (3.5)	0.96 (2)	0.19 (2)		31.27[a]
in water	30.44 (3.5)	1.01 (2)	0.21 (2)		31.66[a]
in water	30.12 (2.5)	1.37 (2)	0.22 (2)		31.71[a]
in water	29.48 (2.5)	1.97 (2)	0.25 (2)		31.70[a]
in water	27.35 (2)	3.52 (1)	0.37 (1)		31.24[a]
With proteolysis in 7 M urea					
80 mg Alcalase 2.4L	29.74 (2)	2.57 (2)	0.18 (1)		32.49[a]
80 mg Auxillase	29.76 (2)	0.93 (2)	0.14 (1)		32.93[a]
240 mg Neutrase	15.29 (4.5)	8.94 (4)	5.11 (3)	0.62 (2)	29.96
240 mg Neutrase	22.39 (3)	6.27(3.5)	1.21 (2)	1.53 (3)	31.36

[a] Same sample of egg yolk.

Table 27.3 shows that about 1% more lipid was extracted with chloroform from egg yolk that had previously been proteolysed. It has not yet been investigated if the lipid is completely extracted in that case. A rather low mean lipid percentage of 33.3 (n = 10, c.v. = 19%) was separated during delipidation from the proteolytically modified protein samples discussed in the previous section. This suggests that the lipids were not completely extracted.

The method, however, enables the proteolytically modified yolk protein to be isolated in such a way that the protein solutions obtained have excellent foaming properties. Thus, extraction with one organic solvent, a prerequisite for reuse in industrial processes, is possible. Unfortunately, extraction with *n*-hexane instead of chloroform, which is probably not safe for food ingredients, did not succeed in sufficient delipidation. After proteolysis of egg yolk protein, only 1-2% of lipid was extracted in 6 h of extraction. This must be due to the lower polarity of *n*-hexane.

References

Bond, J.S. (1989) Commercially available proteases. In: Beynon, R.J. and Bond, J.S. (eds), *Proteolytic Enzymes. A Practical Approach*. IRL Press, Oxford, pp. 232-240.

Burley, R.W. and Sleigh, R.W. (1983) Hydrophobic chromatography of proteins in urea solutions. The separation of apo-proteins from a lipoprotein of avian egg yolk. *Biochem. J.* 209:143-150.

Butler, P.E. (1989) Solubilization of membrane proteins by proteolysis. In: Beynon, R.J. and Bond, J.S. (eds), *Proteolytic Enzymes. A Practical Approach*. IRL Press, Oxford, pp. 193-200.

Cheftel, J.C., Cuq, J.L., and Lorient, D. (1985) Amino-acids, peptides and proteins. In: Fennema, O.R. (ed.), *Food Chemistry*. Marcel Dekker, Inc., New York, pp. 245-369.

Evans, R.J., Bandemer, S.L., Davidson, J.A., Heinlein, K., and Vaghefi, S.S. (1968a) Binding of lipid to protein in the low density lipoprotein from the hen's egg. *Biochim. Biophys. Acta* 164:566-574.

Evans, R.J., Bandemer, S.L., Heinlein, K., and Davidson, J.A. (1968b) Binding of lipid to protein in lipovitellenin from the hen's egg. *Biochemistry* 7: 3095-3102.

Germs, A.C. (1991) Isolation of proteins from egg yolk. In: Oosterwoud, A. and de Vries, A.W. (eds), Quality of poultry products. II Eggs and egg products. *Proceedings 4th European Symposium on the Quality of Eggs and Egg Products*. Spelderholt Centre for Poultry Research and Information Services, Beekbergen, The Netherlands, pp. 243-248.

Greene, R.J.,Jr and Pace, C.N. (1974) Urea and guanidine hydrochloride denaturation of ribonuclease, lysozyme, α-chymotrypsin and ß-lacto-globulin. *J. Biol. Chem.* 249:5388-5393.

Hatta, H., Sin, J.S., and Nakai, S. (1988) Separation of phospholipids from egg yolk and recovery of water-soluble proteins. *J. Food Sci.* 53:425-431.

Herbert, A.B. and Dunnill, P. (1988) Limited modifications of soyaproteins by immobilized subtilisin: comparison of products from different reactor types. *Biotechnol. Bioeng.* 32:475-481.

Janssen, H.J.L. (1971) Methods of determination of the quality of chicken egg white. *Voedingsm. Technol.* 2(36):11-13.

Kwan, L., Li-Chan, E., Helbig, N., and Nakai, S. (1991) Fractionation of water-soluble and -insoluble components from egg yolk with minimum use of organic solvents. *J. Food Sci.* 56:1537-1541.

Larsen, J.E. and Froning, G.W. (1981) Extraction and processing of various components from egg yolk. *Poultry Sci.* 60:160-167.

Long, C. (ed.) (1961) *Biochemists' Handbook.* D. van Norstrand Co. Inc. Princeton, NJ, USA.

Morr, C.V., German, B., Kinsella, J.E., Regenstein, J.M., Van Buren, J.P., Kilara, A., Lewis, B.A., and Mangino, M.E. (1985) A collaborative study to develop a standardized food protein solubility procedure. *J. Food Sci.* 50:1715-1718.

Parkinson, T.L. (1966) The chemical composition of eggs. *J. Sci. Food Agric.* 17:101-111.

Scheumack, D.D. and Burley, R.W. (1988) Separation of lipid-free egg yolk proteins by high-pressure liquid chromatography using solvents containing formic acid. *Anal. Biochem.* 174:548-551.

Tokarska, B. and Clandinin, M.T. (1985) Extraction of egg yolk oil of reduced cholesterol content. *Can. Inst. Food Sci. Technol. J.* 18:256-258.

Chapter Twenty-Eight

Radiation Pasteurisation of Frozen Whole Egg

J. Kijowski, G. Lesnierowski, J. Zabielski,[1] W. Fiszer,[1] and T. Magnuski[1]

Laboratory of Poultry Products Technology and [1]Laboratory of Nuclear Techniques in Agriculture, University of Agriculture, Poznań, Poland

Abstract   Since frozen whole egg magma is frequently contaminated with Enterobacteriaceae, the effect of radiation dose on survival of *Salmonella* sp. and *Escherichia coli* was determined. The decimal reduction doses were found to be 0.39 kGy and 0.52 kGy, respectively. A technologically feasible dose for elimination of those pathogens from unpasteurised frozen whole egg was selected on the basis of the functional, chemical, and sensory features: rheological and baking properties, the free fatty acid and malonaldehyde content, and sensory evaluation of scrambled egg and sponge-cake. No detrimental effect of radiation doses up to 2.5 kGy on those features was found in the frozen (-18°C) whole egg. The sensory quality of sponge-cake was slightly improved in comparison with non-irradiated samples. Since the dose of 2.5 kGy reduced the probability of survival of *Salmonella* sp. by over six orders of magnitude and that of *E. coli* by over four orders, this technique would be a useful tool for improving the hygienic status of frozen whole eggs.
(*Key words:* frozen whole egg, radiation pasteurisation, functionality, sensory quality)

Introduction

Although fresh eggs are usually bacteria-free inside, further industrial processing increases the probability of infection. Since laying hens are frequently *Salmonella* carriers (Granville, 1963; Mossel, 1977), the surface of the egg shell becomes a potential source of contamination. According to Wos (1987),

Salmonella were identified on the surface of up to 3% of the collected eggs.

Both manual and mechanical breaking lead to transmission of bacteria from the shell surface into the egg. For this reason, an essential step in preparation of frozen whole egg magma is thermal pasteurisation. Under commercial conditions, this process requires heating of the material at 66-68°C for 2 min prior to freezing. However, even properly conducted thermal pasteurisation does not ensure that the product is completely pathogen-free. Residual microflora can encounter favourable growing conditions during thawing of the frozen egg and/or during manufacturing of products which do not require further thermal processing, e.g. mayonnaise, salad dressings, sauce, etc.

The potential of ionising radiation for the elimination of pathogenic bacteria from egg products has been studied since 1950. Proctor *et al.* (1953) found that a radiation dose of 2.5 kGy reduced survival of *S. typhimurium* and *S. paratyphi* in thawed whole egg by 7 log cycles ($7 \times D_{10}$). The D_{10} value, i.e. the decimal reduction dose, depends not only on the type of bacterium, but also on the physical state of the product. While investigating the radiation resistance of 18 strains of *Salmonella* in commercially frozen whole egg magma, Comer *et al.* (1963) found the $7 \times D_{10}$ doses varied from 3.6 kGy to 5.4 kGy. Following that, the D_{10} dose values were 0.51 and 0.77 kGy, respectively. Similar findings regarding radiation resistance of *Salmonellae* were reported by Thornley (1963) and Brooks *et al.* (1959). Thieulin *et al.* (1960) recommended 4-5 kGy for the elimination of mesophilic aerobes, coliforms, and *Staphylococcus aureus* from frozen whole egg magma. Since heat reduces the radiation resistance of *Salmonellae*, Licardello (1964) proposed the combined treatment of mild heat with a low dose of irradiation as an alternative to thermal pasteurisation.

Technologically feasible radiation doses, however, cannot be established on the basis of microbial efficiency only. Mossel (1960) found that an off-odour appeared in liquid whole egg treated with 2 kGy. The threshold dose could be even as low as 0.09 kGy (Comer *et al.* 1963). These undesirable changes are due to the radiolysis of chemical components of the material. High water and protein content, and both the content and composition of lipids in the liquid egg create an environment enhancing formation of volatiles that affect sensory properties. Since the pathway of secondary reactions of radiolytic products formed upon the treatment is temperature-related (Taub and Halliday, 1979), lowering the temperature reduces the intensity of these processes. Further drying of the irradiated whole egg magma (Proctor *et al.*, 1953) eliminated the irradiation off-odour. Formulation of dough with irradiated egg products and baking of cakes also mask undesirable changes (Brooks *et al.*, 1959; Niewiarowicz *et al.*, 1980).

In view of the above data, successful application of radiation pasteurisation to frozen egg magma obviously depends on the specific requirements of the final product to be manufactured using the irradiated material. For this reason, a study was undertaken to find a technologically feasible radiation dose for pasteurising frozen whole egg to be applicable in food production.

Materials and Methods

Liquid whole egg (not pasteurised, homogenised) was collected from a commercial processing plant. The material was poured into polyethylene bags under aseptic conditions. The polyethylene bags had been sterilised with gamma rays (25 kGy) prior to use. Sealed bags were frozen until the temperature reached -18°C. The same temperature was applied for storage of the samples.

A Dewar vessel containing a small amount of solidified CO_2 was used for irradiation of the bags. This process was performed in a laboratory-type radiation source PXM-gamma-20, ^{60}Co, at a dose of 0.11 kGy/min. Frozen samples were thawed at ambient temperature prior to analyses until the whole content (200 g) of bags was liquefied.

Standard dilution technique was applied for determination of the number of total aerobic mesophiles. Quantitative determination of *Salmonellae* and *Escherichia coli* were performed according to the procedures recommended by Burbianka *et al.* (1983).

The Instron 1140 universal apparatus was used to determine the rheological properties of the thermal gel prepared from egg magma. Test tubes filled with the liquid material were placed in a thermostat. The gelation temperature of 90°C for 10 min was controlled with a thermistor connected to the Ellab apparatus. Cylindrical gel samples of 1 cm in diameter were compressed twice to reach half of their initial height. Hardness, elasticity, and cohesiveness were calculated from the recorded peaks according to Lyon *et al.* (1980).

A mixture of ethanol and ethylene ether (1:1, v/v) was used to extract lipids from frozen egg magma. The filtered extract was titrated with 0.1 N NaOH in the presence of an indicator. The free fatty acid content was calculated as per cent of oleic acid. The content of malonaldehyde was determined according to a modification of the procedure of Witte *et al.* (1970). The content of malonaldehyde was expressed in mg per 100 g of fat.

For sensory evaluation, two types of standardised products were made. Scrambled egg was prepared by heating liquid whole egg magma in a boiling water bath until the egg white coagulated. After 0.5% salt had been added, the palatability, odour, and colour of warm scrambled egg were evaluated by a panel of five persons experienced in testing irradiated foods. Cake mixture prepared from the treated egg was baked at 160-180°C until the typical, golden colour of cake was reached. Colour, odour, texture, taste, and cake height were assessed. A five-point scale was used to assess these attributes, from 1 (extremely poor) to 5 (extremely good).

Results and Discussion

Liquid whole egg freshly prepared under commercial conditions was highly contaminated with bacteria. In the unpasteurised material, the total number of mesophilic aerobes was 2.3×10^9/g (mean of three determinations), the number of colony forming units of *Salmonella* sp. was 1.5×10^8 per g and that of *E. coli* was 1.0×10^5/g.

The effect of irradiation on microflora can be discussed in terms of a dose-effect relationship. By plotting the log of survival fraction versus dose values (Fig. 28.1), one can obtain the survival curves. Valuable information was derived from the analysis of regression formulas. For *E. coli* the best fitting curve was a straight line:

$$Y = 0.3374 - 1.9328 \ D \qquad (r^2 = 0.9718)$$

where Y is the log of surviving fraction and D the radiation dose in kGy.

Mathematically, the reciprocal of the slope, which was $1/1.9328 = 0.5174$, gives the decimal reduction dose, D_{10}.

The survival of *Salmonella* sp. was calculated in a similar way:

$$Y = -1.1482 - 2.5322 \ D \qquad (r^2 = 0.8275)$$

Thus, the D_{10} value is 0.3949 kGy. In this case, fitting the experimental data to straight line was less accurate ($r = 0.9096$), but still significant ($P < 0.05$). The above calculations are a source of important technological information. Assuming that a dose of 2.5 kGy has to be applied to control microbial quality, the extent of *E. coli* reduction would be 4.8 D_{10}, which is equivalent to a reduction of the initial contamination by a factor of about 63,000. The respective value for *S. enteritidis* is over 2,000,000.

The effect of radiation dose on mesophilic aerobes was less pronounced since the slope of the curve in Fig. 28.1 is less steep. Those findings may lead to the conclusion that irradiation, even within the low dose range, is highly effective in the elimination of most important pathogenic Enterobacteriaceae and less effective for other types of bacteria.

The hardness of thermal gels (Fig. 28.2) appeared to have a tendency to decline slightly with increasing radiation. However, the analysis of variance of data proved the effect was not significant ($F = 0.48$). In contrast, both elasticity and cohesiveness of gels were affected by the dose value.

The effect was not linear, nevertheless statistical significance of those features at $P = 0.05$ ($F = 17.14$ and 8.22) indicated that irradiation affects the formation of cross-links within the gel structure (Taub and Halliday, 1979).

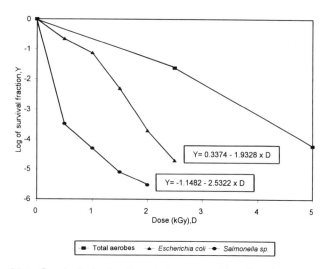

Fig. 28.1. Survival of microflora in frozen and irradiated egg magma.

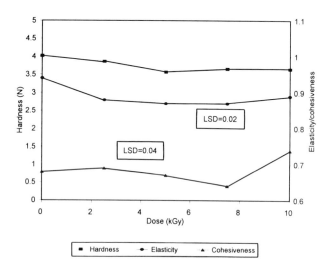

Fig. 28.2. Rheological properties of thermal gels made of irradiated egg magma. Elasticity and cohesiveness were calculated from the first and second bite compression curve. Elasticity was the ratio of the two sections of lines (cm/cm) and cohesiveness was the ratio of the two total areas under the curve; both were therefore unitless values.

In the case of free fatty acid (FFA) content, a slight increasing tendency was observed (Fig. 28.3). The highest FFA content (2.60%) was found in samples irradiated with 10 kGy and the effect was statistically significant in comparison with the non-irradiated sample (2.23%); the lowest significant difference (LSD) was 0.24%. In absolute figures, however, difference between the control and 10 kGy irradiated sample was only 0.37% and therefore the effect on sensory features seems to be questionable. In contrast, radiation treatment influenced significantly the malonaldehyde content. Its increase from 0.35 mg/100g of fat in the control sample, to 1.24 mg/100 g in the 10 kGy irradiated sample (LSD = 0.05) shows the increased intensity of lipid oxidation. A similar tendency was found by Niewiarowicz *et al.* (1980) and the effect was particularly pronounced as frozen egg magma was thawed at 1°C. Apart from irradiation, thawing of material at an elevated temperature probably influenced the final effect. This seems to be a reasonable explanation, because primary radiolytic products that have been trapped and immobilised in the frozen solidified magma may undergo further reactions during thawing at ambient temperatures (Taub and Halliday, 1979). It is not known to what extent these changes are reflected in sensory properties of dishes made from irradiated whole egg.

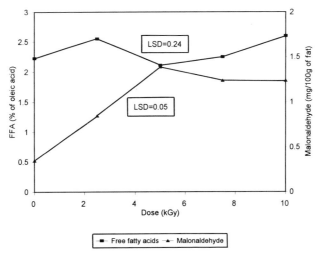

Fig. 28.3. The malonaldehyde and FFA contents of frozen, irradiated whole eggs.

Scrambled egg made of irradiated magma and tested while warm was easily distinguished from the control, non-irradiated sample (Fig. 28.4). This was mainly due to the specific, unpleasant odour which appeared in the samples treated with the lowest dose of 2.5 kGy. Those samples were given odour scores of 2.6 in comparison with 3.4 for the non-irradiated controls (LSD = 0.7). It has to be pointed out that the scrambled egg contained irradiated whole egg only, and that no other ingredients (except NaCl) were added.

Unlike scrambled egg, the sensory properties of sponge cake made of irradiated whole egg magma appeared to be somewhat improved (Fig. 28.5). The scores for colour, odour, texture, and taste of cake were slightly higher than those of the control samples. The difference in cake height was statistically significant at $P = 0.05$ (LSD = 0.3).

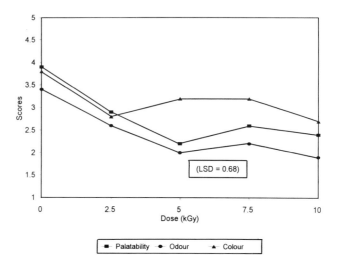

Fig. 28.4. Sensory properties of scrambled eggs made of irradiated magma.

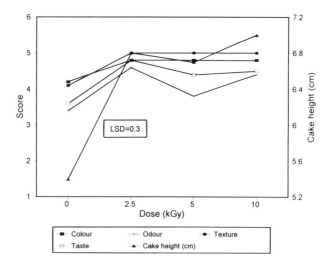

Fig. 28.5. Sensory properties of sponge cake baked with irradiated whole eggs.

At least two reasons may be suggested: the quantity of the irradiated raw material was only about 40% (by weight) in the cake recipe and/or effect of baking in 160-180°C masked the post-irradiation odour of whole egg.

In our previous study, we observed that pathogenic bacteria in frozen or dried egg white could be destroyed by irradiation and that slight changes in the functional properties of the egg white did not adversely affect its technological usefulness or sensory acceptability (Kijowski *et al.*, 1991). Therefore, it may be concluded that frozen and dried egg albumen and frozen whole egg may be suitable for radiation pasteurisation. In our experience, dried whole egg was not suitable for preservation by gamma irradiation. Products made of dried and irradiated whole egg had undesirable taste and odour, and this was particularly pronounced for doses above 2.5 kGy. The intensity of these changes could be reduced to a certain extent through packaging under an inert gas, e.g. nitrogen or carbon dioxide.

Generally, the range of changes observed in the egg products after radiation pasteurisation is dependent on the type of egg product and its physical state (frozen or dried) as well as on the dose applied.

Conclusions

From the standpoint of microbiological safety, the low dose radiation treatment may successfully pasteurise contaminated frozen whole egg. To what extent this procedure could replace thermal pasteurisation remains to be investigated. At least, the radiation treatment might be considered an optional technology in the case of need for product re-pasteurisation.

The experiments described above also prove that processing of frozen and whole egg magma has its technological limitations regarding the use of irradiated raw material. This problem requires further research.

References

Brooks, J., Hannan, R.S., and Hobbs, C. (1959) Irradiation of egg and egg products. *Int. J. Appl. Radiat. Isotopes* 6:49-52.

Burbianka, M., Pliszka, A., and Burzyñska, M. (1983) *Food Microbiology.* PZWL, Warsaw (in Polish).

Comer, A.G., Anderson, G.W., and Garrard, E.H. (1963) Gamma irradiation of *Salmonella* species in frozen whole egg. *Can. J. Microbiol.* 9:321-324.

Granville, A. (1963) The epidemiology of Salmonellosis in relation to its transmission by food and feed products. IAEA Rep. Ser. 22. *Radiation Control of Salmonellae in Food and Feed Products*. Vienna. Austria.

Kijowski, J., Kwasniewska, L., Miszczak, R., Zabielski, J., and Fiszer, W. (1991) Gamma irradiation of frozen and dried egg white - its bacteriological, functional and organoleptic quality. *Proceedings of the 4th European Symposium on the Quality of Egg and Egg Products.* Doorewerth, The Netherlands.

Licardello, J.J. (1964) Effect of temperature on radiosensitivity of *Salmonella typhimurium. J. Food Sci.* 19:469-471.

Lyon, C.E., Lyon, B.G., Davis, C.E., and Townsend, W.E. (1980) Texture profile analysis of patties made from mixed and flaked mechanically deboned poultry meat. *Poultry Sci.* 59:59-69.

Mossel, D.A.A. (1960) The destruction of *Salmonella* bacteria in refrigerated liquid whole egg by gamma irradiation. *Int. J. Appl. Radiat. Isotopes* 9:109-111.

Mossel, D.A.A. (1977) The elimination of enteric bacterial pathogens from food and feed of animal origin by gamma irradiation with particular reference to *Salmonella* radiation. *J. Food Quality* 1:85-92.

Niewiarowicz, A., Fiszer, W., Zabielski, J., and Starega, M. (1980) Radiacyjna pasteryzacja mroªonej masy jajowej. *Medycyna Wet.* 36:365-367.

Proctor, B.E., Joslyn, R.P., Nikerson, J.R.T., and Lockhart, E.E. (1953) Elimination of *Salmonella* in whole egg powder by cathode ray irradiation prior to drying. *Food Technol.* 7:291-293.

Taub, J.A. and Halliday, J.W. (1979) *Chemical Reactions in Proteins Irradiated at Subfreezing Temperatures.* Advanced Chemistry Series 180. American Chemical Society.

Thieulin, G., Brunelet, L., Sarrazin, P., and Basille, D. (1960) La Revue Gen. du Froid 7:654-658. (In: Bednarczyk, W. 1963. *Food Preservation by Ionising Radiation.* WPL, Warsaw. In Polish).

Thornley, M. (1963) *Microbiological Aspects of the Use of Radiation for the Elimination of Salmonellae from Foods and Feeding Stuffs.* IAEA Techn. Rep. Series 22. Vienna. Austria.

Witte, V.C., Krause, G.F., and Bailey, M.E. (1970) A new extraction method for determining 2-thiobarbituric acid values of pork and beef during storage. *J. Food Sci.* 35:582-585.

Wos, Z. (1987) *Occupancy of Salmonellae in the Hen and Duck Egg.* Technical Report of Poultry and Development Center, Poznañ (in Polish).

Chapter Twenty-Nine

Fatty Acid Modification of Yolk Lipids and Cholesterol-Lowering Eggs

Z. Jiang[1] and J.S. Sim

Department of Animal Science, University of Alberta, Edmonton, Alberta, Canada T6G 2P5 ([1]Present address: Shur-Gain, 2700 Matheson Blvd. E., Suite 600, East Tower, Mississauga, Ontario, Canada L4W 4V9)

Abstract The importance of dietary ω3 polyunsaturated fatty acids (ω3 PUFA) in cardiovascular diseases has been established only in the past decade. Flax seeds are rich in the parent ω3 PUFA, α-linolenic acid (LnA), which can be converted in both animals and humans to longer chain ω3 PUFA such as eicosapentaenoic (EPA) and docosahexaenoic acids (DHA). Feeding laying hens diets containing full-fat flax seeds enriched chicken eggs with ω3 PUFA. The major proportion of the ω3 PUFA incorporated into the yolk was LNA, although a considerable amount of longer chain ω3 PUFA was also deposited into the yolk lipids, particular into yolk phosphatidylethanolamine. Subsequent feeding of the ω3 yolk powders to rats reduced plasma total cholesterol, enriched plasma lipids with ω3 PUFA, and reduced prostaglandin E_2 synthesis by skeletal muscle, when compared with the feeding of either control eggs or eggs enriched with ω6 PUFA. The inclusion of flax seeds in laying hen diets did not adversely affect the performance of laying hens in terms of hen-day production or mortality, or the internal qualities of the egg in terms of the Haugh units, specific gravity, or yolk index. Sensory studies, however, revealed that eggs from hens fed flax seeds exhibited a fishy flavour and were scored lower in the preference study than the control, indicating the need for further research.
(Key words: ω3 fatty acids, flax seeds, eggs, plasma cholesterol, triglyceride, rat)

Introduction

The chicken egg, viewed by many as one of nature's most perfect foods, has been a staple food item for humans since the dawn of civilisation. It provides the most complete, therefore the highest quality proteins, all necessary vitamins except vitamin C, and minerals (Moreng and Avens, 1985; Shrimpton, 1987). The per capita consumption of the egg, however, has been declining steadily over the past four decades in Western countries. Changing life style and more choices of food products are two contributing factors, but by far the major cause is consumers' increasing health consciousness over their diet and the controversial relationship between dietary cholesterol and coronary heart disease. Due to its relatively high cholesterol level and great culinary advantage, the egg has been used as the exclusive source of dietary cholesterol in both animal and human experiments to study the effect of dietary cholesterol. Consequently, the term 'egg' nowadays has become a synonym of 'cholesterol' which, in turn, implies heart disease, to many laymen and medical professionals alike.

The relationship between dietary cholesterol and coronary heart disease, however, is far from being settled (Texon, 1989; Stehbens, 1989; Steinberg, 1989), although it has been extensively examined over the past four decades in experimental animals as well as in observational epidemiological and clinical studies. At the same time, many attempts have been made to reduce egg cholesterol content but with only little practical success (Naber, 1983; Noble, 1987).

An alternative way to reduce the cholesterolaemic effects of eggs is by altering the fatty acid composition of the yolk. The cholesterol-lowering effects of ω6 polyunsaturated fatty acids (PUFA), mainly linoleic acid (LA), have been known for decades, and the hypocholesterolaemic properties of monounsaturated fatty acids (MUFA), such as oleic acid (OA), have also been recognised (Grundy, 1986). The past decade has witnessed the growing interests in another family of PUFA, namely ω3 PUFA, such as α-linolenic acid (LnA), eicosapentaenoic acid (EPA), and docosahexaenoic acid (DHA). The ameliorative effects against atherosclerosis and other diseases and the essentiality for normal function and development of these ω3 PUFA have been appreciated only since the later 1970s (Dyerberg et al., 1978; Dyerberg and Bang, 1979). The beneficial effects of dietary ω3 PUFA include, among others, the reduction of the concentration of plasma triglycerides (TG) (Connor, 1986; Harris, 1989), blood pressure (Knapp, 1989), platelet aggregation (Von Shacky et al., 1985), and tumour growth (Karmali et al., 1984). The ω3 PUFA exhibit their beneficial effects mainly through their influences on the quantity and quality of eicosanoids produced, and on the metabolism of lipoproteins (Kinsella et al., 1990). Thus the cholesterolaemic properties of the chicken egg might be altered by incorporating either ω3 or ω6 PUFA or MUFA into the egg yolk.

It is known that the fatty acid composition of chicken eggs can be altered easily through dietary manipulation (Cruickshank, 1934). In a series of studies carried out in our laboratory, we have demonstrated that ω3 or ω6 PUFA or ω9 MUFA can be readily incorporated into the yolk lipids by inclusion of full-fat oil seeds in the diet of laying hens. The effects of yolk fatty acid modification on the cholesterolaemic effects of eggs in the experimental animals, and on the internal and sensory qualities of eggs have also been examined and are presented herein.

Fatty Acid Modification of Egg Yolk Lipids

Full-Fat Oil Seeds and Laying Hen Diet

Full-fat oil seeds, namely flax (Hn-3) and regular high linoleic acid (Hn-6) or high oleic acid sunflower (Hn-9) seeds, were selected as the dietary source of ω3 PUFA, ω6 PUFA and ω9 MUFA, respectively, for the laying hen rations.

Table 29.1. Nutrient composition of laying hen diets.

Ingredients and analyses	Laying hen diet (%)			
	Hn-9	Hn-3	Hn-6	Control
Wheat	61.3	66.9	58.7	72.4
Soybean meal	8.0	5.3	7.6	11.8
Flax seed	0	15.0	0	0
Sunflower seed (ω9)	18.0	0	0	0
Sunflower seed (ω6)	0	0	21.0	0
Fat mixture	0	0	0	3.0
Limestone	8.3	8.3	8.3	8.3
Calcium phosphate	1.9	2.0	1.9	2.0
Salt	0.3	0.3	0.3	0.3
DL-methionine	0.1	0.1	0.1	0.1
Layer premix[a]	2.1	2.1	2.1	2.1
Calculated analyses:				
Crude protein (%)	15.0	15.0	15.0	15.0
Metabolisable energy (kcal/kg)	2732	2728	2700	2751

[a] The following quatities were supplied per kilogram of diet: vitamin A, 8000 IU; cholecalciferol, 1200 ICU; vitamin E, 5 IU; riboflavin, 4 mg; calcium pantothenate, 6 mg; niacin, 15 mg; vitamin B_{12}, 10 µg; choline chloride, 100 mg; biotin, 100 µg; selenium, 0.1 mg; DL-methionine, 500 mg; manganese sulphate, 0.4 g; zinc oxide, 0.1 g.

They were chosen because i) the full-fat seeds protected the oils from auto-oxidation; ii) these full-fat seeds have been shown to be rich sources of dietary protein and metabolisable energy; and iii) these seeds are readily produced in abundance in Canada (Sim, 1990). The full-fat seeds were incorporated into

laying hen diets at 15-21% which would contribute about 5% of fat to the diet
(Table 29.1) (Jiang et al., 1991).

Changes in Fatty Acid Composition of Yolk Total Lipids

Dietary treatment had no effect on the total lipid content of the egg yolk (33.6%
(w/w) of fresh yolk). The fatty acid composition of yolk total lipids, however,
readily responded to the changes in dietary fat (Table 29.2).

Table 29.2. Major fatty acid composition of yolk total lipids for eggs collected from
hens fed full-fat seed rations.

Fatty acid	Laying hen diet[1] (% of total fatty acids[2])				SEM
	Hn-9	Hn-3	Hn-6	Control	
C16:0	22.3[b]	25.6[a]	23.2[b]	25.3[a]	0.34
C18:0	7.7[b]	9.6[a]	9.9[a]	9.4[a]	0.33
C16:1 (ω7 & ω9)	3.1[b]	3.1[b]	2.3[c]	3.9[a]	0.15
C18:1 (ω7 & ω9)	52.3[a]	39.0[c]	32.1[u]	44.7[b]	0.60
C18:2 (ω6)	11.1[b]	12.1[b]	28.5[a]	11.6[b]	0.68
C18:3 (ω3)	0.2[b]	5.8[a]	0.3[b]	0.9[b]	0.30
C20:4 (ω6)	2.2[b]	1.3[c]	2.7[a]	2.0[b]	0.13
C20:5 (ω3)	0[b]	0.3[a]	0[b]	0[b]	0.02
C22:5 (ω3)	0[b]	0.3[a]	0.1[b]	0.2[b]	0.05
C22:6 (ω3)	0.8[c]	2.7[a]	0.5[c]	1.5[b]	0.12
Σ Saturated	30.2[c]	35.5[a]	33.4[b]	35.1[a]	0.46
Σ MUFA[3]	55.4[a]	42.2[c]	34.5[d]	48.6[b]	0.64
Σ PUFA[4] (ω6)	13.4[b]	13.4[b]	31.2[a]	13.6[b]	0.66
Σ PUFA (ω3)	1.1[c]	9.1[a]	0.9[c]	2.6[b]	0.26
$\Sigma(\omega$6$)/\Sigma(\omega$3$)$	12.7[b]	1.5[d]	34.3[a]	5.3[c]	1.03

[a-d] Means in a row with no common superscripts differ significantly ($P < 0.05$,
$n = 6$).
[1] The laying hen diets contained high oleic acid sunflower seed (Hn-9), full-fat flax
seed (Hn-3), and high linoleic acid sunflower (regular seed, Hn-6).
[2] Each value was the mean of six eggs.
[3] MUFA = monounsaturated fatty acids.
[4] PUFA = polyunsaturated fatty acids.

When the birds were fed experimental diets containing flax or regular or
high OA sunflower seeds, the contents of LnA, LA and OA, respectively,
increased rapidly and reached a plateau within 2 weeks (Sim, 1990). The major
ω3 PUFA incorporated into yolk total lipids was α-LnA, but a considerable

amount of longer chain ω3 PUFA, namely EPA, DPA (docosapentaenoic acid), and DHA, was also found in the yolk upon flax seed feeding. For instance, the LnA content in yolks from hens fed a flax seed diet increased by 6.4 times over the control. The level of ω6 arachidonic acid (AA), a metabolite of LA, was significantly reduced in the yolk upon flax seed feeding. When laying hens were fed diets containing regular LA sunflower seeds, the LA content of yolks increased by 2.5 times over the control (Table 29.2). On the other hand, the ability of laying hens to increase OA in the yolk seemed limited. An increase of 17% ($P < 0.05$) of OA was found in yolk total lipids, even though the OA content in the high OA sunflower seed diet was twice that of the control diet. The ratio of ω6 to ω3 PUFA in yolk total lipids was significantly reduced by flax seed feeding (1.5) but increased by regular sunflower seed feeding (34.3) when compared with the control yolks (5.3). Due to the use of a mixture of animal tallow and vegetable oil in the control diet, the control eggs in our experiments contained more ω3 PUFA than commercially produced eggs. The ratio of ω6:ω3 in commercial eggs is about 19, much higher than that in our control eggs (Sim *et al.*, 1991; Simopoulos, 1991).

Changes in Fatty Acid Composition of Various Yolk Lipid Classes

Our next project was carried out to investigate the distribution of ω3 and ω6 PUFA in various yolk lipid classes. Yolk total lipids were separated into triglycerides (TG), phosphatidylcholine (PC), and phosphatidylethanolamine (PE) by a thin-layer chromatographic technique (Jiang *et al.*, 1991).

The enrichment of LnA upon feeding flax seed was mainly in yolk TG with only moderate increases in PC and PE (Jiang *et al.*, 1991). In contrast, EPA, DPA, and DHA, the longer chain metabolites of LnA, were exclusively deposited into yolk phospholipids, particularly in the yolk PE fraction. The contents of EPA, DPA, and DHA in PE were three to seven times those in PC. The arachidonic acid content in yolk phospholipids was significantly reduced in ω3 PUFA-enriched eggs.

This study therefore clearly demonstrated that the changes of fatty acids induced by dietary treatment were not uniformly distributed among all lipid classes of egg yolks. The longer chain ω3 PUFA were preferentially deposited into the phospholipid fraction, particularly into the PE fraction of the yolk. Negative relationships ($P < 0.05$) between AA and LnA, and between AA and longer chain ω3 fatty acids were observed in yolk total lipids, TG, PC, and PE. It has been known that the enzymatic pathway for the synthesis of AA from LA is shared by the ω3 fatty acids (Brenner, 1981), and LnA inhibits the Δ6 desaturase and thereby reduces the conversion of LA to AA. The higher contents of longer chain ω3 fatty acids, such as EPA, DPA, and DHA, might also hinder the incorporation of AA into PC and PE and thus reduce AA content in yolk phospholipids.

Cholesterolaemic Effects of Fatty Acid Modified Eggs in Rats

We then proceeded to investigate the effects of yolk fatty acid modification on the cholesterolaemic and lipid modulating properties of the egg. Young female Sprague-Dawley rats were chosen as the experimental model due to their long established suitability for the study of cholesterol metabolism. Egg yolk powder was prepared and incorporated at 15% into semi-synthetic rat diets (Jiang and Sim, 1992). Two separate experiments were carried out. In Experiment 1, the dietary effects of fatty acid modified eggs on plasma and liver cholesterol levels were examined (Jiang and Sim, 1992). It was found that feeding ω3 PUFA-enriched yolk powder to rats for a period of 4 weeks reduced plasma and liver cholesterol significantly when compared with the control yolk powder (Fig. 29.1).

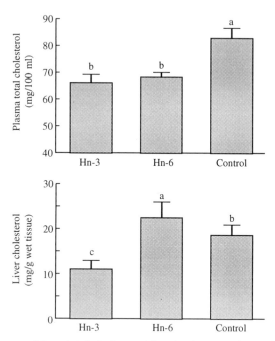

Fig. 29.1. Plasma and liver total cholesterol levels of rats at the end of the 28 day feeding trial (Experiment 1, mean ± SEM, n = 7). The test diets used yolk powders prepared from eggs laid by laying hens fed diets containing 10% flax seed (Hn-3), 12% sunflower seed (Hn-6), or animal tallow (Control). Bars without a common letter differ significantly (*P* < 0.05). Taken from Jiang and Sim (1992) with permission.

Feeding yolk powder enriched with ω6 PUFA also reduced plasma total cholesterol, but elevated liver cholesterol content significantly. Furthermore, feeding ω3 PUFA eggs to rats not only reduced the plasma and tissue cholesterol contents, but also enriched plasma and tissue lipids with ω3 PUFA (Table 29.3).

In the phospholipids fraction of rat liver lipids, the longer chain ω3 PUFA, such as EPA and DHA, increased significantly while the ω6 AA decreased in those fed ω3 eggs.

In Experiment 2, we confirmed the findings of Experiment 1 that feeding ω3 eggs to rats reduced plasma total cholesterol (Fig. 29.2). In addition, feeding eggs enriched with either ω3 or ω6 PUFA had no effect on plasma high density lipoprotein cholesterol levels. The elevation of liver cholesterol levels by dietary ω6 PUFA-enriched eggs was observed again in Experiment 2, indicating differential dietary effects of ω3 or ω6 PUFA on cholesterol metabolism in rats.

Table 29.3. Major fatty acids of liver phosphatidylethanolamine fraction of rats in Experiment 2.

	Test diet[1] (% of total fatty acids)		
	Hn-3	Hn-6	Hn-9
C16:0	15.3 ± 0.8[a]	15.3 ± 0.4[a]	15.4 ± 0.6[a]
C18:0	27.8 ± 1.1[a]	28.4 ± 1.1[a]	29.0 ± 0.7[a]
C16:1	1.3 ± 0.1[a]	1.0 ± 0.1[a]	1.0 ± 0.1[a]
C18:1	12.7 ± 1.2[a]	11.5 ± 1.0[a]	14.2 ± 1.6[b]
C18:2 ω6	5.9 ± 0.2[a]	4.7 ± 0.0[a]	5.0 ± 0.4[a]
C18:3 ω3	1.2 ± 0.2[a]	0.9 ± 0.1[a]	1.1 ± 0.2[a]
C20:4 ω6	19.4 ± 0.8[b]	24.6 ± 1.1[a]	24.3 ± 1.8[a]
C20:5 ω3	2.4 ± 0.1[a]	0.1[b]	0[b]
C22:6 ω3	10.4 ± 0.6[a]	3.7 ± 0.6[b]	6.4 ± 1.0[b]
Σ MUFA	4.0	12.5	15.3
Σ ω6 PUFA	25.3	29.3	29.3
Σ ω3 PUFA	14.7	4.7	7.5
Σ(ω6):Σ(ω3) ratio	1.7	6.2	3.9

[a,b] Means in a row without common letters differ significantly ($P < 0.05$).
[1] The rat test diets contained yolk powders prepared from eggs laid by laying hens fed diets containing full-fat flax seed (Hn-3), regular high LA sunflower seed (Hn-6), or high OA sunflower seed (Hn-9). Data are presented as mean ± SEM, n = 5 pooled samples of two each. MUFA, monounsaturated fatty acids; PUFA, polyunsaturated fatty acids.

We also examined the influence of consuming these fatty acid modified eggs on prostaglandin E_2 (PGE2) production by rat skeletal muscle. The production of PGE2 by rat epitrochlearis muscle was significantly reduced in those fed ω3 eggs when compared with those fed regular eggs (Fig. 29.2). The reduction of PGE2 synthesis could be attributed to a lower ω6:ω3 ratio in tissue phospholipids of rats fed ω3 eggs (Table 29.3). Chronic excessive production of PGE2 has been implied in several degenerative vascular diseases (Simopolous, 1991).

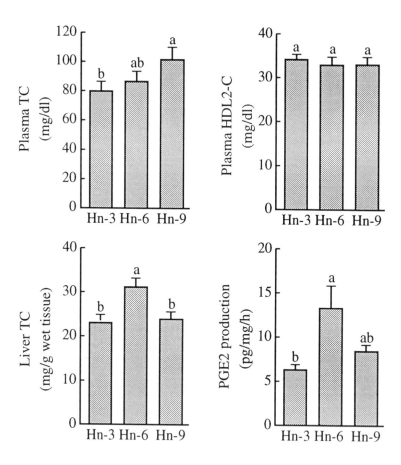

Fig. 29.2. Plasma total (TC), high-density lipoprotein (HDL_2-C) cholesterol, liver cholesterol levels, and the production of prostaglandin E_2 (PGE2) by the epitrochlearis muscle of rats at the end of 31 day feeding trial (Experiment 2, mean ± SEM, n = 10). The test diets used yolk powders prepared from eggs laid by laying hens fed diets containing 15% flax seed (Hn-3), 18% regular high LA sunflower seed (Hn-6), or 21% high OA sunflower seed (Hn-9). Bars without a common letter differ significantly ($P < 0.05$).

In summary, enriching chicken eggs with ω3 PUFA reduced the cholesterol-aemic properties of the egg in rats. Considerable amounts of ω3 PUFA, including the longer chain EPA, DPA, and DHA, were incorporated into tissue lipids, which in turn, resulted in favourable changes in tissue prostaglandin production.

Impact of Fatty Acid Modification on Egg Internal and Sensory Qualities

The ω3 PUFA-enriched eggs may be more appealing to today's health conscious public. Consumer acceptance of a food item, however, also depends on the overall qualities of the item. Therefore, our next study was carried out to examine the effects of enriching eggs with ω3 or ω6 PUFA on the internal and sensory qualities of hard-cooked eggs (Jiang *et al.*, 1992).

The effects of feeding various oil seeds to laying hens on the quality of egg albumen were examined by means of measuring the Haugh units of eggs during the 6 week storage period (Jiang *et al.*, 1992). Overall, enriching eggs with either ω3 or ω6 PUFA did not adversely affect egg Haugh units during storage.

Yolk colour is an important quality trait in influencing consumer acceptance (Hunton, 1987). We found that feeding full-fat flax seed to laying hens resulted in darker ($P < 0.05$) yolks than other dietary regimens (Jiang *et al.*, 1992). Although having no influence on the nutritive value of eggs, darker yolks might be more appealing to consumers of certain ethnic groups. Storage for up to 6 weeks at 4°C did not significantly change yolk fatty acid profiles.

One of the most important quality parameters in determining consumer acceptance of any food item is the sensory characteristics. Two independent studies were carried out at the University of Alberta, Edmonton, and the Food Processing Development Centre, Leduc, Alberta, Canada. As indicated in Fig. 29.3, studies at both the University and the Food Centre generated similar preference patterns. Eggs from hens fed the Hn-3 diet scored significantly lower in preference evaluation than others while those from the sunflower seed diets were not affected when compared with the control. About one-third of the evaluations in both studies indicated a fishy or fish-product related flavour, such as 'cod liver oil-like odour' and 'tuna flavour', for eggs from the Hn-3 diet (Table 29.4).

The cause or causes of fishy flavours in eggs from laying hens fed flax seed remain to be determined. It was speculated that the fishy flavour was due to the presence of lipid oxidation products (Jiang *et al.*, 1992). It has been reported that the rancidity products of PUFA cause a fishy flavour in some food products (Saxby, 1982). In the present study, as eggs were stored at 4°C for 7 days before sensory evaluations were carried out, there might be very little, if any, oxidation of PUFA in yolks during the first week post-harvest. The oxidative rancidity was more likely to occur in the Hn-3 diet during the preparation and feeding of the diet. These oxidation products might find their way into the yolk when the diet was fed to laying hens. As only a trace amount of lipid oxidation products are needed to cause a fishy flavour in food products (Saxby, 1982), the fishy taint could be the result of rancidity of the Hn-3 diet. More recent studies from our laboratory demonstrated that this fishy flavour was indeed eliminated by adding natural antioxidants to the laying hen diets (unpublished results).

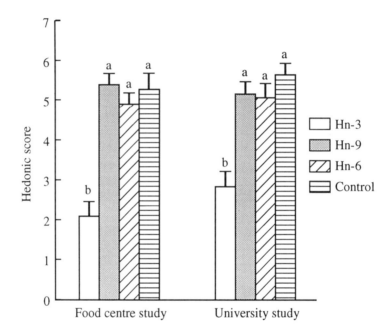

Fig. 29.3. Effects of dietary full-fat oil seeds on the Hedonic scores of eggs (mean $\pm$ SEM, n = 36 and 28 for Food Centre study and University study, respectively). A Hedonic score of 0 represents an extreme dislike, and 9 an extreme like. Bars in the same graph with no common letters differ significantly ($P < 0.05$). The laying hen diets contained either 15% flax seed (Hn-3), 18% high OA sunflower seed (Hn-9), 21% regular high LA sunflower seed (Hn-6), or 3% animal tallow (Control). Taken from Jiang *et al.* (1992) with permission.

In brief, fatty acid modification of yolk lipids did not affect egg internal quality in terms of specific gravity and Haugh units. Yolk colour was darkened by feeding flax seed. Furthermore, feeding flax seed in the present study resulted in a fishy flavour in some eggs which was attributed to the presence of lipid oxidation products in laying hen diet, and it was possible to overcome this problem by adding antioxidants to the diet.

Table 29.4. Number and percentage of evaluations that reported a fishy flavour in eggs from laying hens fed diets containing full-fat oil seeds.

Study by		Laying hen diet[a]			
		Hn-3	Hn-9	Hn-6	Control
Food Centre	number	13	1	0	2
(n = 36)	percentage	36	3	0	6
University	number	10	0	0	0
(n = 28)	percentage	36	0	0	0

[a] The laying hen diets contained either 15% flax seed (Hn-3), 18% high OA sunflower seed (Hn-9), 21% regular high LA sunflower seed (Hn-6), or 3% animal tallow (Control).

References

Brenner, R.R. (1981) Nutritional and hormonal factors influencing desaturation of essential fatty acids. *Prog. Lipid Res.* 20:41-47.

Connor, W.E. (1986). Hypolipidemic effects of dietary ω-3 fatty acids in normal and hyperlipidemic humans: effectiveness and mechanisms. In: Simopoulos, A.P., Kifer, R.R. and Martin, R.E. (eds), *Health Effects of Polyunsaturated Fatty Acids in Seafoods*. Academic Press, New York, NY, pp. 173-195.

Cruickshank, E.M. (1934) Studies in fat metabolism in the fowl. 1. The composition of the egg fat and depot fat of the fowl as affected by the ingestion of large amounts of different fats. *Biochem. J.* 28:965-977.

Dyerberg, J. and Bang, H.O. (1979) Haemostatic function and platelet polyunsaturated fatty acids in Eskimos. *Lancet* ii:433-435.

Dyerberg, J., Bang, H.O., and Stofferson, E. (1978) Eicosapentaenoic acid and prevention of thrombosis and atherosclerosis. *Lancet* ii:117-119.

Grundy, S.M. (1986) Comparison of monounsaturated fatty acids and carbohydrates for lowering plasma cholesterol. *N. Engl. J. Med.* 314:745-748.

Harris, W.S. (1989) Fish oils and plasma lipid and lipoprotein metabolism in humans: critical review. *J. Lipid Res.* 30:785-807.

Hunton, P. (1987) Laboratory evaluations of egg quality. In: Wells, R.G. and Belyavin, C.G. (eds), *Egg Quality - Current Problems and Recent Advances*. Butterworths Co. Ltd, London, UK, pp. 87-102.

Jiang, Z. and Sim, J.S. (1992) Effects of dietary n-3 PUFA-enriched chicken eggs on plasma and tissue cholesterol and fatty acid composition of rats. *Lipids* 27:279-282.

Jiang, Z., Ahn, D.U., and Sim, J.S. (1991) Effects of feeding flax and two types on sunflower seeds on fatty acid composition of yolk lipid classes. *Poultry Sci.* 70:2467-2475.

Jiang, Z., Ahn, D.U., Ladner, L., and Sim, J.S. (1992) Influence of feeding full-fat flax and sunflower seeds on internal and sensory qualities of eggs. *Poultry Sci.* 71:378-382.

Karmali, R.A., March, J., and Fuchs, C. (1984) Effect of omega-3 fatty acids on growth of rat mammary tumor. *J. Natl Cancer Inst.* 73:457-461.

Kinsella, J.E., Lokesh, B., and Stone, R.A. (1990) Dietary n-3 polyunsaturated fatty acids and amelioration of cardiovascular disease: possible mechanisms. *Am. J. Clin. Nutr.* 52:1-28.

Knapp, H.R. (1989) Omega-3 fatty acids, endogenous prostaglandins and blood pressure regulation in humans. *Nutr. Rev.* 47:301-313.

Moreng, R.E., and Avens, J.S. (eds) (1985) In: *Poultry Science and Production.* Reston Publishing Company, Inc., Reston, VA, pp. 15-46.

Naber, E.C. (1983) Nutrient and drug effects on cholesterol metabolism in the laying hen. *Fed. Proc.* 42:2486-2493.

Noble, R.C. (1987) Egg lipids. In: Wells, R.G. and Belyavin, C.G. (eds), *Egg Quality - Current Problems and Recent Advances.* Butterworths Co. Ltd, London, UK, pp. 159-177.

Saxby, M.J. (1982). Taints and off flavors in foods. In: Morton, I.D. and Macleod, A.J. (eds), *Food Flavors. Part A. Introduction.* Elsevier Scientific Publishing Company, Amsterdam, The Netherlands, pp. 439-457.

Shrimpton, D.H. (1987). The nutritive value of eggs and their dietary significance. In: Wells, R.G. and Belyavin, C.G. (eds), *Egg Quality - Current Problems and Recent Advances.* Butterworth & Co. Ltd, London, UK, pp. 11-25.

Sim, J.S. (1990) Flax seed as a high energy/protein/omega-3 fatty acid feed ingredient for poultry. In: *Proceedings of the 53rd Flax Institute of the United States,* University of North Dakota, Fargo, ND, pp. 65-71.

Sim, J.S., Chung, S.O., Borgersen, P.A., Jiang, Z., and Chorney, W.P. (1991) Cholesterol and fatty acid compositions of commercially produced Alberta eggs (a provincial-wide survey study). In: Baracos, V. and Price, M. (eds) *70th Feeders' Day Report,* University of Alberta, pp. 20-21.

Simopoulos, A.P. (1990) Genetics and nutrition: or what your genes can tell you about nutrition. In: Simopoulos, A.P. and Childs, B. (eds), *Genetic Variation and Nutrition.* World Review of Nutrition and Diet. Karger, New York, pp. 25-34.

Simopoulos, A.P. (1991) Omega-3 fatty acids in health and disease and in growth and development. *Am. J. Clin. Nutr.* 54:438-463.

Stehbens, W.E. (1989) The controversial role of dietary cholesterol and hyper-cholesterolemia in coronary heart disease and atherogenesis. *Pathology* 21:213-221.

Steinberg, D. (1989) The cholesterol controversy is over. Why did it take so long? *Circulation* 80:1070-1078.

Texon, M. (1989) The cholesterol-heart disease hypothesis (critique) - time to change course? *Bull. N. Y. Acad. Med.* 65:836-841.

Von Shacky, C., Fischer, S., and Weber, P.C. (1985) Long term effects of dietary marine omega-3 fatty acids upon plasma and cellular lipids, platelet function and eicosanoid formation in humans. *J. Clin. Invest.* 76:1626-1631.

Chapter Thirty

High Linolenic Acid Eggs and Their Influence on Blood Lipids in Humans

L.K. Ferrier, S. Leeson, B.J. Holub, L. Caston, and E.J. Squires

Department of Animal and Poultry Science, University of Guelph, Guelph, Ontario, Canada N1G 2W1

Abstract Linolenic acid (LnA) increased about 30-fold and docosahexaenoic acid (DHA) increased nearly 4-fold in eggs when hens were fed a diet containing 20% flax seeds. Total polyunsaturated fatty acids nearly doubled, increasing to 22% from 12% of the total fatty acids. The ratio of ω6:ω3 fatty acids decreased from 25:1 in control to 1.4:1 in modified eggs. Five volunteers consumed either four control eggs, four modified eggs, or no eggs per day for 2 week periods. Changes in lipid distribution in plasma and ω3 fatty acid levels in blood platelet phospholipids were monitored to determine the effect of eating the modified eggs. Blood platelet DHA levels increased by about 55%, from 0.85 to 1.3% of phospholipid fatty acids, in response to consumption of modified (relative to control) eggs whereas no change in LnA or EPA (eicosapentaenoic acid) was seen. DHA has been found to reduce blood platelet aggregatability and, potentially, the risk of heart disease. High density lipoprotein-cholesterol levels increased by about 15% (relative to no eggs) when either regular or modified eggs were consumed but plasma cholesterol showed no change. Plasma triglycerides were reduced when modified, but not regular, eggs were eaten. Blood plasma LnA and platelet phospholipid EPA did not change significantly. Increases have been shown in animal studies. Thus, more research in this area is warranted. Fresh, modified eggs were described as having a trace of off-flavour described as 'stale or rancid'. Flavour scores did not change significantly during 8 weeks' storage at 4°C or 4 weeks at room temperature. These modified eggs offer a possible means of delivering very high levels of dietary ω3 fatty acids including DHA and LnA.
(*Key words:* high linolenic acid eggs, ω3 fatty acids, human blood lipids, docosahexaenoic acid, linolenic acid, cholesterol)

Introduction

Omega-3 fatty acids in eggs have been increased to over 10% of the fatty acids in egg yolk by feeding flax seeds to laying hens (Yu and Sim, 1987; Caston and Leeson, 1990). Linolenic acid (LnA, 18:3ω3) accounted for the majority of the increase in ω3 fatty acids. One modified egg may contain over 400 mg of LnA, as compared with 16 mg for the controls.

Omega-3 fatty acids are now regarded as essential in the diet for brain function and visual acuity in humans based on research conducted and on the Canadian Nutritional Recommendations (Health and Welfare Canada, 1990). The importance of ω3 fatty acids was first clearly demonstrated by a case of deficiency reported by Holman *et al.* (1982). Increased LnA intake (as linseed oil) has been associated with lower morbidity and mortality from heart disease, and with reduced platelet adhesiveness associated with diabetes and atherosclerosis (Owren *et al.*, 1964). While diets enriched in linoleic acid (ω6) or LnA (ω3) can lower blood cholesterol levels (Cunnane *et al.*, 1989; MacDonald *et al.*, 1989), the LnA can also provide for the entry of the ω3, eicosapentaenoic acid (EPA) into blood platelet phospholipids in humans via LnA metabolism to EPA (Weaver *et al.*, 1990). These ω3 fatty acids may slow the rate of blood clotting, thereby reducing the risk of cardiovascular disease (Holub, 1988).

Another ω3 fatty acid, docosahexaenoic acid (DHA, 22:6ω3), may be more potent than EPA in reducing the aggregation rate of platelets (Gaudette and Holub, 1991). DHA may be a critical ω3 fatty acid derived from LnA via metabolism for brain and retinal function (Bjerve *et al.*, 1988). DHA has been found to reduce platelet aggregatability, platelet adhesiveness, and blood triglyceride (TG) levels and thus may reduce the risk of cardiovascular disease (Gaudette and Holub, 1990, 1991).

Eggs high in ω3 fatty acids (ω3 eggs) are highly desirable in light of recent recommendations that Canadian diets contain an ω6:ω3 fatty acid ratio of about 6:1 (Health and Welfare Canada, 1990). The current Canadian diet is estimated to contain an ω6:ω3 ratio of 14:1 because of the high proportion of ω6 fatty acids in most of the foods we consume (including eggs). Since there is normally no appreciable accumulation of LnA in animal tissue (meat), ω3 eggs could be a valuable source of this essential nutrient. The modified eggs can deliver as much as 40% of the daily recommended ω3 fatty acid intake for humans per egg and very high levels of LnA.

The objective of this study was to measure changes in the ω3 fatty acid distribution of blood platelets as well as favourable changes in blood lipids in response to consumption of high ω3 fatty acid eggs. The platelet phospholipid fatty acids of interest were mainly DHA, EPA, and LnA. The plasma lipids of concern were TG, total cholesterol, and high density lipoprotein (HDL)-cholesterol.

Materials and Methods

Egg Production

Sixty commercial strain Single Comb White Leghorn hens, 30 weeks old, were allocated to individual cages in an environmentally controlled room. Hens were housed and fed according to Caston and Leeson (1990). High LnA eggs were produced by feeding 30 hens a diet containing 20% (w/w) flax seeds. Control hens were fed a standard layer diet to produce normal eggs. Diets were fed for 32 days before the nutritional trial began. Eggs were collected daily and weighed. Eggs were selected at random for nutritional trials, analyses, storage or taste testing during the course of the study.

Composition of Eggs

Fatty acids and cholesterol in eggs were measured biweekly. Lipids were extracted from 1 g samples of egg yolk using chloroform:methanol according to Bligh and Dyer (1959). The cholesterol was analysed with HPLC using the method of Chung *et al.* (1991). Fatty acids were analysed using GLC according to the method of Holub and Skeaff (1988). Oxidation in lipids from egg yolk was measured with thiobarbituric acid as described by Squires (1989).

Feeding and Nutrition Trial

Four modified or four control eggs per day or no eggs were fed for 2 week periods to five healthy male volunteers (aged 35-52; mean weight 79 kg) to test the effects of the eggs on blood cholesterol, HDL-cholesterol, plasma TG levels, and platelet phospholipid fatty acids. Subjects did not eat eggs during weeks 1, 2, 5, and 6. Control eggs were consumed in weeks 3 and 4 and modified eggs during weeks 7 and 8. Subjects were instructed to refrain from eating eggs, Canola oil, fish, aspirin, and other medication during the experimental period of 8 weeks. Alcoholic beverages were limited to two per day. Other dietary habits and normal physical exercise levels were to be maintained. Otherwise diets were *ad libitum.* None of the volunteers smoked. Eggs were cooked as desired by the volunteer since cooking method does not appear to alter the fatty acid composition (Van Elswyk *et al.*, 1990). Typically, volunteers ate two eggs for breakfast and two for dinner. Diets were restricted in that no fish, fish oils, Canola oil, or other high $\omega 3$ fatty acid foods were to be consumed.

Flavour

Changes in flavour were monitored using experienced taste panellists and by measuring malonyldialdehyde (MDA) in extracted egg lipids. Scrambled eggs were tasted fresh (at room temperature) on three occasions and after storage for

2, 4, or 8 weeks at 4°C and for 2 and 4 weeks at room temperature to determine if detectable changes in egg flavour occurred during storage. Panellists were trained by tasting scrambled eggs, exposing them to flavours and aromas they might encounter, discussing and reaching consensus on terminology and meanings to be used. This took three, 1 h sessions. The rating scale used was based on 5 = the flavour of reference (supermarket) eggs. Samples could be rated as having more or less of a flavour using the scale 'trace, slight, moderate or high'. Trace was defined as there being some uncertainty about the presence of the off-flavour. Slight was used to describe a low level of the off-flavour where the panellist was sure of its presence. For statistical analysis, ratings were changed to numerical values where trace = 1, slight = 2, etc. These values were added to or subtracted from 5 depending on whether the flavours were considered desirable or undesirable by panellists. Statistical analysis (ANOVA) was performed using the Systat program (Wilkinson, 1986).

Plasma Lipid Analysis

Fasting blood samples were drawn from the antecubital vein of each subject on day 0 and weekly thereafter. Siliconised vacutainer tubes (Becton-Dickinson and Co., Rutherford, NJ) were used. Blood was incubated at room temperature (20-24°C) for 5 h to allow clotting to occur and then centrifuged at 350 g to obtain serum. Plasma TG and total cholesterol concentrations were measured using a DACOS^R chemistry analyser using a modification of the Fossati and Prencipe (1982) and McGowan *et al.* (1983) methods.

HDL-cholesterol was isolated by precipitation of all other lipoproteins from plasma using an HDL-cholesterol reagent (Boehringer Mannheim Canada, Dorval, Quebec) and quantified enzymatically using a Cholesterol High Performance CHOD-PAP kit (Boehringer Mannheim Canada).

Platelet Fatty Acid Analysis

Whole blood, drawn into siliconised vacutainer tubes as described above, was centrifuged at 250 g for 7 min at room temperature (20-24°C) to obtain platelet-rich plasma (PRP). 100 mM Na$_2$EDTA, at a concentration of 60 µl/ml PRP, was added and the suspension was refrigerated at 4°C for 1 h. The PRP was then centrifuged at 2000 g for 15 min at 4°C in order to obtain a platelet pellet. Platelets were resuspended in 2 mM Na$_2$EDTA, 0.15 M NaCl, 0.02 M Tris-HCl buffer (pH 7.4), washed and resuspended in 134 mM NaCl, 15 mM Tris-HCl, 5 mM glucose buffer (pH 7.4). Lipids were extracted from washed platelet suspensions by the method of Bligh and Dyer (1959), applied to glass plates coated with a thin layer of silica gel 60 (E. Merck, BDH Chemicals, Canada), and separated into their various lipid components using TLC on silica gel 60 with heptane/isopropyl ether/acetic acid (60:40:3, by volume) as the developing mixture. The lipid classes were detected under ultraviolet light after spraying with 2,7-dichlorofluorescein and exposing to ammonia vapour. The total phospholipid band was then scraped from the plate, transmethylated in the presence of 6% (by volume) H_2SO_4 in methanol and known amounts of an internal standard, monopentadecanoin (Nu Chek Prep. Inc. Elysian, MN) for 1.5-3 h at 80°C. Fatty acid methyl esters derived from the phospholipid fraction were analysed isothermally on a Hewlett-Packard 5890 GLC (Hewlett Packard, Palo Alto, CA) equipped with a DB-225 megabore column (Chromatographic Specialties, Inc., Brockville, Ontario).

Fatty acids were identified by comparison of their retention times with those of known standards (Nu Chek Prep. Inc.).

Statistical Analysis

All plasma lipid and platelet fatty acid data were analysed by a paired Student's *t*-test.

Results and Discussion

In this study, the effect of consumption of LnA-enriched eggs on plasma lipid profiles and on the distribution of platelet phospholipid fatty acids was studied. Enrichment of LnA in eggs was accomplished by feeding ground flax seeds, an abundant source of LnA.

All ω3 fatty acids increased in modified eggs whereas all ω6 fatty acids except linoleic decreased compared with control eggs (Table 30.1).

Table 30.1. Distribution of polyunsaturated fatty acids in egg phospholipids for modified and control eggs.

Fatty acid	% w/w of yolk lipid	
	ω3 eggs	Control eggs
Linoleic acid	11.7	9.5
LnA	8.2	0.3
Arachidonic acid	0.5	0.9
20:3ω3	0.1	0.0
EPA	0.1	0.0
22:5ω6	0.7	1.2
DHA	0.7	0.2
Total ω3	9.1	0.5
Total ω6	12.9	11.6
Ratio ω6:ω3	1.4:1	24.6:1

The total ω3 fatty acids increased to about 9% from 0.5%. LnA increased to 7-10% from 0.3% of the phospholipid fatty acids, an increase of about 30-fold. DHA increased to 0.7% from 0.2% of the egg phospholipid fatty acids. This may be due to desaturation and elongation of LnA within the hen (Tinoco, 1982) since flax seed is devoid of long chain fatty acids. EPA increased to 0.9% of the phospholipid fatty acids in modified eggs from 0.0% in controls. LnA is a known precursor of both EPA and DHA in blood (Gaudette and Holub, 1991).

The ratio of ω6 to ω3 fatty acids decreased dramatically, from 25:1 in control to 1.4:1 in modified eggs (Table 30.1). Total polyunsaturated fatty acids nearly doubled, thus increasing to 22% from 12% of the total fatty acids in modified and control eggs respectively. The saturated fatty acids decreased to 41% from 51% of the fatty acids. These changes in egg composition all are expected to reduce the risk of heart disease.

Egg cholesterol increased in the LnA egg, to about 212 mg/egg compared with 182 mg/egg for control although the change was not statistically significant. Hargis *et al.* (1991) and Farrell (1991) noted that cholesterol remained the same when fish oil was included in the diet of laying hens. However, Adams *et al.* (1988) observed higher egg cholesterol levels after feeding menhaden oil.

Consuming four modified eggs per day resulted in a 55% increase (from 0.85 to 1.3%) in the DHA in blood platelet phospholipids compared with eating control eggs (Fig. 30.1).

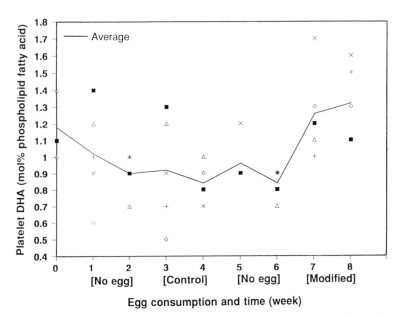

Fig. 30.1. Change in platelet phospholipid DHA. Line shows mean for five subjects. Symbols show individual levels during the experiment. Increase in DHA on consumption of modified eggs (from week 6 to weeks 7-8) was significant (*P* < 0.05).

There are few reports of increased platelet phospholipid DHA without a rise in platelet phospholipid LnA. Consumption of fish oil will result in this effect (Gaudette and Holub, 1991). This is the first report of a change in platelet phospholipid DHA in humans with no change in platelet phospholipid LnA or EPA (see below) due to consumption of eggs high in ω3 fatty acids. Consumption of either LnA or DHA in eggs may account for the increased DHA in platelets but DHA is the more probable source. The increase in DHA without an increase in EPA or LnA is intriguing both scientifically and for special food uses because it may provide a means of increasing DHA in food used in clinical feeding of, for example, premature infants who are undergoing rapid brain development.

Plasma TG decreased when the modified eggs were eaten (Fig. 30.2). It is now becoming recognised that plasma TG is a significant predictor of coronary heart disease.

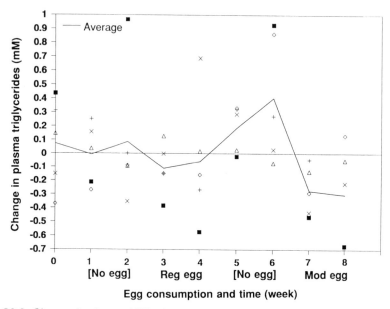

Fig. 30.2. Change in plasma HDL-cholesterol. Symbols show changes from each subject's mean value during the experimental period. Line shows mean for five subjects.

Plasma HDL-cholesterol levels increased by about 15% (as compared with an egg-free diet) when either regular or modified eggs were consumed (Fig. 30.3). This suggests that the ingestion of LnA eggs may possible offer some protection against heart disease. Ingestion of control eggs also raised total cholesterol whereas consumption of LnA eggs resulted in higher LDL-cholesterol levels. Increases in total cholesterol and HDL-cholesterol have been positively and inversely correlated, respectively, with the risk of heart disease (Kannel et al., 1979; Gordon et al., 1977).

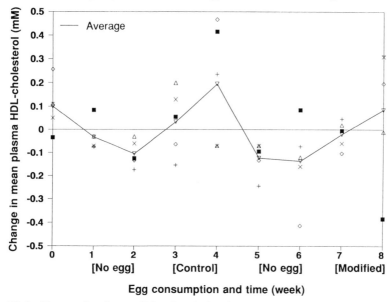

Fig. 30.3. Change in plasma TGs. Symbols show changes from each subject's mean value during the experimental period. Line shows the mean for five subjects.

Platelet phospholipid LnA and platelet phospholipid EPA did not change significantly. An increase in platelet LnA was predicted based on animal studies and because LnA intakes were very high (~150% of the recommended daily intake of nutrients from the eggs alone). Others have observed a decrease in plasma cholesterol in test animals fed diets high in ω3 fatty acids (Yu and Sim, 1987; Hargis *et al.*, 1991; Sim *et al.*, 1991). Therefore, further study is warranted.

Taste panellists described fresh, modified eggs as having a trace of off-flavour described as 'stale or rancid' by some panellists. The terms fishy or painty were not used. The difference was small but statistically significant (scores were 5.0 for control and 4.4 for modified eggs (Table 30.2). The mean score did not change significantly during 8 weeks of storage at 4°C or 4 weeks at room temperature. No oxidation of yolk lipids was found by analysis for MDA in egg yolk lipids during storage at 4°C for 5 weeks (Fig. 30.4). However, at 22°C, MDA was detected at 4 weeks. The detection of an off-flavour by taste panellists is a concern which merits further investigation as to the actual cause. Other workers have reported that high LnA eggs had objectionable or different flavours (Yu and Sim, 1987; Van Elswyk *et al.*, 1990). The method of cooking (boiling or scrambling) may be an important factor affecting taste.

Egg production was normal during the course of the study. Yolks from modified eggs were pale due to replacement of corn in the diet with flax. The colour may be corrected with alfalfa meal or dietary pigments in the feed.

Table 30.2. Mean sensory scores for high LnA eggs.

Eggs	Storage temp.(°C)	Sensory score[a] Age of egg (days)			
		0 days	14 days	28 days	56 days
Control	4	5.0[b] ± 0.3	4.8 ± 0.4	4.7 ± 0.1	4.9 ± 0.3
High LnA	4	4.4 ± 0.4	3.9 ± 0.1	5.0 ± 0.2	4.2 ± 0.2
Control	22	5.0 ± 0.3	5.2 ± 0.2	5.3 ± 0.5	
High LnA	22	4.4 ± 0.4	4.3 ± 0.1	4.1 ± 0.0	

[a] Mean ± SD. No. of panellists = 11-14.
[b] Reference sample set = 5, 3 = slight off-flavour, 4 = trace off-flavour,
6 = trace improvement in flavour.
Mean for control adjusted to 5.0. All other means adjusted by same amount.

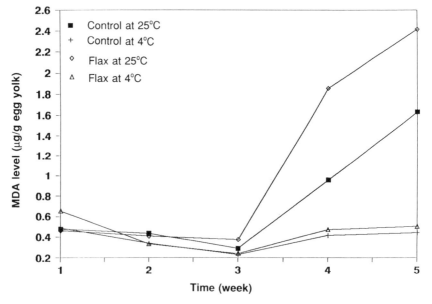

Fig. 30.4. Oxidation of fatty acids (MDA) in egg yolk lipids in ω3 eggs and control eggs during storage at 4°C and room temperature.

Discussion

Omega-3 eggs offer a possible means of delivering very high levels of both LnA and DHA, as much as 40% of the recommended daily ω3 fatty acid intake per egg. Eggs which are high in ω3 fatty acids may be highly desirable to today's health conscious consumers. These value-added eggs could provide a niche market for processed foods containing ω3 eggs, where the ω3 enrichment is declared on the label and in advertising. In our opinion, these eggs have the potential to overshadow by far the perceived negative aspects of cholesterol in eggs. In addition, there may be a spin-off effect in increased consumption of regular eggs. Another important spin-off could be value-added speciality products, such as egg oil, phospholipids, and lecithin enriched in ω3 fatty acids, which could be extracted from the LnA egg.

Speciality eggs may be considered a premium product, especially if they have a perceived health benefit. For example, 'cholesterol reducing' eggs sell for 50-100% above normal egg prices in the USA. Due to a worldwide interest in ω3 eggs, the possibilities for export are high for companies that are market leaders. We estimate the value of the modified eggs to Canadian farmers to be over $5 million per year if 2% of the current Canadian production is converted to high LnA eggs and they sell at the farm gate for about 25 cents extra per dozen over cost. We estimate the additional costs of production, handling, and packaging of the modified eggs to be less than 8 cents/dozen. The retail price of these eggs would probably increase significantly (e.g. by 75 cents/dozen) when the public became aware of the modified eggs.

It is important to note that feeding flax to hens to produce high ω3 fatty acid eggs is unlikely to trigger opposition by Health and Welfare Canada or the US FDA because the feed is a natural grain and contains a high proportion of desirable polyunsaturated fatty acids. Furthermore, altering fatty acid composition in egg yolk in a manner considered desirable by nutritionists and the Canadian government is also unlikely to trigger concern by regulatory agencies unlike the situation recently seen with the marketing of 'cholesterol reducing' (high iodine) eggs (Muirhead, 1990).

Conclusion

Very high levels of the nutritionally essential ω3 fatty acids were delivered in these modified eggs. Their consumption resulted in beneficial changes in plasma and platelet lipids. We also report the first observed rise in plasma DHA in humans due to consumption of non-fish food which is high in ω3 fatty acids. The increase in platelet DHA without an increase in EPA or LnA is intriguing both scientifically and for special food uses because it may provide a means of increasing DHA in blood of premature infants who are undergoing rapid brain development. The detection of an off-flavour merits further investigation. These data should be confirmed in a larger experiment. Value-added speciality products, with possible niche applications, could be extracted from the ω3 eggs.

References

Adams, R.L., Pratte, D.E., Lin, J.H., and Stadelman, W.J. (1989) Introduction of omega-3 polyunsaturated fatty acids into eggs. *Poultry Sci.* 68 (Suppl. 1):166.

Bjerve, K.S., Mostad, I.L., and Thoresen, L. (1988) Alpha-linolenic acid deficiency in patients on long-term gastric-tube feeding: estimation of linolenic acid and long-chain fatty unsaturated n-3 fatty acid requirement in man. *Am. J. Clin. Nutr.* 45:66-77.

Bligh, E.G. and Dyer, W.J. (1959) A rapid method of total lipid extraction and purification. *Can. J. Biochem. Physiol.* 37:911-917.

Caston, L. and Leeson, S. (1990) Research note: dietary flax and egg composition. *Poultry Sci.* 69:1617-1620.

Chung, S.L., Ferrier, L.K., and Squires, E.J. (1991) Survey of cholesterol level of commercial eggs produced on Canadian farms. *Can. J. Animal Sci.* 71: 205-209.

Cunnane, S.C., Jenkins, D.J.A., Ganguli, S., Armstrong, J.K., and Wolever,T.M.S. (1989) Flax consumption by humans increases plasma and red cell omega-3 fatty acids and decreases serum cholesterol. (Abstract) *J. Am. Oil Chem. Soc.* 66:438.

Farrell, D.J. (1991) The enrichment of the hen's egg with omega-3 fatty acids and their increase in the plasma of humans eating these eggs. *Poultry Sci.* 70 (Suppl. 1):40.

Fossati, P. and Prencipe, L. (1982) Serum triglyceride determined colorimetrically with an enzyme that produces hydrogen peroxide. *Clin. Chem.* 28:2077-2088.

Gaudette, D.C. and Holub, B.J. (1990) Albumin-bound docosahexaenoic acid and collagen-induced human platelet reactivity. *Lipids* 25:166-169.

Gaudette, D.C. and Holub, B.J. (1991) Docosahexaenoic acid (DHA) and human platelet reactivity. *J. Nutr. Biochem.* 2:116-121.

Gordon, T., Casfelli, M.P., Hjotland, M.C., Kennel, W.B., and Dawber, T.R. (1977) High density lipoprotein as a protective factor against coronary heart disease. The Framingham study. *Am J. Med.* 62:707- 714.

Health and Welfare Canada (1990) *Nutrition Recommendations.* Ministry of Supply and Services, Ottawa, Canada.

Hargis, P.S., van Elswyk, M.E., and Hargis, B.M. (1991) Dietary modification of yolk lipid with menhaden oil. *Poultry Sci.* 70:874-883.

Holman, R.T., Johnson, S.B., and Hatch, T.F. (1982) A case of human linolenic acid deficiency involving neurological abnormalities. *Am. J. Clin. Nutr.* 35:617-623.

Holub, B.J. (1988) Dietary fish oils containing eicosapentaenoic acid and the prevention of atherosclerosis and thrombosis. *Can. Med. Assn. J.* 139:377-381.

Holub, B.J. and Skeaff, C.M. (1988) Nutritional regulation of cellular phosphatidylinositol. *Methods Enzymol.* 141:234-244.

Kannel, W.B., Castelli, W.P., and Gordon T. (1979) Cholesterol in the prediction of atherosclerotic disease. New perspectives based on the Framingham study. *Ann. Intern. Med.* 90:85-91.

MacDonald, B.E., Gerrard, J.M., Bruce, V.M., and Corner, E.J. (1989) Comparison of the effect of canola and sunflower oil on plasma lipids and lipoproteins and on *in vivo* thromboxane A_2 and prostacyclin production in healthy young men. *J. Clin. Nutr.* 50:1382-1388.

McGowan, M.W., Artiss, J.D., Strandbergh, D.R., and Zak, B. (1983) A peroxidase-coupled method for the colorimetric determination of serum triglycerides. *Clin. Chem.* 29:538-542.

Muirhead, S. (1990) Development of cholesterol-reducing egg still on track. *Feedstuffs*, 29 Jan., p. 7.

Owren, P.A., Hellem, A.J., and Ödergaard, A. (1964) Linolenic acid for the prevention of thrombosis and myocardial infarction. *Lancet* ii:975-979.

Sim, J.S., Barbour, G.W., and Jiang, Z. (1991) Modulation of egg yolk fatty acids and its effect on plasma and liver cholesterol levels in the rat. *Poultry Sci.* 70 (Suppl. 1):110.

Squires, E.J. (1989) High performance liquid chromatographic analysis of the malondialdehyde content of chicken liver. *Poultry Sci.* 69:1371-1376.

Tinoco, J. (1982) Dietary requirements and functions of α-linolenic acid in animals. *Prog. Lipid Res.* 21:1-45.

Van Elswyk, M.E., Hargis, B.M., Williams, J.D., Sams, A.R., and Hargis, P.S. (1990) Omega-3 fatty acid enriched eggs and human health. *FASEB J.* 4: A932.

Weaver, B.J., Corner, E.J., Bruce, V.M., McDonald, B.E., and Hollub, B.J. (1990) Dietary canola oil: effect on the accumulation eicosapentaenoic acid in the alkenylacyl fraction of human platelet ethanolamine phosphoglyceride. *Am. J. Clin. Nutr.* 51:594-598.

Wilkinson, L. (1986) *SYSTAT: The System for Statistics.* SYSTAT Inc., Evanston, IL.

Yu, M.M. and Sim, J.S. (1987) Biological incorporation of n-3 fatty acids into chicken eggs. *Poultry Sci.* 66 (Suppl. 1):195.

Chapter Thirty-One

Altering Fatty Acid and Cholesterol Contents of Eggs for Human Consumption

T.M. Shafey and B.E. Cham[1]

University of Queensland, Gatton College, Department of Animal Production, Lawes, Queensland 4243, Australia and [1]University of Queensland, Department of Medicine, Royal Brisbane Hospital, Brisbane, Queensland 4029, Australia

Abstract Dietary and genetic factors are known to modify yolk cholesterol (CHL) and fatty acid concentrations. In a comparison of different grains, substituting triticale for 100% wheat improved the ratio of yolk polyunsaturated to saturated fatty acids (P:S). The substitution of oat for 50% wheat in the diet reduced yolk CHL content. However, dietary factors which may affect egg CHL content have no effect on daily CHL output. The degree of unsaturation of yolk can be altered by increasing the level of unsaturated fats in the diet. The addition of 4% fish oil to a wheat-based diet increased yolk $\omega 3$ fatty acids. Differences between strains of laying hens were mainly due to differences in responses in egg traits and not to responses in daily CHL output. CHL content of the egg varies with age of the hen, so a reduction in egg CHL can be achieved by choosing laying strains that are highly productive at a young age. It was concluded that an egg yolk with a lower CHL content and a higher than 'normal' P:S ratio may be of benefit to man.
(*Key words:* grain type, soy oil, fish oil, strain, age, egg cholesterol, egg fatty acids)

Introduction

The public awareness of the relationship between dietary lipid and the incidence of coronary heart disease (CHD) has increased in recent years. The diet of western society is rich in lipid (40% energy), about 40-50% of which is saturated and the ratio of polyunsaturated:saturated fatty acids (P:S) is about 0.40 (English, 1987). Current health recommendations have encouraged individuals to reduce the consumption of total lipid, saturated fatty acids, and cholesterol (CHL) and to increase the proportion of monounsaturated and polyunsaturated fatty acids in their diets (Walsh *et al.*, 1975).

Consumers' attitudes towards lipids in general have changed their attitudes towards egg consumption because of fears that yolk CHL would raise their blood CHL levels. The average egg of 50 g contains approximately 5 g lipid, about 4.2% (213 mg) of which is CHL, and the P:S ratio of yolk lipid is about 0.59 (USDA, 1989). In a review, Moore (1987) has stated that reducing or increasing egg consumption does not significantly affect the blood CHL levels in normal people. He has concluded that the type of fat intake is more important in influencing blood CHL. The higher the level of ingested unsaturated fatty acids, the lower the CHL level in the blood. Eggs are considered to be less atherogenic than other animal products, despite the high level of CHL present in the eggs. In comparing the P:S ratio of red meat and eggs, Sinclair and O'Dea (1987a,b) have concluded that beef and lamb are more atherogenic than eggs because of their P:S ratio of 0.19 and 0.05, respectively. The effect of substituting unsaturated for saturated fatty acids on blood CHL was predicted by Keys *et al.* (1957). These authors suggested that each gram of saturated fat ingested had approximately twice the effect on raising blood CHL as did a gram of unsaturated fat.

Work in Australia by McDonald and Shafey (1989, 1990), Shafey and Dingle (1991, 1992), and Farrell and Gibson (1991) has examined a range of genetic and non-genetic factors. It has indicated that the CHL and fatty acid content of eggs can be modified to enhance the health standard of eggs and consequently improve the image of egg and egg products.

The purpose of this paper is to demonstrate the manner by which research findings can be used to enhance the health standard of eggs and to help in marketing of eggs and egg products. Our approach has been to try not only to lower yolk CHL but also to alter yolk fatty acids. It is not intended here to review the effects of different factors on egg yolk lipid. Factors influencing the composition of the hen's egg were reviewed by Stadelman and Pratt (1989).

The experiments described in this report were carried out with three different strains of laying hens. Birds were housed in flat deck cages and fed the experimental diets for 11-12 weeks. Only a brief description of each experiment is supplied.

Diet and Yolk Lipid

Dietary factors have been shown to modify yolk CHL and fatty acid concentrations (Hargis 1988; Stadelman and Pratt 1989). It has been known for many years that the degree of unsaturation of yolk can be altered by dietary means. Increasing dietary level of unsaturated fats increases the degree of unsaturation of yolk, mainly by altering the levels of oleic and linoleic acids. Researchers at the University of Queensland, Gatton College, have been examining the effect of including wheat or triticale as a major grain in the diet, the addition of oat as a source of fibre, and the use of vegetable oil on egg production and yolk lipid. An attempt was made to enrich yolk lipid with ω3 fatty acids by the addition of fish oil in the diet.

In comparing different cereal grains (wheat, triticale, or oat), the substitution of triticale for wheat improved the yolk P:S ratio and produced small non-significant reduction in yolk CHL content. In contrast, the addition of 2% soy oil to a triticale-based laying hen diet has been shown to give eggs healthy combination of low CHL and a high P:S ratio (Table 31.1). Substituting oats for 50% wheat in the diet ($P < 0.07$) reduced yolk CHL content (Table 31.2).

Table 31.1. Effects of type of grain and soy oil on yolk CHL and fatty acid concentrations of egg yolks.

Type of grain	Yolk weight (g)	Rate of laying[a]	CHL concn. (mg/g yolk)	CHL content (mg/g yolk)	CHL output (mg/day)	Linoleic acid 18:2%[b]	18:1 to 18:2 ratio[c]
Wheat	15.6	0.91	11.7	182.4	165.9	12.2	3.18
Triticale	15.5	0.89**	11.6	179.2	160.1	13.6	2.49**
Control	15.5	0.89	11.8	186.4	162.8	11.6	3.43
Added oil	15.6	0.91**	11.5	181.5	163.2	14.2**	2.24**
SEM[d]	0.3	0.01	0.5	10.2	4.5	1.0	0.14
DF[e]	51	22	51	51	51	51	51

[a] Eggs/hen/day.
[b] % of total methyl esters of yolk lipid.
[c] The proportion of monounsaturated (oleic) to polyunsaturated (linoleic) fatty acid in yolk lipid.
[d] Standard error of means.
[e] Degrees of freedom.
** ($P < 0.01$).

The addition of fish oil has been shown to enrich egg yolk lipid with ω fatty acids (Table 31.3). The addition of 4% fish oil to a wheat-based diet increased yolk ω3 (linolenic plus docosahexaenoic plus eisosapentaenoic) fatty acids, whilst yolk ω6 (linoleic plus arachidonic) fatty acids were decreased by dietary treatment. The ratio of ω6 to ω3 fatty acids was reduced from 15 to 3 by the inclusion of fish oil in the diet. Dietary treatment did not alter yolk CHL. The beneficial effects of ω3 fatty acids in cholesterol metabolism and cardiovascular diseases in humans may be of interest to consumers. However, the use of high levels of fish oil in poultry diets may taint eggs. The addition of 4% fish oil to the diet did not affect the flavour or taste of the eggs.

Table 31.2. Effects of oats on yolk CHL and fatty acid concentrations of egg yolks.

Type of grain	Yolk weight (g)	Rate of laying[a]	CHL concn. (mg/g yolk)	CHL content (mg/ yolk)	CHL output (mg/day)	Linoleic acid 18:2%[b]	18:1 to 18:2 ratio[c]
Wheat	17.7	0.82	12.0	212.6	176.3	11.56	3.16
Oats[f]	17.3	0.83	11.4	197.1	163.7	12.17	3.01
SE[d]	0.3	0.03	0.3	7.5	10.7	0.33	0.11
DF[e]	49	23	24	24	24	24	24

[a,b,c,d,e] See Table 31.1.
[f] Substituting whole oats for 50% wheat.
* ($P < 0.05$).

Numerous studies have attempted to lower egg yolk CHL by dietary and genetic means (Hargis, 1988). It is clear that manipulating yolk lipid by available means does not provide a significant reduction in yolk CHL. In a review, Washburn (1979) concluded that the inability to reduce significantly egg yolk CHL is indicative of the natural selection pressures for a stable nutrient level in the egg. Connor *et al.* (1969) have shown that 90% of the CHL in the brain of the chicken embryo is synthesised, the remainder of body CHL comes from the yolk of the egg. CHL is an essential nutrient for the developing embryo and consequently might affect hatching of eggs. However, egg yolk CHL can be reduced by up to 10% by choosing a suitable source of fibre.

It appears that dietary factors which may affect yolk CHL content have no effect on daily yolk CHL output. CHL content of egg is influenced by the effect of dietary factors on egg production traits. Dietary factors which may affect egg production traits may also affect egg yolk CHL content. The enrichment of egg yolk lipid with ω3 fatty acids may be used as a marketing tool.

Table 31.3. Effects of fish oil on yolk CHL and fatty acid concentrations of egg yolks.

Treatment	Yolk weight (g)	Rate of lay[1]	CHL concentration (mg/g yolk)	CHL content (mg/yolk)	CHL output (mg/day)	Fatty acid %[2] ω6 PUFA[3] 18:2	20:4	ω3 PUFA[6] 18:3	20:5	22:6	ω6 PUFA / ω3 PUFA ratio
Control	15.9	0.87	13.4	212.9	183.4	11.38	1.29	.28	0.05	0.49	15.45
Added oil[7]	16.1	0.86	13.1	210.8	181.5	11.24	0.33**	0.45**	0.32**	3.03**	3.07**
SE[4]	0.4	0.01	0.4	8.8	4.9	0.87	0.38	0.05	0.1	0.9	3.9
DF[5]	143	59	49	49		39					

1,2,4,5, ** See Table 31.1.

[3] ω6 Polyunsaturated fatty acids = linoleic acid (18:2) plus arachidonic acid (20:4).

[6] ω3 Polyunsaturated fatty acids = linolenic acid (18:3) plus docosahexaenoic acid (22:6) plus eicosapentaenoic acid (20:5).

[7] 4% fish oil.

Genetic Aspects of Yolk Lipid

There is very little available comparative strain information on egg yolk lipid. A summary of comparison of egg yolk CHL and the correlations between egg yolk CHL and egg production parameters of three Australian strains is shown in Table 31.4.

Table 31.4. The effects of strain of laying hen on egg yolk CHL and the correlations between egg yolk CHL and egg production parameters.

	CHL concentration (mg/g yolk)			CHL content (mg/yolk)			Daily CHL output (mg/day)		
Strain	1	2	3	1	2	3	1	2	3
Egg yolk	11.8[b]	11.9[b]	12.6[a]	190.8[b]	208.1[a]	211.9[a]	169.6	172.3	177.2
SE[1]		0.2			7.6			6.9	

The correlations between egg yolk CHL and egg production parameters:[2]

Egg weight	0.10	0.06	0.45	0.42	0.61	0.60	0.28	0.53	0.51
Yolk weight	-0.10	-0.01	0.46	0.63	0.63	0.86	0.68	0.50	0.70
Laying rate	-0.20	-0.41	-0.56	0.06	-0.54	-0.64	0.53	-0.01	-0.07

[a,b] See Table 31.1. [1] Standard error of means with 135 degrees of freedom.
[2] For the correlation coefficient (n = 54) to be significant, the following limit values should be used as references: 0.26 ($P = 0.05$), 0.34 ($P = 0.01$), and 0.43 ($P = 0.001$).

Genotypes differed in yolk CHL concentration and yolk CHL content. However, differences between strains were not significant when egg CHL was calculated as daily CHL output.

Significant correlations were found between yolk CHL concentration and yolk CHL content ($0.74, P < 0.001$), between yolk weight and yolk CHL content ($0.73, P < 0.001$), between egg weight and yolk weight or yolk CHL content (0.67 or $0.62, P < 0.01$), and between daily yolk CHL output and egg weight, yolk weight, yolk CHL concentration, and yolk CHL content ($0.45, 0.64, 0.59$, and $0.84, P < 0.01$). Rate of lay was negatively correlated with egg weight, yolk CHL concentration, and yolk CHL content ($-0.36, -0.35$, and $-0.39, P < 0.01$). Significant genotype differences occurred between correlated egg traits (Table 31.4).

Results from this study have demonstrated that a change in yolk CHL was negatively associated with a change in rate of egg production. However, when the total daily CHL was calculated, the negative correlation between yolk CHL and egg numbers resulted in no difference in total daily CHL output among strains of birds. These findings suggest that factors which may directly affect the production of egg and its components have an indirect effect on egg CHL but not daily yolk CHL output.

The non-significant difference in daily CHL output among strains of birds despite the significant differences in egg and yolk weights, egg production (Table 31.5), and CHL concentration and content of the yolk (Table 31.4) may suggest that the yolk CHL is not influenced by total CHL output. However, yolk CHL is affected by other egg traits which are controlled by the genetic background of the hen. There was no difference among strains in the ability to produce CHL. However, the ability of hens to produce eggs controlled the amount of CHL found in these eggs. Highly productive strains of birds deposit less CHL per egg than less productive ones.

Table 31.5. The effects of strain of laying hens on egg weight, yolk weight, and rate of laying.

Strain	Egg weight (g)	Yolk weight (g)	Rate of laying (egg/hen/day)
S1	54.2[b]	16.2[b]	0.89[a]
S2	58.2[a]	17.5[a]	0.83[b]
S3	58.2[a]	16.8[ab]	0.84[b]
SE[1](79)	0.5	0.5	0.01

[a,b] Means within columns followed by different superscripts are significantly different ($P < 0.05$).
[1] See Table 31.1.

The inconsistent correlations found between yolk CHL and different egg traits among the three strains reflect the differences of genetic makeup of these birds (Table 31.4). Yolk CHL content was negatively correlated with egg production in strains 2 and 3 but not in strain 1. The differences in yolk CHL content found in eggs produced by strains 1 and 2 were attributed to differences in yolk weights, but differences in yolk CHL content in strains 1 and 3 may be attributed to differences in yolk weight and yolk CHL concentration. The correlations between yolk CHL (mg/g yolk or mg/yolk), and rate of laying, egg weight, and yolk weight suggested that the lower yolk CHL produced by strain 1 birds was associated with lower egg and yolk weights and a higher rate of laying when compared with higher egg and yolk weights and lower rate of laying of strain 3 birds. Selection for higher egg weight may be expected to increase yolk CHL content of individual eggs. Selection for higher egg production may be expected to decrease yolk CHL content of individual eggs.

Effect of Age of the Hen on Yolk Lipid

The relationship between egg yolk lipid and age of the hen has been reported by Marion *et al.* (1966) and Menge *et al.* (1974) who found that yolk lipid and CHL concentrations increased with age of the hens. The relationship between age and yolk CHL is given in Fig. 31.1.

Significant correlations were found between age and yolk CHL concentration, daily yolk CHL output, egg weight, yolk weight, and yolk CHL content (0.30, 0.36, 0.48, 0.54, and 0.57, $P < 0.01$). Rate of lay was negatively correlated with age of the bird (-0.4, $P < 0.01$). Egg and yolk produced at 56 weeks of age were approximately 8.7 and 10.5% heavier than those produced at 30 weeks of age, whereas yolk CHL content produced over the same period was approximately 19% higher than that produced at the younger age.

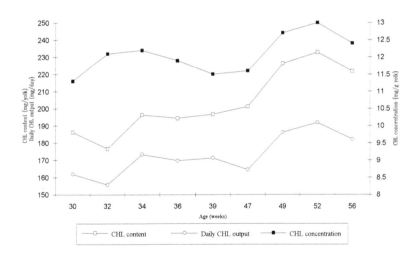

Fig. 31.1. The effects of age on egg yolk CHL.

However, when CHL concentration was calculated on the basis of egg weight (mg cholesterol/g egg weight), the increase in CHL over the same period was lessened to 9.2%. It appears that the increase in CHL content was approximately equivalent to the increase in egg and yolk weights. The finding that the smaller eggs contained significantly less CHL than larger eggs, could be used to promote the consumption of smaller eggs. Consumer demands for healthier eggs and a lower CHL intake could be met by marketing the small eggs as lower CHL eggs.

Management and Yolk Lipid

Management practices relative to age and strain of the laying hens may provide an opportunity for manipulating egg yolk CHL. Age and strain of bird significantly influenced egg production, egg weight, yolk weight, and consequently yolk CHL content. Younger hens produced more smaller eggs with less deposition of CHL in each egg. Lower egg CHL can be achieved by choosing laying strains that are highly productive at a young age. Considerations have to be made when altering egg yolk lipid by dietary means such as consumer acceptance, economical effect, and storability of the eggs.

The Relationship of Diet to Blood CHL and the Influence of Coronary Heart Disease (CHD)

Comparisons of different populations have demonstrated a relationship between the quantity of saturated dietary fat and both the concentration of serum CHL and the frequency of CHD. The independent contribution of dietary CHL to these correlations has yet to be demonstrated; however, death rates from CHD are positively related to intake of saturated fatty acids and negatively related to intake of monounsaturated fatty acids. Total fat and other dietary variables do not appear to be significantly related to CHD (Keys *et al.*, 1986). The relationship between changes in dietary fats and CHL and changes in serum CHL was described by Keys (1984) in the following equation:

$$\text{Change in serum CHL (mg/dl)} = 2.7 \, \Delta S - 1.35 \, \Delta P + 1.5 \, \sqrt{C}$$

in which ΔS and ΔP are the variations in intake of saturated and polyunsaturated fat, respectively, expressed as percent of total energy intake, and C is the quantity of dietary CHL in mg/4280 kJ per day. From this equation it can be seen that the changes in serum CHL are more pronounced by the variations in P and S rather than C in the diet. For example this equation predicts that reduction in saturated fat intake from 17 to 6% of total energy intake will lead to a mean 0.725 mmol/l (28 mg/dl) fall in serum CHL. Whereas, reducing CHL intake from 240 to 80 mg/4280 kJ leads to a more modest 0.259 mmol/l (10 mg/dl) average reduction.

It is clear that the most detrimental dietary component affecting plasma CHL is the intake of saturated fatty acids. Diets high in saturated fat will increase plasma CHL level. Lowering total fat, and particularly the amount of saturated fat, is the most important dietary modification to be recommended. Reducing dietary CHL is less important. Although the contribution of the egg to daily fat intake of human is not quantitatively significant, increasing the degree of unsaturation of eggs would improve the quality of fat intake. The enrichment of eggs with long chain polyunsaturated fatty acids has been shown to be beneficial to man (Farrell and Gibson, 1991).

At this stage it would be prudent to suggest that eggs may raise blood CHL levels; however, the specific effect of dietary CHL is smaller than that of dietary saturated fat. Considerable inter-individual variation in responsiveness to dietary CHL is well recognised (Beynen and Katan, 1985; McNamara *et al.*, 1987). A reduction in the intake of egg yolks may be warranted in the treatment of hypercholesterolaemia, but such dietary measures should not have priority over reductions in the intake of saturated fat. Therefore, an egg yolk with a lower CHL content and a higher than 'normal' P:S ratio may be of benefit to man. The decline in shell egg consumption may be reversed by encouraging the development of innovative quality egg-based products and by the introduction of health promotion campaigns.

Acknowledgements

Dr J.G. Dingle is thanked for his valuable advice. The technical assistance of F. Gorbacz, P. Kalinowski, and R. Englebright is gratefully acknowledged. This study was funded by the Egg Industry Research and Development Council.

References

Beynen, A.C. and Katan, M.B. (1985) Reproducibility of the variations between humans in the response of serum cholesterol to cessation of egg consumption. *Atherosclerosis* 57:19-31.

Connor, W.E., Johnston, R., and Lin, D.S. (1969) Metabolism of cholesterol in the tissues and blood of the chick embryo. *J. Lipid Res.* 10:388-394.

English, R. (1987) *Towards Better Nutrition for Australians: Report of Nutrition.* Taskforce of the Better Health Commission, Commonwealth Department of Health. Australian Government Publishing Services, Canberra, p. 51.

Farrell, D.J. and Gibson, R.A. (1991) The enrichment of eggs with omega-3 fatty acids and their effects in humans. In: Farrell, D.J. (ed.), *Recent Advances of Animal Nutrition in Australia.* University of New England Armidale, Australia, pp. 256-270.

Hargis, P.S. (1988) Modifying egg yolk cholesterol in domestic fowl, a review. *World Poultry Sci. J.* 44:17-29.

Keys, A. (1984) Serum cholesterol response to dietary cholesterol. *Amer. J. Clin. Nutr.* 40:351-359.

Keys, A., Anderson, J.T., and Grande, F. (1957) Prediction of serum-cholesterol responses of man to changes in fats in the diets. *Lancet* ii:959-966.

Keys, A., Menotti, A., Karvonen, M.J., Aravanis, C., Blackburn, H., Buzina, R., Djordjevic, B.S., Dontas, A.S., Fidnaza, F., Keys, M.H., Kromhout, D., Nedeljkovk, S., Punsar, S., Seccareccia, F., and Toshima, H. (1986) The diet and 15-year death rate in the seven countries study. *Amer. J. Epidemiol.* 124:903-915.

McDonald, M.W. and Shafey, T.M. (1989) Nutrition of the hen and egg cholesterol. In: Egg Industry Research Council (ed.), *Cholesterol in Eggs Seminar*. Egg Industries Research Council, Sydney, Australia, pp. 30-34.

McDonald, M.W. and Shafey, T.M. (1990) Cholesterol in poultry production. In: Mannion, R.F., Moffatt, B.W., Loveday, M., Runge, G.A., Obst, J., Stewart, G.D., Munt, R., Houweling, J., McGoldrick, S., Peiker, T., Mason, B., Aspinall, J., Sansom, G., Clark, M., and Milnc, F.N. (eds), *Poultry Information Exchange*. Department of Primary Industries, Brisbane, pp. 123-133.

McNamara, D.J., Kolb, R., Parker, T.S., Batwin, H., Samouel, P., Brown, C.D., and Aherns, E.H., Jr (1987) Heterogeneity of cholesterol homeostasis in man. Response to changes in dietary fat quality and cholesterol quantity. *J. Clin. Invest.* 79:1729-1739.

Marion, J.E., Woodroff, J.G., and Tindell, D. (1966) Physical and chemical properties of eggs as affected by breeding and age of hens. *Poultry Sci.* 45:1189-1195.

Menge, H., Littlefield, L.H., Frobish, L.T., and Weinland, B.T. (1974) Effect of cellulose and cholesterol on blood and yolk lipids and reproductive efficiency of the hen. *J. Nutr.* 104:1554-1556.

Moore, J.H. (1987) Biochemical aspects of the relationships between dietary cholesterol, blood cholesterol and ischaemic heart disease. In: Wells, R.G. and Belyavin, C.G. (eds), *Egg Quality - Current Problems and Recent Advances*. Butterworths, London, pp. 27-56.

Shafey, T.M. and Dingle, J.G. (1991) Management and egg yolk cholesterol. *Survival in a Competitive Market*. Department of Primary Industry, Queensland, Australia, pp. 99-101.

Shafey, T.M. and Dingle, J.G. (1992) Factors affecting egg fatty acid and cholesterol content. In: Johnson, R.J. Annison, E.F., Balnave, D., Bryden, W.L., Farrell, D.J., Fraser, D.R., Prowse, S., Pym, R.A.E., and Sheldon, B.L. (eds), *Poultry Research Husbandry Foundation Symposium*. University of Sydney, Australia, pp. 79-83.

Sinclair, A.J. and O'Dea, K. (1987a) The lipid levels and fatty acid compositions of the lean portions of Australian beef and lamb. *Food Technol. Australia* 39:228-231.

Sinclair, A.J. and O'Dea, K. (1987b) The lipid levels and fatty acid compositions of the lean portions of pork, chicken and rabbit meats. *Food Technol. Australia* 39:232-233.

Stadelman, W.J. and Pratt, D.E. (1989) Factors influencing composition of the hen's egg. *World Poultry Sci. J.* 45:247-266.

USDA (1989) Composition of foods, dairy and egg products - raw, processed, prepared. *Agricultural Handbook* No. 8-1 (Suppl.). USDA, ARS, Washington, DC.

Walsh, R.J., Day, M.F., Fenner, F.J., McCall, M., Saint, E.G., Scott, T.W., Tracey, M.V., and Underwood, E.J. (1975) *Diet and Coronary Heart Disease*. Report no. 18, March 1975. Australian Academy of Science, Canberra.

Washburn, K.W. (1979) Genetic variation in the chemical composition of the egg. *Poultry Sci.* 58:529-535.

Chapter Thirty-Two

The Fortification of Hens' Eggs with ω3 Long Chain Fatty Acids and Their Effect in Humans

D.J. Farrell

Department of Agriculture, University of Queensland, St Lucia, Qld 4072, Australia

Abstract Of experiments conducted on diets with added fish or vegetable oil a mixture of oils, trial 1 showed that cod liver, rapeseed (Canola), or linseed o at 70 g/kg diet gave substantial fortification of the ω3 polyunsaturated fatty acic (PUFA) in egg yolk. Greatest enrichment was in egg yolk from hens fed linsec oil but this was mainly as α-linolenic acid (LnA). Fatty acid profiles of plasm from starved hens showed similar ω3 PUFA profiles to those in egg yolk. I both cases there were only small amounts of eicosapentaenoic acid (EPA) preser in plasma or egg yolk but substantial concentrations of docosahexaenoic aci (DHA) on the experimental diets. A sensory evaluation by 35 subjects could n distinguish between egg types in several categories. When 44 humans consume two fortified eggs daily for 7 weeks and two eggs daily from hens on commercial diet for 2 weeks before and 4 weeks after the 7 week period, the were no changes in plasma cholesterol or triglycerides, while high densii lipoproteins (HDL) decreased as did white blood cell counts. Plasma from th fasted subjects after consuming fortified eggs for 7 weeks showed substanti enrichment of ω3 PUFA particularly those eating eggs from hens given the cc liver oil and Canola oil diets. EPA in human plasma was higher than in egg yo and hen plasma but LnA was much lower. In trial 2, cod liver or mackerel o with combinations of Canola and linseed oil were added to layer diets (60 g c 40 g/kg). It was shown that ω3 PUFA could be manipulated in the fortified egg Mackerel oil was slightly superior to cod liver oil giving higher enrichments c EPA, DHA, and total ω3 PUFA when added at 60 g/kg diet.
(*Key words:* ω3 long chain fatty acids, egg lipid fortification, huma consumption, plasma enrichment, plasma lipid fractions)

Introduction

There is much interest in the therapeutic value for humans of the ω3 long chain polyunsaturated fatty acids (PUFA). Their increasing importance has led to substantial research (Galli and Simopoulos, 1989; Simopoulos *et al.*, 1991a) and the topic has been reviewed recently by Sinclair (1991). The main source of ω3 PUFA for humans is fish. Marine phytoplankton has high amounts of ω3 PUFA and it is in this way that they get into the food chain. There is considerable variation in the oil content of different species of fish and in the ω3 PUFA in fish oil (Evans *et al.*, 1986; Fogerty and Sovoronos, 1987). Some vegetable oils such as linseed, rapeseed (Canola), and soybean contain varying amounts of α-linolenic acid (LnA). This ω3 PUFA may be less important than the longer chain ω3 PUFA for human health.

It is important that major sources of ω3 PUFA other than fish are freely available. Even though regular consumption of only small amounts of fish can dramatically reduce in humans the incidence of death from cardiovascular disease (Kromhout *et al.*, 1985; Burr and Fehily, 1991; Dolecek and Grandits, 1991), fish is not always liked, sometimes not available, may be contaminated with heavy metals or pesticide residues, and also may be expensive.

The purpose of the experiment to be reported here was to enrich hens' eggs with ω3 PUFA by dietary manipulation and to observe changes in different parameters, including plasma lipid fractions, when these eggs were eaten by volunteers.

Materials and Methods

Laying Hens

Seventy-two cross-bred layers in single cages were assigned to four dietary treatments. The basal diet was formulated to meet commercial specifications (Table 32.1). Cod liver oil, Canola, linseed, or no oil was added to the basal diet at 70 g/kg. The diets were fed over a 12 week period and the usual production parameters recorded. After 10 days on the experimental diets, eggs were collected and stored at 5°C for subsequent distribution to the volunteers. At the end of the experiment, hens were starved for 24 h and 5 ml of blood was drawn from a wing vein.

In a second preliminary experiment, groups of four hens were offered a commercial diet containing either cod liver oil or mackerel oil in various amounts, or in combination with mackerel oil with additions of linseed oil or rapeseed (Canola) oil. Eggs were collected after 19 days and prepared for fatty acid analysis. The object here was to determine the amounts and proportions of ω3 PUFA that could be incorporated in egg yolk by dietary manipulation.

Table 32.1. Composition of layer diets (kg/tonne).

	Basal	Experimental
Wheat	689	443
Wheat pollard	65	167
Soybean meal	98	162
Meat and bone meal	60	60
Limestone	81	92
DL-methionine	1.5	1.7
Lysine monoHCl	0.7	---
Salt	2.6	3
Vitamin-mineral premix[a]	2	2
Ethoxyquin (g)	---	125
Oil[b]	---	70

[a] Standard layer premix
[b] Cod liver, Canola, or linseed

Volunteers

Forty-four apparently healthy free-living adult subjects of both sexes were recruited. They were asked not to alter their life style or dietary habits. They were allocated at random to one of four treatment groups represented by the different egg types. For the first 2 weeks all volunteers were given eggs from a commercial flock fed on a standard layer diet. For the next 7 weeks volunteers were provided with coded eggs from hens on one of the four treatments. They consumed two eggs daily, preferably scrambled or soft-boiled.

Taste Panel

A sensory evaluation of the egg types was undertaken by a trained taste panel of 35 persons in a purpose-designed room. Eggs were prepared, scrambled, served, and scored, and data analysed according to standard procedures (Chesterman *et al.*, 1989).

Measurements

At the commencement of the experiment and after 2, 5, 7, 9, and 13 weeks volunteers fasted overnight and between 07.30 and 09.30 h were weighed, a blood sample (5 ml) was drawn, and blood pressure was measured using an automatic digital blood pressure monitor (Omron HEM-703-C). Sufficient eggs were then distributed until the next measurement period.

Analytical Methods

Some of the heparinised blood was set aside for white blood cell (WBC) counts. The remainder was centrifuged and the plasma removed for analysis for cholesterol, triglycerides, and high-density lipoproteins (HDL) using a spectrophhotometric analyser (The Cobas Bio, Roche Diagnostics, F. Hoffman-la Roche, Basle, Switzerland) and the appropriate diagnostic kit.

For fatty acid analysis, egg yolk was separated from the albumen in the whole, preweighed egg and freeze dried. Wet and freeze-dried yolk weights were recorded. Lipids were extracted from the yolk and following transmethylation of the total lipids (Lepage and Roy, 1986), the fatty acid methylesters were separated by gas chromatrography using a 50 m capillary column coated with SP-2340 (Supelco, Inc.). Plasma fatty acids and various oils were determined similarly. Cholesterol was measured in egg lipid using the colorimetric method and test kit of Boehringer Mannheim (cat. no. 139050).

Data were analysed using standard statistical procedures. Where appropriate, variation in initial values was adjusted using covariance analysis.

Results

Fatty Acids in Oils

The fatty acid profile of Canola, linseed, mackerel, and cod liver oils are given in Table 32.2. The two vegetable oils contained substantial amounts (19%) of 18:2 ω6 (linoleic acid, LA), while fish oils contained only small amounts. Both vegetable oils contained substantial quantities of 18:3 ω3 LnA but no 20:5 ω3 eicosapentaenoic acid (EPA) nor 22:6 ω3 docosahexaenoic acid (DHA). Mackerel and cod liver oils had similar amounts of EPA, but mackerel oil contained twice as much DHA and much less LnA.

Laying Hens

Previously we reported on layer performance (Farrell and Gibson, 1990). The addition of the various oils to the basal diet generally enhanced production parameters relative to the diet without oil. We also found that it was necessary to feed similar diets for more than 5 days before enrichment of eggs was at a maximum (Farrell and Gibson, 1990).

Mean cholesterol (mg/egg) was similar after 7 days (186) and 63 days (187). But when expressed as g/100 g dry yolk it was significantly ($P < 0.01$) reduced at 63 days (2.22 versus 1.99). There was also an effect of diet ($P < 0.05$). Eggs from hens on the basal and cod liver oil diets had higher ($P < 0.05$) mean cholesterol concentrations (2.22, 2.21) than eggs from hens on the Canola and linseed oil diets (2.02, 1.98).

Table 32.2. Fatty acid (%) profile of vegetable and fish oils.

Fatty acid	Canola	Linseed	Mackerel	Cod liver
14.0	0.0	0.0	5.8	8.0
14.1	0.0	0.0	0.3	0.3
15.0	0.0	0.0	0.1	0.5
16.0	3.8	6.1	15.3	16.8
16.1	0.0	0.0	7.6	7.2
17.0	0.0	0.0	0.3	0.3
17.1	0.0	0.0	2.2	1.5
16.4	0.0	0.0	0.1	0.2
18.2	2.1	3.1	4.0	2.6
18.1	62.6	17.3	17.3	22.0
18.2 (ω6)	19.5	19.1	1.3	4.9
18.3 (ω6)	0.0	0.0	0.6	0.3
18.3 (ω3)	10.9	53.0	0.3	2.1
18.4 (ω3)	0.0	0.0	0.7	2.4
20.0	0.2	1.5	0.2	0.1
20.1 (ω9)	0.0	0.2	3.3	4.2
20.2 (ω6)	0.0	0.0	0.2	0.1
20.3 (ω6)	0.0	0.0	0.2	0.1
20.4 (ω6)	0.0	0.0	1.9	1.1
20.4 (ω3)	0.0	0.0	0.8	0.7
20.5 (ω3)	0.0	0.0	11.7	10.4
22.0	0.2	0.0	0.0	0.0
22.1 (ω11)	0.0	0.0	3.6	3.2
22.1 (ω9)	0.0	0.0	0.7	0.9
22.4 (ω6)	0.0	0.0	0.2	0.0
22.5 (ω6)	0.0	0.0	0.3	0.0
22.5 (ω3)	0.0	0.0	3.2	1.1
22.6 (ω3)	0.0	0.0	13.3	6.0
Total sat.	6.3	10.7	26.2	28.3
Monounsat.	62.6	17.5	34.1	39.4
Total ω6	19.5	19.2	3.7	6.6
Total ω3	10.9	52.9	29.9	22.6
ω6+ω3	30.4	72.1	33.6	29.2
Ratio ω6:ω3	1.8	0.3	0.1	0.2

The effects of feeding hens diets containing different amounts of cod liver oil and mackerel oil, or mackerel oil in combination with linseed oil or Canola oil on egg yolk fatty acid profile and the ratio of ω6:ω3 fatty acids are given in Table 32.3.

In all cases egg lipid contained large amounts of palmitic acid (16:0) and oleic acid (18:1), both saturated fatty acids. Only when 2% linseed oil was included in the hens' diet was there a significant content (>3%) of LnA in egg lipid. Interestingly the combination of Canola and linseed oils elevated DHA concentration in yolk lipid relative to the commercial diet. Mackerel oil was superior to cod liver oil in enriching eggs with EPA and DHA and gave a lower ratio of ω6:ω3 fatty acids.

Table 32.3. The effects of oil type and amount in layer diets on the fatty acid content of egg yolk lipid.

Diet oil	(%)	Fatty acid (%)							Ratio
		16:0	18:1	18:2 ω6	18:3 ω3	20.5 ω3	22:6 ω3	Total ω3	ω6:ω3
Commercial		25.1	46.7	12.7	0.2	0.2	0.6	1.2	13.0
Cod liver	6	26.9	42.5	11.3	0.6	0.9	4.3	6.3	1.9
Mackerel	6	27.8	38.5	9.9	0.3	1.1	5.8	7.9	1.4
Mackerel	4	27.2	40.1	12.5	0.4	0.7	4.4	6.0	2.3
Mackerel	2	27.9	40.5	11.8	0.4	0.4	3.4	4.6	2.8
Linseed/ mackerel	2 2	24.0	41.1	12.5	3.2	0.6	3.5	7.8	1.7
Canola/ mackerel	2 2	25.6	42.3	13.1	0.8	0.4	3.1	4.6	3.1
Canola/ linseed	2 2	22.2	46.9	14.1	3.3	0.3	2.0	5.9	2.6
SEM		0.43	0.60	0.26	0.23	0.06	0.27	0.36	0.65

Even though fish oils contain considerable concentrations of EPA (Table 32.2), only very small amounts were incorporated in egg lipid particularly when compared with DHA (Table 32.3).

Shown in Fig. 32.1a are the concentrations of ω3 PUFA in the plasma of starved hens bled after 14 weeks on the experimental diets with 7% added or no oil. Linseed oil is rich (>50%) in LnA (Table 32.2), consequently there was significant transfer from diet to blood. Similarly, cod liver oil is rich in DHA and concentrations in plasma were highest in hens on this diet. Surprisingly, Canola and linseed oils gave substantial transfer of DHA although some of this came from the basal (control) diet. EPA was present in plasma in only very small amounts.

Although linseed oil in the diet provided the highest total concentration of ω3 PUFA and the lowest ratio of ω6:ω3, cod liver oil probably provided the most appropriate balance. The ratio of ω6:ω3 PUFA was 9.7, 2.2, 4.2, and 1.6 in plasma of hens on the control, cod liver oil, Canola, and linseed oil-based diets respectively.

Shown in Fig. 32.1b are ω3 PUFA concentrations in egg yolk after 21 days
on the four diets. Clearly there was substantial enrichment, particularly of LnA
in eggs from hens on the linseed oil diet. Cod liver oil provided small but
significant increases in EPA and DHA as did Canola and linseed oil for DHA.
Measurements after 42 days on the four diets showed similar concentrations of
the ω3 PUFA in egg yolk (Fig. 32.1c) indicating that after 21 days on the
experimental diets ω3 PUFA were at a maximum and proportions had stabilised.

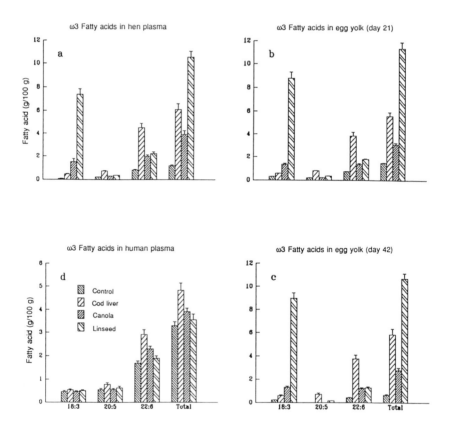

Fig. 32.1. Concentrations of individual and total ω3 long-chain polyunsaturated fatty
acids in (a) hen plasma, egg yolk after 21 days (b) or 42 days (c), and (d) human
plasma when volunteers had consumed for 7 weeks eggs from hens on diets with
or without added oil. Bars represent SEM.

Taste Panel

Results of the sensory evaluation of eggs are given in Table 32.4. Only texture was different. Relative to eggs from the control birds, texture was less acceptable ($P < 0.05$) in eggs from hens on the fish and Canola oil diets. The overall acceptability of egg types was the same.

Table 32.4. Sensory evaluation (n = 35) of enriched eggs from hens offered diets with either no oil or 70 g/kg of three different types of oil.

	Flavour[1]	Taste[2]	Colour[2]	Texture[3]	Overall acceptability[2]
Control	5.7	4.9	4.2	7.1[ab4]	4.9
Fish oil	7.0	6.3	6.1	4.7[b]	6.3
Canola oil	6.2	4.8	4.6	4.8[b]	5.4
Linseed oil	6.3	3.6	5.0	6.5[ab]	5.0
Pooled SEM	0.52	0.53	0.55	0.55	0.51

[1] From 0 (strong) to 12 (weak).
[2] From 0 (liked a lot) to 12 (strongly disliked).
[3] From 0 (tough) to 12 (tender).
[4] Values with the same or no superscript are not significantly different ($P > 0.05$).

Human Measurements

The results are shown in Table 32.5. Compared with the values at the commencement of the study (week 0), cholesterol and triglycerides showed no change, while HDL and WBC declined at week 2 and remained low to the end of the experiment (week 13). Both systolic and diastolic blood pressure also decreased at week 2 compared with week 0, and were significantly ($P < 0.05$) reduced by week 13.

In order to determine effects on volunteers of the enriched eggs *per se* on the different parameters, values at week 2 (before subjects were given the fortified eggs) were adjusted by covariance analysis and data for weeks 5, 7, and 9 were analysed separately (Table 32.5). It was found that HDL decreased and low density lipoproteins (LDL, calculated by difference) increased ($P < 0.05$) on the fish oil treatment. Dietary treatment had no effect on other measures (Table 32.5).

The concentrations of ω3 PUFA in human plasma taken after volunteers had consumed enriched eggs for 7 weeks, are given in Fig. 32.1d. There was significant enrichment of EPA and DHA in plasma of volunteers eating eggs from hens on the fish oil diet and to a lesser extent those from hens on the Canola oil diet relative to controls.

Table 32.5. Effects of dietary treatment (weeks 5, 7, and 9 only) and time (weeks 0-13) on changes in blood pressure and blood parameters including HDL, LDL and triglycerides (TG) of 11 human subjects per treatment eating two eggs per day.

	Diet oil				
	Control	Fish	Canola	Linseed	Significance[a]
Cholesterol (mM)	4.9	5.0	4.8	4.9	NS
HDL (mM)	1.1	1.0	1.1	1.0	*
LDL (mM)	3.8	4.0	3.7	3.8	*
TG (mM)	1.1	1.4	1.3	1.2	NS
White cells (10³/cm³)	7.9	7.7	7.5	7.6	NS
Blood pressure (mm)					
systolic	121	121	121	122	NS
diastolic	71	71	69	72	NS
Body weight (kg)	72	71	72	71	NS

White cells $(10^3/cm^3)$

Blood pressure (mm)

Weeks on experiment
(Combined dietary treatments)

	0	2	5	7	9	13[b]	LSD ($P < 0.05$)
Cholesterol (mM)	5.2	4.8	4.7	4.8	5.1	4.8	0.36
HDL (mM)	1.5	1.0	1.1	1.0	1.0	1.0	0.15
LDL (mM)	3.6	3.8	3.6	3.8	4.1	3.7	0.52
TG (mM)	1.3	1.2	1.3	1.2	1.2	1.4	0.36
White cells (10³/cm³)	11.5	8.4	8.1	7.1	8.0	7.7	0.94
Blood pressure (mm)							
systolic	122	124	122	122	121	117	5.6
diastolic	71	78	71	72	71	72	4.2
Body weight (kg)	72	72	72	72	71	71	4.9

[a] NS = not significant, * $P < 0.05$.
[b] Subjects off experiment for 4 weeks.

The increase in total ω3 PUFA was substantial compared with the controls except for those consuming the linseed oil eggs. Despite the very high enrichment of LnA in eggs from hens on the linseed oil-based diet, this ω3 PUFA was not elevated in human plasma.

Discussion

Recent research has demonstrated the enrichment of broiler meat with ω3 PUFA using flax seed (Ajuyah *et al.*, 1991) or redfish meat (Hulan *et al.*, 1988, 1989) in the diet and of duck meat using different oils in the diet (Farrell, 1991). Fortification of egg lipid with ω3 PUFA through the dietary addition of 3% menhaden oil or 15% flax seeds (linseed) was demonstrated by Hargis *et al.* (1991) and Jiang *et al.* (1991) respectively. The diet of Hargis *et al.* (1991) contained 7.2% DHA and only 0.8% EPA when expressed as % of total dietary fatty acids. Enrichment of the yolk with DHA was about 180 mg and about 25 mg for EPA and LnA.

The manipulation of egg lipid composition is well documented (Noble *et al.*, 1990). Karunajeewa and Tham (1987) showed that dietary levels of several PUFA, particularly LA and arachidonic acid, were reflected in the lipid in eggs from hens offered experimental diets containing oat groats, rice pollard, or corn oil.

It is of interest that significant quantities of DHA in egg yolk can result from the addition of linseed oil to the diet (Fig. 32.1b and c). But EPA is incorporated in significant amounts only when fish oil is added to the hen's diet. Previously we fed hens on a diet containing 4 or 8 g of a specially blended fish oil (Max EPA, R.P. Scherer, Florida) which contains 12% DHA and 18% EPA (Gibson and Farrell, 1990). Eggs from these hens contained about 2.4% DHA and 1.8% EPA of the total fatty acids.

The comparatively large concentrations of DHA relative to EPA in egg yolk, even on the cod liver oil-based (Fig. 32.1b and c) diet suggest that the hen probably converts some of the dietary EPA to DHA. The concentration of EPA in plasma of starved hens was low on all diets (Fig. 32.1a).

Despite the lack of DHA or EPA in linseed and Canola oil (Table 32.2), there were significant amounts of DHA in hen blood plasma. This suggests that the hen can convert some LnA to DHA by carbon chain elongation and desaturation. In rats the conversion of LnA to EPA and DHA takes place rapidly but in humans it is thought to occur slowly or in Eskimos possibly not at all (Sinclair, 1991).

The manipulation of ω3 PUFA in egg lipid can be achieved with the appropriate mix of oils in the diet (Table 32.3). Mackerel oil is slightly richer in EPA than cod liver oil (Table 32.2) and these differences were reflected in small changes in ω3 PUFA in egg yolk lipid. Again, despite only small amounts of DHA in the basal diet and none in linseed and Canola oil, hens were able to deposit significant amounts of DHA in egg lipid.

Fish oils differ greatly in their PUFA contents. For example pollock oil and tuna oil contain 17.3 and 7.5% of EPA, and 4.6 and 21.6% of DHA respectively (Childs *et al.*, 1990). Clearly the source of fish oil is extremely important when attempting to achieve the correct mix of ω3 PUFA in eggs.

R.A. Gibson (personal communication) found that LA competes for all ω3

PUFA incorporation. He was able to enrich egg lipid with 8% DHA, but the fatty acids in his layer diet contained only 5% LA. Sinclair (1991) has also pointed out the adverse effects of LA on ω3 PUFA incorporation as well as its effect in reducing the rate of conversion of LnA to EPA and DHA. Simopoulos and Salem (1989) observed that in Greece, hens consuming large amounts of purslane, a green leafy vegetable rich in ω3 PUFA, laid eggs that contained much higher amounts of the important ω3 PUFA than did commercial eggs from a supermarket. From their data it can be calculated that the content of ω3 PUFA was about 106 mg in a 60 g egg compared with only 10 mg in the supermarket egg. Their enriched eggs contained substantially smaller amounts than those shown in Table 32.3 and Fig. 32.1b and c.

Although the taste panel was unable to differentiate between egg types (Table 32.4), it is well recognised that giving poultry diets with significant amounts of fish meal results in 'tainted' eggs or meat often described as having a 'fishy flavour'. It is probable that much of the taint originates from the fish meal and not from the fish oil. The difference in texture between eggs from hens on the fish oil and Canola oil diets detected by the taste panel (Table 32.4) may be due to the higher contents of the longer ω3 PUFA in these fortified eggs. The scrambled eggs were often kept warm in an oven for a few minutes to await arrival of panel members and these egg types may have become less acceptable in texture as a consequence.

Eating two eggs daily for 9 weeks did not elicit any significant change in cholesterol or triglyceride levels in plasma of volunteers (Table 32.5). This observation on cholesterol is in keeping with measurements made by O'Dea and Sinclair (1991) who found that blood cholesterol levels of humans consuming two eggs daily were not elevated. The nature and amount of total fat in the diet determine to a large extent the effect of eggs on blood cholesterol as do other factors (Reiser, 1988). However interpretation of data relating to diet and disease should be treated with caution (Blaxter and Webster, 1991). Oh *et al.* (1988) reported an increase in plasma cholesterol of humans consuming four regular eggs daily for 4 weeks but no change when these eggs were fortified with ω3 PUFA. Their layer diets contained either 5 or 10% of Max EPA fish oil.

In the present study subjects were asked not to alter their dietary habits but no survey was undertaken to establish this nor to quantify their nutrient intake during the experimental period. On the other hand the important and beneficial HDL were significantly reduced after 2 weeks of consuming two eggs daily. They were also reduced by a small but significant amount in humans consuming eggs from hens on the fish and linseed oil diets. LDL on the other hand tended to be elevated in plasma of subjects consuming eggs from hens on the diet with fish oil.

Oh *et al.* (1988) observed a significant decrease in systolic and diastolic blood pressure of subjects consuming four ω3 PUFA enriched eggs daily for 4 weeks. Although we observed a decline in blood pressure of subjects over the 13 week experimental period (Table 32.5) this could not be attributed to any dietary treatment.

We are unaware of previous reports of the transfer of the ω3 PUFA from diet to hen through egg yolk to humans. Plasma levels found here (Fig. 32.1d) are from subjects who had fasted (for about 12 h) and are therefore unlikely to represent chylomicron PUFA of food origin. De Lany *et al.* (1990) gave human subjects either 5 g or 20 g of fish daily and showed elevated levels of EPA and DHA in serum phospholipids particularly at 20 g per day of fish oil. However, the latter amount is possibly unphysiological and although it demonstrates a point, in practice such large amounts of fish oil are unlikely to be consumed regularly from natural food sources.

The relative importance of the ω3 PUFA in the prevention of cardiovascular disease is well known (Dryberg and Jorgensen, 1982). EPA is recognised as the key ω3 PUFA and not much emphasis has been placed on the others. Gaudette and Holub (1991) have recently compared the effects of DHA and EPA on platelet function. They concluded that DHA was probably just as effective as EPA in inhibiting platelet reactivity, probably by decreasing the formation of thromboxane A2 (Sinclair, 1991). The aggregation of platelets in blood vessels is a major factor involved in initiating thrombus formation (Dryberg and Jorgensen, 1982). As mentioned, the importance of LnA as a processor of EPA and DHA is uncertain in humans but our data suggest that conversion of LnA also occurs in the hen (Fig. 32.1a) but largely to DHA.

Despite the substantial evidence that ω3 PUFA play a significant role in the prevention of several diseases, including coronary heart disease (Simopoulos *et al.*, 1991a), the Food and Drug Administration of the USA has concluded after reviewing the data that 'this evidence does not provide a basis upon which to authorise such a health claim' (Anonymous, 1991).

The importance of DHA in neural and retinal tissue of infants and particularly its rapid accumulation in brain cell membranes during the last intra-uterine trimester is recognised (Carlson and Salem, 1991). There exists the possibility of inadequate accumulation of DHA in the premature infant. Visual function may be impaired and learning difficulties may develop if a DHA source is not provided for these infants. In normal infants there is a linear increase in the accumulation of the ω3 PUFA and other PUFA in the whole cerebrum of infants to about 2 years of age (Martinez, 1991). Human milk contains small but significant amounts of DHA while infant formula usually contains negligible quantities (Carlson and Salem, 1991). Fortification of the infant diet with fish oil may be inappropriate; undesirable contaminants of the oil may be transmitted to the infant. It is possible to oversupply DHA and EPA in fish oil causing adverse effects (Carlson and Salem, 1991).

A major advantage of providing ω3 PUFA in eggs rather than in fish is that the egg is not only a cheap and complete source of essential nutrients but the hen acts as a 'biological sieve'. Eggs are normally free of undesirable contaminants and the content and mix of the ω3 PUFA can be manipulated.

The question of human requirements for the ω3 PUFA is difficult to determine. Clearly requirement will depend on the mix of ω3 PUFA and there may be a minimum requirement for EPA and DHA for example. Simopoulos *et al.* (1991b) suggested that a daily dietary intake of 0.5-1.0 g of n-3 PUFA was needed to reduce the risk from cardiovascular disease.

Pederson (1991) proposed a recommended dietary intake (RDI) of at least 0.5% of energy intake. Assuming a daily energy intake of 7.5 MJ, this would amount to about 1 g of these fatty acids per day.

Bjerve (1991) has given more precise estimates for LnA in the absence of dietary EPA and DHA of 0.86-1.2 g/day. Optimal requirements for EPA and DHA are 350-400 mg with a minimal requirement of 100-200 mg/day. Bjerve (1991) suggested that LnA was two to three times less effective than EPA and DHA in curing and preventing clinical symptoms of ω3 PUFA deficiency. This RDI is a minimum and in the future may well increase to 2 g per day. If these ω3 PUFA are to be provided in fortified eggs the data in Table 32.3 can be used as the basis for preliminary calculations. A mixture of 2% mackerel oil and 2% Canola oil gave a total of 7.5% ω3 PUFA with less than half as LnA. Eggs contain about 10% lipid (Noble *et al.*, 1990), thus one fortified 60 g egg would provide about 0.5 g of ω3 PUFA which is 50% of RDI and at what extra cost to the producer? Conversion of food to egg mass is about 2.5:1. A mixture of 2% fish oil and 2% Canola oil at about US$380 per metric tonne (mt) would add an additional $15-16 per mt of layer feed or 40¢ per kg of eggs (assuming no other costs); for a dozen 60 g eggs this would be about 25¢. This is an extremely modest sum particularly when compared with the purchase of fish oil in capsule form. A 2 g capsule of fish oil which may retail at 40¢ is equivalent to $200,000 per mt fish oil. The fortified egg would probably be a little more expensive per gram of ω3 PUFA than the capsule but likely safer for human consumption in terms of avoiding potential hypervitaminosis and undesirable contaminants. Furthermore, there may be an additional bonus of providing substantial quantities of high-quality protein and other important nutrients without changing plasma cholesterol in the majority of consumers.

Acknowledgements

I wish to thank the 44 volunteers who participated in this study. Dr R.A. Gibson of Flinders Medical Centre, Adelaide undertook many of the fatty acid analyses. I thank Maryanne Betts, Carrol Quilkey, and John Roberts for skilled technical assistance, Evan Thomson for technical and computing assistance, and Ruth Fox for typing the manuscript. I acknowledge the help of Dr G. Skurray and his staff at the University of Western Sydney, Hawkesbury for conducting the sensory evaluation of the eggs. The Egg Industry Research and Development Council and The University of New England generously provided financial support.

References

Ajuyah, A.O., Lee, K.H., Hardin, R.T., and Sim, J.S. (1991) Changes in the yield and in the fatty acid composition of whole carcass and selected meat portions of broiler chickens fed full fat oil seeds. *Poultry Sci.* 70: 2304-2314.

Anonymous (1991) Food labelling health claims and label statements: omega 3 fatty acids and coronary heart disease. Proposed rule. *Federal Register* 56:No. 299, p. 60663.

Bjerve, K.S. (1991) n-3 fatty acid deficiency in man: implications for the requirements of alpha-linolenic acid and long chain n-3 fatty acids. In: Simopoulos, A.P., Kifer, R.R., Martin, R.E., and Barlow, S.M. (eds), *Health Effects of Omega 3 Polyunsaturated Fatty Acids in Seafoods.* Karger, Basel, pp. 133-142.

Blaxter, K.L. and Webster, A.J.F. (1991) Animal production and food: real problems and paranoia. *Anim. Prod.* 53:261-269.

Burr, M.L. and Fehily, A.M. (1991) Fatty fish and heart diesease: a randomized controlled trial. In: Simopoulos, A.P., Kifer, R.R., Martin, R.E., and Barlow, S.M. (eds), *Health Effects of Omega 3 Polyunsaturated Fatty Acids in Seafoods.* Karger, Basel, pp. 306-312.

Carlson, S.E. and Salem, N. (1991) Essentiality of n-3 fatty acids in growth and development of infants. In: Simopoulos, A.P., Kifer, R.R., Martin, R.E., and Barlow, S.M. (eds), *Health Effects of Omega 3 Polyunsaturated Fatty Acids in Seafoods.* Karger, Basel, pp. 74-86.

Chesterman, C., Durham, R., Newell, G., and D'Mello, T. (1989) *Sensory Evaluation Manual.* Department of Food Technology, University of Sydney, Hawkesbury, NSW 2753, Australia.

Childs, M.T., King, I.B., and Knopp, R.H. (1990) Divergent lipoprotein responses to fish oils with various ratios of eicosapentaenoic acid and docsahexaenoic acid. *Am. J. Clin. Nutr.* 52:623-629.

De Lany, J.P., Vivian, V.M., Snook, J.T., and Anderson, P.A. (1990) Effects of fish oil on serum lipids in man during a controlled feeding trial. *Am. J. Clin. Nutr.* 52:477-485.

Dolecek, T.A. and Grandits, G. (1991) Dietary polyunsaturated fatty acids and mortality in the multiple risk intervention trial (MRFIT). In: Simopoulos, A.P., Kifer, R.R., Martin, R.E., and Barlow, S.M. (eds), *Health Effects of Omega 3 Polyunsaturated Fatty Acids in Seafoods.* Karger, Basel, pp. 205-216.

Dryberg, J. and Jorgensen, K.A. (1982) Marine oils and thrombogenesis. *Prog. Lipid Res.* 21:255-269.

Evans, A.J., Fogerty, A.C., and Sainsbury, K.J. (1986) The fatty acid composition of fish from the North West Shelf of Australia. *CSIRO Food Res. Q.* 46:40-45.

Farrell, D.J. (1991) Manipulation of growth, carcass composition and fatty acid content of meat-type ducks using short term feed restriction and dietary additions. *J. Anim. Physiol. Anim. Nutr.* 65:146-153.

Farrell, D.J. and Gibson, R.A. (1990) Manipulation of the composition of lipid in egg and poultry meat. In: Smith, W.C. (ed.), *Proceedings of the Inaugural Massey Pig and Poultry Symposium.* Massey University, New Zealand, pp. 164-179.

Fogerty, A.C. and Sovoronos, D. (1987) Fatty acids in canned fish. *CSIRO Food Res. Q.* 47:12-21.

Gaudette, D.C. and Holub, B.J. (1991) Docosahexaenic acid (DHA) and human platelet reactivity. *J. Nutr. Biochem.* 2:116-121.

Galli, C. and Simopoulos, A.P. (eds) (1989) *Dietary Omega-3 and Omega-6 Fatty Acids. Biological Effects and Nutritional Essentiality.* Plenum Press, New York.

Gibson, R.A. and Farrell, D.J. (1991) Egg yolk as a dietary source of n-3 poly-unsaturated fatty acids for infants. In: Simopoulos, A.P., Kifer, R.R., Martin, R.E., and Barlow, S.M. (eds), *Health Effects of Omega 3 Polyunsaturated Fatty Acids in Seafoods.* Karger, Basel, pp. 545-549.

Hargis, P.S., Van Elswyk, M.E., and Hargis, B.M. (1991) Dietary modification of yolk lipid with menhaden oil. *Poultry Sci.* 70:874-883.

Hulan, H.W., Ackman, R.G., Ratnayake, W.M.N., and Proudfoot, F.G. (1988) Omega-3 fatty acid levels and performance of boiler chickens fed red-fish meal or redfish oil. *Can. J. Anim. Sci.* 68:533-547.

Hulan, H.W., Ackman, R.G., Ratnayake, W.M.N., and Proudfoot, F.G. (1989) Omega-3 fatty acid levels and general performance of commercial broilers fed practical levels of redfish meal. *Poultry Sci.* 68:153-162.

Jiang, Z., Ahn, D.U., and Sim, J.S. (1991) Effects of feeding flax and two types of sunflower seeds on fatty acid compositions of yolk lipid classes. *Poultry Sci.* 70:2467:2475.

Karunajeewa, H. and Tham, S.H. (1987) Dietary manipulation of the poly-unsaturated fatty acids content in egg yolks. *Proc. Nutr. Soc. Aust.* 12: 113-116.

Kromhout, D., Bosschieter, E.B., and de Lezenne-Coulander, C. (1985) The inverse relation between fish consumption and 20-year mortality from coronary heart disease. *N. Engl. J. Med.* 312:1205-1209.

Lepage, G. and Roy, G.C. (1986) Direct esterification of all classes of lipids in a one-step reaction. *J. Lipid Res.* 27:114-120.

Martinez, M. (1991) Developmental profiles of polyunsaturated fatty acids in the brain of normal infants and patients with peroxisomal diseases: severe deficiency of docosahexaenoic acid in Zellweger's and pseudo Zellweger's syndromes. In: Simopoulos, A.P., Kifer, R.R., Martin, R.E., and Barlow, S.M. (eds), *Health Effects of Omega 3 Polyunsaturated Fatty Acids in Seafoods.* Karger, Basel, pp. 87-102.

Noble, R.C., Cocchi, M. and Turchetto, E. (1990) Egg fat - a case for concern? *World Poultry Sci. J.* 46:109-118.

O'Dea, K. and Sinclair, A.J. (1991) Eggs: implications for human health.

Proceedings of the Australian Poultry Science Symposium. University of Sydney, NSW, pp. 23-28.

Oh, S.Y., Hsieh, T., Ryue, J., and Bell, D.E. (1988) Effects of dietary eggs enriched with omega-3 fatty acids on plasma cholesterol, lipoprotein composition and blood pressure in humans. *FASEB J.* 15:A425.

Pederson, J.I. (1991) Nordic recommended dietary allowances for n-3 and n-6 fatty acids. In: Simopoulos, A.P., Kifer, R.R., Martin, R.E., and Barlow, S.M. (eds), *Health Effects of Omega 3 Polyunsaturated Fatty Acids in Seafoods.* Karger, Basel, pp. 161-164.

Reiser, R. (1988) Egg cholesterol and human health. In: *Proceedings of the World's Poultry Science Congress.* Sept. 4-9, 1988. Japan Poultry Science Association, Nagoya, pp. 16-22.

Simopoulos, A.P. and Salem, N. (1989) n-3 Fatty acids in chicks from range-fed Greek chickens. *N. Engl. J. Med.* 321:1412

Simopoulos, A.P., Kifer, R.R., Martin, R.E., and Barlow, S.M. (eds) (1991a) *Health Effects of Omega 3 Polyunsaturated Fatty Acids in Seafoods.* Karger, Basel.

Simopoulos, A.P., Kifer, R.R., Martin, R.E., and Barlow, S.M. (1991b) Conference statement. In: Simopoulos, A.P., Kifer, R.R., Martin, R.E., and Barlow, S.M. (eds), *Health Effects of Omega 3 Polyunsaturated Fatty Acids in Seafoods.* Karger, Basel, p. 15.

Sinclair, A.J. (1991) The good oil: omega 3 polyunsaturated fatty acids. *Today's Life Sci.* 3:18-27.

Chapter Thirty-Three

Omega-3 Fatty Acid Enriched Eggs as a Source of Long Chain ω3 Fatty Acids for the Developing Infant

G. Cherian and J.S. Sim

Department of Animal Science, University of Alberta, Edmonton, Alberta, Canada T6G 2P5

Abstract Fats in commercial formulas are derived from vegetable oil rich in C18:2ω6 or C18:3ω3 and do not contain long chain polyunsaturated fatty acids (PUFA). Therefore, the adequacy of infant formula in meeting the infant's ω3 fatty acid requirement is questionable. In the present study, we enriched eggs with ω3 fatty acids. The resulting eggs contained 690 mg of ω3 fatty acid including 165 mg of longer chain ω3 fatty acid (C20:5ω3, C22:5ω3, and C22:6 ω3). Addition of enriched eggs in the diet of nursing women resulted in the deposition of total ω3 fatty acids at 3.5% compared with 1.9% for the pre-test milk. The total long chain fatty acids (C20:5ω3, C22:5ω3, and C22:6ω3) comprised 1.0% compared with 0.4% in the pre-test milk. The result of this study indicates the importance of dietary ω3 fatty acids in modulating the ω3 fatty acid content of breast milk. Comparison of the fatty acid composition of enriched egg yolk and breast milk showed similarities in fatty acid composition. Thus, with proper manipulation of the hens' diet, egg yolk lipids could serve as a source of PUFA for the formula-fed infant or the oils derived from egg lipids would be a natural lipid alternative for infant formula preparation.
(Key words: docosahexaenoic acid, egg, infant formula, breast milk)

Introduction

Linoleic (C18:2ω6) and linolenic acid (C18:3ω3) are recognised as essential fatty acids for humans. These fatty acids are the precursors of longer chain ω6 or ω3 polyunsaturated fatty acids (PUFA) (Rosenthal, 1987) which are essential for the synthesis of structural lipids and prostaglandins. In human infants, most C22:6ω3 accumulates during the last intra-uterine trimester and during early stages of life suggesting that ω3 fatty acids may be indispensable for neural development of the newborn (Neuringer and Connor, 1986). During these periods, infants consume ω6 and ω3 fatty acids both as C18:2ω6 and C18:3ω3 unless fed on breast milk. However, breast milk reflects the mother's diet (Henderson *et al.*, 1992) and because modern diets are low in PUFA (Simopoulos, 1991), it is assumed that infants at present are receiving less long chain ω3 or ω6 fatty acids from breast milk. The present paper focuses on the importance of longer chain C20 and C22 PUFA on growth and development of nervous system, role of maternal diet in providing long chain ω3 fatty acids to the developing brain, and preliminary evidence suggesting that the dietary consumption of ω3 fatty acid enriched eggs can supply the desired levels of long chain ω3 PUFA to growing infants.

Development of the Central Nervous System

The period when the brain grows at a high rate, exceeding other organs, is known as the brain growth spurt (Dobbing and Sands, 1979). In general, the growth spurt includes dendritic development, formation of synaptic contacts, and proliferation of myelin-forming cells. At this time, the brain is more vulnerable to nutritional insult than any other stages in development. If one stage is perturbed or missed, the chances of recuperation are reduced. In humans, the brain growth spurt occurs during the last trimester of pregnancy and first 18 months of life after birth (Dobbing and Sands, 1979).

Omega-3 Fatty Acids in the Central Nervous System

The nervous system is the organ with the greatest concentration of lipids after adipose tissue. These lipids are structural and are high in PUFA of the ω6 and ω3 series (Sastry, 1985). Docosahexaenoic acid (DHA, C22:6ω3) is the predominant ω3 fatty acid and C20:4ω6 is the major ω6 fatty acid in the central nervous system (CNS) of mammals and birds. Long chain C20 and C22 PUFA control the composition of membranes, thereby affecting their enzymatic activity, transport of nutrients, membrane fluidity, activity of membrane bound enzymes such as Na^+, K^+ ATPase and 5′-nucleotidase (Bourre *et al.*, 1991).

Accumulation of DHA during Development

In human brain, substantial quantities of DHA are deposited in the CNS during the brain growth spurt. Information on the fatty acid composition of human brain was derived by analyses of autopsy tissue of fetuses from 26 to 44 weeks of gestation and newborn to 13 week old infants. These investigations indicated that during the third trimester, approximately 41 mg of ω6 and 14.5 mg of ω3 fatty acid accrue in the fetal brain per week (Clandinin *et al.*, 1980a,b; Dobbing and Sands, 1979).

Functional Effects of ω3 Fatty Acid Deficiency

Studies conducted on rodents reported that restriction of ω3 fatty acids interferes with normal visual function and learning difficulties in the offspring (Walker, 1967; Lamptey and Walker, 1976). These researchers evaluated several aspects of behaviour in rats fed safflower oil for two generations. Levels of DHA in pup brain tissue phospholipids were reduced up to 90% compared with the controls fed soy oil, resulting in reduced exploratory behaviour and difficulties with black/white discrimination. These results were in agreement with those of Enslen *et al.* (1991), Benolken *et al.* (1973), and Yamamoto *et al.* (1987) who observed lower exploratory behaviour and reduced colour discrimination in rat pups from ω3 fatty acid deficient dams. Connor *et al.* (1991) reported that ω3 fatty acid deficiency in rhesus monkey resulted in impaired vision, abnormal electro-retinograms, and polydypsia. In primates, the abnormal electroretinograms induced by ω3 fatty acid deficiency were irreversible. Deficiency of ω3 fatty acids has also been reported to alter anaesthetic and ethanol tolerance (Lefkowith *et al.* 1986; Haycock and Evers, 1988) and increased mortality to neurotoxin (Bourre *et al.*, 1989). Deficiency continued through early post-natal life may accentuate delays in development of certain motor skills (Morgan *et al.*, 1981; Ruthrich *et al.*, 1984). Improved visual acuity in breast-fed versus formula-fed infants has been related to the increased dietary availability of ω3 fatty acids through breast milk. Increasing the DHA content of infant formulas resulted in improved visual acuity similar to that of breast milk fed infants (Carlson *et al*, 1992). Maternal supplementation of ω3 fatty acids has been reported to increase eye-opening and sensory motor development in mouse pups (Wainwright *et al.*, 1991). The fatty acid changes found in the CNS of essential fatty acid deficient animals include an increase in C20:3ω9 and C22:3ω9 and a decrease in ω6 and ω3 PUFA predominantly C20:4ω6 and C22:6ω3.

Sources of Essential Fatty Acids during Brain Development

During development, fetal accretion of DHA may result from maternal sources through the placenta. However, after birth the infant receives DHA through milk or by endogenous synthesis from precursor C18:3ω3. Studies conducted on neonatal piglets (Foot *et al.*, 1990) and clinical trials in human infants have revealed lower DHA in the brain and tissue lipids when C18:3ω3 was the only source of ω3 fatty acid in the diet. This limited synthesis has been attributed to the inefficiency of microsomal desaturases in converting C18 fatty acids to C22 PUFA (Putnam *et al.*, 1982; Carlson *et al.*, 1986; Clandinin *et al.*, 1989).

Linolenic acid, the precursor of DHA, goes through a series of enzymatic desaturation and chain elongation reaction to give rise to endogenous formation of C20 and C22 fatty acids such as C20:5ω3, C22:5ω3, and DHA (Brenner, 1974) (Fig. 33.1).

The rate limiting enzyme in this pathway is Δ6 desaturase (Stoeffel, 1961). As the enzyme activity is limiting during this period, optimal postnatal accretion may depend on dietary supply of preformed DHA to the newborn.

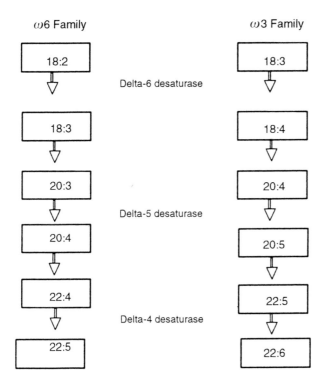

Fig. 33.1. Schematic diagram of the major pathways of ω6 and ω3 fatty acid metabolism.

Supply of ω3 PUFA for the Infant

Breast Milk

For the infant, milk provides the optimum single source of nutrients. Human breast milk contains C20 and C22 PUFA. However, the PUFA levels in milk cannot be used as an absolute model to justify the need of the infant for C20 and

C22 PUFA because the milk fatty acid levels are dependent on the maternal diet. Levels of DHA have been reported to be higher in the breast milk of women consuming an omnivorous rather than a vegan diet, and in the red blood cell lipids of nursing infants fed by omnivorous rather than vegan women (Simopoulos, 1991). A positive relationship between maternal diet ω3 fatty acid, milk DHA, and the levels of DHA in the piglet brain has been reported recently (Arbuckle and Innis, 1992). These findings substantiate the dependency of milk ω3 fatty acids on maternal diet and reveal the dietary impact on infant tissue levels of DHA.

Infant Formula

An alternative way of nourishing the newborn is through commercially available infant formulas. We analysed the fatty acid composition of some of the common formulas available on the market; the results are shown in Table 33.1.

The fats in the formulas are derived from one or more vegetable oils, usually coconut, soy, safflower, or corn oil. All vegetable oils differ from human milk in such a way that they do not contain fatty acids such as C20:4ω6 or DHA, and the C18:3ω3 content of corn oil varies from 0.8 to 1.2%, appreciably lower than that in human milk. Thus, the formulas marketed do not have any ω6 or ω3 PUFA longer than C18. Studies carried out on low-birth-weight infants and term infants has established that erythrocyte phospholipid fatty acids from formula-fed infants were depleted in long chain PUFA as compared with infants fed breast milk (Putnam *et al.*, 1982; Carlson *et al.*, 1986). Similarly, infant formulas with low levels of C18:3ω3 and no DHA have been reported to limit accretion of ω3 PUFA to the developing brain (Hrboticky *et al.*, 1990). Because of the lack of long chain PUFA in formula, and limited tissue reserves at birth in pre-term infants, the requirements for long chain PUFA of pre-term infants fed on formula is high. Thus, there is a good reason to question and evaluate the adequacy of today's infant formula and design an improved fatty acid balance to meet the infant's C20 and C22 PUFA requirement.

Alteration of Brain Tissue DHA: Role of Maternal Diet

In our laboratory, we have investigated the effect of maternal diets high in ω3 or ω6 essential fatty acid on the PUFA metabolism of the brain tissue of developing progeny using eggs and hatched chicks. This animal model which is isolated from maternal influence provides a unique opportunity to study the accretion of fatty acids in the brain during development. Laying hens were fed diets high in ω9 (Hn-9), ω3 (Hn-3), or ω6 (Hn-6) fatty acids. Fertilised eggs were incubated and the fatty acid compositions of the hatched chick tissues were analysed. The brain tissue of chicks hatched from the ω3 fatty acid enriched group had significantly higher total ω3 and DHA compared with those hatched from the ω9 or ω6 fatty acid enriched group (Fig. 33.2).

Table 33.1. Fatty acids in the breast milk of women consuming ω3 fatty acid enriched egg compared with fatty acids in infant formula or egg yolk lipids.

Fatty acid	Breast milk	Infant formula	Egg lipids
8:0	0.6	3.5	0.0
10:0	1.1	2.8	0.0
12:0	4.5	21.6	0.0
14:0	6.0	7.7	0.1
16.0	17.4	12.0	23.9
18:0	8.2	5.0	10.2
20:0	0.5	0.0	0.0
16:1	2.3	0.3	2.6
18:1	41.6	19.3	35.6
20:1	1.1	0.0	0.2
18:2 ω6	9.7	24.3	12.9
20:4 ω6	0.6	0.0	1.4
20.2 ω6	0.1	0.0	0.0
20:3 ω6	0.2	0.0	0.0
22:4 ω6	0.1	0.0	0.0
22:5 ω6	0.1	0.0	0.0
18:3 ω3	2.5	3.4	10.3
20:5 ω3	0.3	0.0	0.2
22:5 ω3	0.2	0.0	0.5
22:6 ω3	0.6	0.0	2.5
Total SAT	38.3	52.8	34.2
Total MUFA	45.0	19.6	38.4
Total ω6	10.8	24.3	14.3
Total ω3	3.6	3.4	13.5
Total LC ω6	1.1	0.0	1.4
Total LC ω3	1.1	0.0	3.3
Total ω6/ω3	3.0	7.2	1.1

SAT = saturated fatty acids, MUFA = monounsaturated fatty acids.
LC ω6, LC ω3 = longer chain ω6 or ω3 fatty acids.

The incorporation of long chain ω3 fatty acids such as C20:5ω3, C22:5ω3, and C22:6ω3 was mainly in the phosphatidylethanolamine fraction of the chick brain (Cherian and Sim, 1992a). The fatty acid composition of the hatched chick liver, plasma, and whole body were significantly altered by maternal fat composition (Cherian and Sim, 1992b). As incubation proceeded, the concentration of C20:4ω6 and C22:6ω3 in the yolk sac phospholipids diminished, suggesting that the developing chick embryo preferentially utilised phospholipids rich in C20 and C22 PUFA.

Results from this study indicate that changes in dietary source of fat during development of brain tissue can affect the fatty acid composition. Thus, despite the obvious developmental differences between mammals and avians, the subsequent usage and metabolism of PUFA is similar. This is evidenced by accretion and preferential uptake of C20 and C22 PUFA by the chick brain during the last week of incubation, suggesting that the egg and the hatched chick is a unique research model in studying the effect of maternal diet on the metabolism of PUFA in the brain.

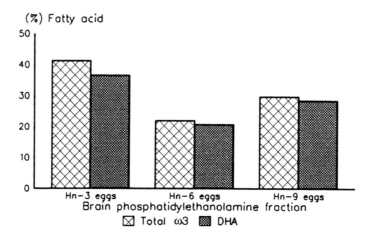

Fig. 33.2. Total ω3 and DHA content in the brain tissue of chicks hatched from eggs enriched with ω3, ω6, or ω9 fatty acids.

Egg as a Source of ω3 PUFA

Omega-3 Fatty Acid Enriched Egg Consumption and Breast Milk Fatty Acids

Eggs enriched with ω3 fatty acids (Hn-3 eggs) were produced by feeding diets containing flax seeds to laying hens. The resulting eggs provided up to 690 mg of ω3 fatty acid, of which 163 mg was longer chain ω3 fatty acids such as eicosapentaenoic acid (C20:5ω3), docosapentaenoic acid (C22:5ω3) and DHA compared with 87 mg of total ω3 fatty acid in a supermarket egg (Fig. 33.3). We were, therefore, interested in investigating the effect of ω3 fatty enriched eggs on the breast milk ω3 fatty acids in nursing women.

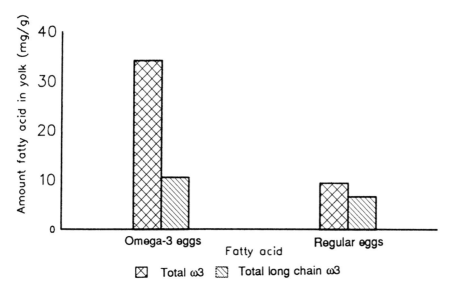

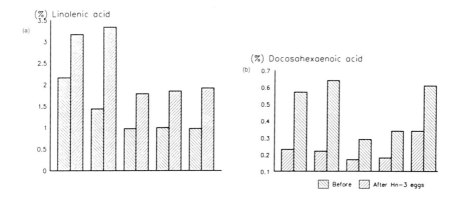

Fig. 33.3. Total ω3 and total long chain ω3 fatty acid content (mg/g) of yolk in ω3 eggs and regular supermarket eggs.

Nursing women were recruited and instructed to continue their usual diets supplemented with two 'designer eggs' per day and limit the intake of seafood for the duration of the study (4 weeks). Breast milk samples were collected once every week and analysed. Addition of two eggs daily to the diet resulted in a significant increase in the amount of ω3 fatty acids in the breast milk lipids (Fig. 33.4a and b).

Fig. 33.4. Effect of ω3 fatty acid enriched egg consumption on the amount of (a) linolenic acid (18:3ω3) and (b) DHA in human breast milk.

The major ω3 fatty acid in the breast milk was C18:3ω3 at 2.5%. However, the longer chain ω3 fatty acids such as C20:5ω3, C22:5ω3, and C22:6ω3 constituted 1.0% compared with 0.4% for the pre-test milk. Addition of 'designer eggs' to the diet did not change the ω6 content of milk lipids. The total ω6 fatty acid content in the milk lipids was 10.8% with C18:2ω6 being the major ω6 fatty acid. The total longer chain C20 ω6 fatty acids constituted 1.1%. The increase in ω3 fatty acids decreased the ω6:ω3 ratio of breast milk to 3.0 from 6.2 for the pre-test milk. The mean intake of breast milk by a one month old infant has been reported to be 794 ml (Boersma *et al.*, 1991). If so, the infants nursed from women consuming 'designer eggs' could have up to 280 mg of long chain ω3 fatty acids such as 20:5ω3, 22:5ω3 and 22:6ω3. Thus, diets supplemented with 'designer eggs' automatically provide the necessary amount of DHA to the developing infant through breast milk. These observed differences in the ω3 PUFA content of the breast milk could have implications for the development of the lipid structures of the CNS and possibly neurological development of the infant. More work in this direction is required.

'Designer Eggs' as Weaning Food for Infants

After three months of age, infants are often given supplementary food. Jackson and Gibson (1989) pointed out that although foods such as meat, egg yolks, or canned baby foods contain PUFA, large amounts of food are necessary to match the intake of long chain PUFA from breast milk. Addition of one medium-sized egg yolk can provide up to 165 mg of long chain ω3 PUFA. Further, the designer egg along with abundance of amino acids, vitamins, and minerals would be a useful supplemental or weaning food.

The Designer Egg and Infant Food Industry

Although there has been a significant increase in breast-feeding in the last decade, infant formulas remain the major source of nutrition for many infants during the first 12 months of life. At present, commercially available infant formulas do not contain the longer chain ω3 or ω6 fatty acids. Thus providing a formula with fatty acid composition that ensures optimal structural-lipid accretion is of major concern. This could be achieved by properly designing the formula with added longer chain PUFA in a manner and compositions that would reflect the fatty acid composition of breast milk. The fatty acid composition of the ω3 fatty acid enriched egg yolk lipids and breast milk lipids shows similarities in its saturated, MUFA and PUFA (Table 33.1). Thus, if egg yolk lipids extracted from ω3 fatty acid enriched egg are ideal as a natural lipid alternative, their incorporation into formula preparations may achieve an improved fatty acid balance.

Acknowledgements

The authors wish to acknowledge the financial assistance from the Natural Sciences and Engineering Research Council of Canada and Flax Council of Canada. The authors thank the nursing women who participated in the study.

References

Arbuckle, L.D. and Innis, S.M. (1992) Docosahexaenoic acid in developing brain and retina of piglets feed high or low α-linolenate formula with and without fish oil. *Lipids* 27:89-93

Benolken, R.M., Anderson, R.E., and Wheeler, T.G. (1973) Membrane fatty acids associated with the electrical response in visual excitation. *Science* 182:1253-1254.

Boersma, E.R., Offringa, P.R., Muskiet, F.A.J., and Chase, W.M. (1991) Vitamin E, lipid fractions and fatty acid compositions of colostrum, transitional milk, and mature milk: an international comparative study. *Am. J. Clin. Nutr.* 53:1197-1204.

Brenner, R.R. (1974) The oxidative desaturation of unsaturated fatty acids in animals. *Mol. Cell. Biochem.* 3:41-52.

Bourre, J.M., François, M., Youyou, A., Dumont, O., Piciotti, M., Pascal, G., and Durand, G. (1989) The effects of dietary α-linolenic acid on the composition of nerve membranes, enzymatic activity, amplitude of electrophysiological parameters, resistance to poisons and performance of learning tasks in rats. *J. Nutr.* 119:1880-1892.

Bourre, J.M., Dumont, O., Piciotti, M., Clement, M., Chaudiere, J., Bonneil, M., Nalbone, G., Lafont, H., Pascal, G., and Durand, G. (1991) Essentiality of ω3 fatty acids for brain structure and function. In: Simopoulos, A.P, Keifer, R.R., Martin, R.E., and Barlow, S.M. (eds), *Health Effects of ω3 Polyunsaturated Fatty Acids in Sea Foods.* World Rev. Nutr. Diet. 66:103-118.

Carlson, S.E., Rhodes, P.G., and Ferguson, M.G. (1986) Docosahexaenoic acid status of preterm infants at birth and following feeding with human milk or formula. *Am. J. Clin. Nutr.* 44:798-804.

Carlson, S.E., Cooke, R.J., Rhodes, P.G., Peebles, J.M., and Werkman, S.H. (1992) Effect of vegetable and marine oils in preterm infant formulas on blood arachidonic acid and docosahexaenoic acids. *J. Pediatr.* 120:S159-167.

Cherian, G. and Sim, J.S. (1992a). Preferential accumulation of n-3 fatty acids in the brain of chicks from eggs enriched with n-3 fatty acids. *Poultry Sci.* 71:1658-1668.

Cherian, G. and Sim, J.S. (1992b) Omega-3 fatty acid and cholesterol content of newly hatched chicks from α-linolenic acid enriched eggs. *Lipids* 27:706-710.

Clandinin, M.T., Chappell, J.E., and Leong, S. (1980a) Intrauterine fatty acid accretion rates in human brain: implications for fatty acid requirements. *Early Hum. Dev.* 4:121-129.

Clandinin, M.T., Chappell, J.E., Heim, T.S., Sawyer, P.R., and Chance, G.W. (1980b) Extrauterine fatty acid accretion in infant brain implications for fatty acid requirements. *Early Hum. Dev.* 4:131-138.

Clandinin, M.T., Chappell, J.E., and Van Aerde, J.E.E. (1989) Requirements of newborn infants for long chain polyunsaturated fatty acids. *Acta. Pediatr. Scand.* Suppl. 351:63-71.

Connor, W.E., Neuringer, M., and Reisbick, S. (1991) Essentiality of ω3 fatty acids: evidence from the primate model and implications for human nutrition. In: Simopoulos, A.P, Keifer, R.R., Martin, R.E., and Barlow, S.M. (eds), *Health Effects of ω3 Polyunsaturated Fatty Acids in Sea Foods.* World Rev. Nutr. Diet. 66:118-132.

Dobbing, J. and Sands, J. (1979) Comparative aspects of brain growth spurt. *Early Hum. Dev.* 3:79-83.

Enslen, M., Milon, H., and Malone, A. (1991) Effect of low intake of n-3 fatty acids during development on brain phospholipid fatty acid composition and exploratory behaviour in rats. *Lipids* 26:203-208.

Foot, K.D., Hrboticky, N., Mackinnon, M.J., and Innis, S.M. (1990) Brain synaptosomal, liver, plasma, and red blood cell lipids in piglets fed exclusively on a vegetable-oil-containing formula with and without fish-oil supplements. *Am. J. Clin. Nutr.* 51:1001-1006.

Haycock, J.C. and Evers, A.S. (1988) Altered phosphoinositide fatty acid composition, mass and metabolism in brain essential fatty acid deficiency. *Biochim. Biophys. Acta* 960:54-60

Henderson, R.A., Jensen, R.G., Lammi-Keefe, C.J., Ferris, A.M., and Dardick, K.R. (1992) Effect of fish oil on the fatty acid composition of human milk and maternal and infant erythrocytes. *Lipids* 27:863-869.

Hrboticky, N., Mackinnon, M.J., and Innis, S.M. (1990) Effect of a vegetable oil formula rich in linoleic acid on tissue fatty acid accretion in the brain, liver, plasma, and erythrocytes of infant piglets. *Am. J. Clin. Nutr.* 51:173-182.

Jackson, K.A. and Gibson, R.A. (1989) Weaning foods cannot replace breast milk as sources of long-chain polyunsaturated fatty acids. *Am. J. Clin. Nutr.* 50:980-982.

Lamptey, M.S. and Walker, B.L. (1976) A possible essential role for dietary linolenic acid in the development of the young rat. *J. Nutr.* 106:86-93.

Lefkowith, J.B., Evers, A.S., Elliot, W.J., and Needleman, P. (1986) Essential fatty acid deficiency: a new look at an old problem. *Prostaglandin Leukotr. Essent. Fatty Acids* 23:123-127.

Morgan, B.L.G., Oppenheimer, J., and Winick, M. (1981) Effects of essential fatty acid deficiency during late gestation on brain N-acetyl neuraminic acid metabolism and behaviour in the progeny. *Br. J. Nutr.* 46:223-230.

Neuringer, M.E. and Connor, W.E. (1986) N-3 fatty acids in the brain and retina: evidence for their essentiality. *Nutr. Rev.* 44:285-294.

Putnam, J.C., Carlson, S.E., and DeVoe, P.W. (1982) The effect of variations in dietary fatty acids on the fatty acid compositions of erythrocyte phosphatidylcholine and phosphatidylethanolamine in human infants. *Am. J. Clin. Nutr.* 36:106-111.

Rosenthal, M.D. (1987) Fatty acid metabolism in isolated mammalian cells. *Prog. Lipid Res.* 26:87.

Ruthrich, H.L., Hoffman, P., Matthies, H., and Forster, W. (1984) Perinatal linoleate deprivation impairs learning and memory in adult rats. *Behav. Neural Biol.* 40:205-212.

Sastry, P.S. (1985) Lipids of nervous tissue: comparison and metabolism. *Prog. Lipid Res.* 24:69-176.

Simopoulos, A.P. (1991) Omega-3 fatty acids in health and disease and in growth and development. *Am. J. Clin. Nutr.* 54:438-463.

Stoeffel, W. (1961) Biosynthesis of polyenoic fatty acids. *Biochem. Biophys. Res. Commun.* 6:270-273.

Wainwright, P.E., Huang, Y.S., Bulman-Fleming, B., Mills, D.E., Redden, D.E., and McCutcheon, D. (1991) The role of n-3 essential fatty acids in the brain and behavioral development. A cross-fostering study in the mouse. *Lipids* 26:37-45.

Walker, B.L. (1967) Maternal diet and brain fatty acids in young rat. *Lipids* 2:497-500.

Yamamoto, N., Saitoh, M., Moriuchi, A., and Nomura, M. (1987) Effect of dietary linolenate/linoleate balance on brain lipid compositions and learning ability of rats. *J. Lipid Res.* 28:144-151.

Chapter Thirty-Four

Consumption of ω3 PUFA Enriched Eggs and Changes of Plasma Lipids in Human Subjects

J.S. Sim and Z. Jiang[1]

Department of Animal Science, University of Alberta Edmonton, Alberta, Canada T6G 2P5 ([1]Present address: Shur-Gain, 2700 Matheson Blvd. E., Suite 600, East Tower, Mississauga, Ontario, Canada L4W 4V9)

Abstract The purpose of this article is to report the changes in plasma lipids in humans upon consumption of ω3 PUFA enriched eggs. In trial 1, 23 male university students consumed two ω3 or regular eggs per day for a period of 18 days. Plasma cholesterol and fatty acid profiles were determined before and after the test period. In those who consumed ω3 PUFA enriched eggs, plasma total cholesterol levels were virtually unchanged, and plasma triglycerides were significantly reduced. No reduction of plasma total cholesterol or triglycerides was observed in those who consumed regular eggs. In trial 2, 23 volunteers consumed two or three ω3 eggs daily for 30 days. Both plasma total cholesterol and triglyceride levels were reduced. In a third trial, a group of nine volunteers consumed two regular eggs for the first 2 weeks and then two ω3 eggs per day for another 2 weeks. Plasma total cholesterol and triglycerides were slightly increased after consumption of regular eggs and decreased after consumption of ω3 eggs. These results indicated that the cholesterolaemic effects of chicken eggs in humans can be reduced by incorporating ω3 fatty acids into the egg.
(*Key words:* ω3 fatty acids, flax seeds, eggs, plasma cholesterol, triglyceride)

Introduction

Many attempts have been made to reduce the cholesterol content of chicken eggs through dietary and drug treatment but without any practical success (Noble, 1987). It is known, however, that the fatty acid composition of the egg yolk lipids can be altered by changing the type of dietary fat of laying hens (Cruickshank, 1934), and that the type of dietary fat affects cholesterol metabolism. For instance, the importance of ω3 polyunsaturated fatty acids (PUFA) in the prevention and amelioration of atherosclerosis has gained attention during the past decade (Kinsella *et al.*, 1990; Simopoulos, 1991). Several studies have demonstrated that the much needed ω3 fatty acids could be incorporated into the egg yolk by feeding laying hens diets containing fish oil, flax and canola seeds and oils (Yu and Sim, 1987; Hargis *et al.*, 1991; Jiang *et al.*, 1991; see also Chapter 32). Full-fat flax seeds are used in our experiments as a dietary source of ω3 fatty acids because these seeds are also high in dietary protein and metabolisable energy (Sim, 1990).

Previous studies from our laboratory demonstrated that the ω3 PUFA enriched eggs were less cholesterolaemic than regular eggs in rats (Jiang and Sim, 1992). Feeding ω3 eggs to rats lowered plasma total cholesterol and increased the ω3 fatty acid levels in tissue lipids. The production of prostaglandin E_2 by rat epitrochlearis muscle was also significantly reduced in rats fed ω3 eggs when compared with those fed regular eggs, indicating that the profile of eicosanoids produced by animal tissue could be modulated through dietary treatment (Jiang, 1992).

This paper presents the results of three studies on the effects of ω3 egg consumption in human subjects.

Materials and Methods

In Experiment 1, 23 male students were recruited from the University of Alberta. They were divided randomly into two groups of 11 and 12 each. They consumed either two ω3 PUFA enriched or two regular eggs a day in addition to their habitual diet for a period of 18 days. Before and after the test period, their plasma lipids were analysed by an independent clinical laboratory (Dr T.A. Kasper and associates, Edmonton).

In Experiment 2, 22 volunteers of both genders (15 male and seven female) consumed two or three ω3 PUFA enriched eggs a day in addition to their habitual diets for a period of 30 days. Plasma lipids were measured before and after the trial.

In Experiment 3, a group of nine volunteers consumed two regular eggs each day for 2 weeks and then two ω3 eggs for another 2 weeks. Plasma lipids were determined before, and 2 and 4 weeks after dietary treatment.

Results

Experiment 1

The changes in plasma total cholesterol (TC) and triglyceride (TG) levels of healthy volunteers (n = 11 and 12 for ω3 egg and regular egg groups, respectively) are shown in Fig. 34.1. Mean plasma TC increased by 5.1% in those who consumed two regular eggs each day. However, for those who consumed ω3 eggs, plasma TC was virtually unchanged. Mean plasma TG levels were significantly ($P < 0.05$) decreased in those who consumed ω3 eggs, but not affected ($P > 0.05$) in those who consumed regular eggs.

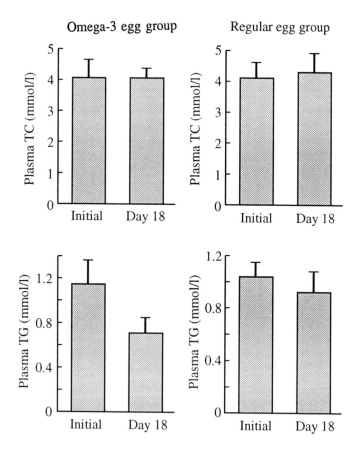

Fig. 34.1. Plasma total cholesterol (TC) and triglyceride (TG) levels in human subjects who consumed two ω3 PUFA enriched or regular eggs a day with their habitual diets for a period of 18 days. Day 18 values with * are significantly different from initial values ($P < 0.05$, Experiment 1).

Experiment 2

On average, plasma TC levels were reduced by 5.3%, and plasma TG decreased by 9.1% after 30 days of consumption of ω3 eggs (Fig. 34.2).

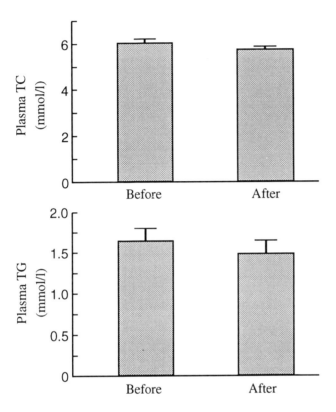

Fig. 34.2. Changes of plasma total cholesterol (TC) and triglyceride (TG) levels in human subjects who consumed two or three ω3 PUFA enriched eggs a day for a period of 30 days (Experiment 2).

Experiment 3

Figure 34.3 shows the changes in plasma TC and TG in those who consumed two regular eggs per day for the first 2 weeks and two ω3 eggs per day for another 2 weeks. Large variations in individual response to egg intake were observed. On average, plasma TC increased slightly (+2.5%) after regular egg consumption and decreased slightly (-1.2%) after ω3 egg consumption. Plasma TG was increased by 6.1% during the first 2 weeks (regular eggs) and was reduced by 12.5% during the second 2 week period (ω3 eggs).

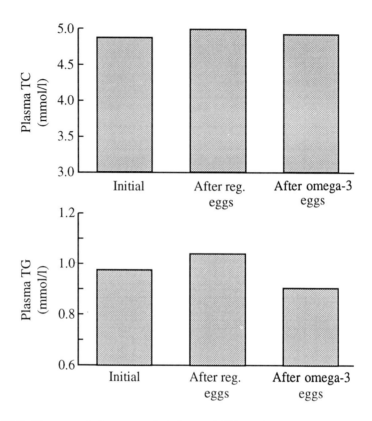

Fig. 34.3. Changes of plasma total cholesterol (TC) and triglyceride (TG) levels in human subjects who consumed two regular eggs a day for 2 weeks and then two ω3 PUFA enriched eggs a day for another 2 weeks (Experiment 3).

Discussion

These studies demonstrated that enriching eggs with ω3 PUFA changes the cholesterolaemic property of chicken eggs. Feeding laying hens diets containing full-fat flax seeds enriched eggs mainly with α-linoleic acid (LnA), although considerable amounts of longer chain ω3 PUFA such as EPA and DHA were also incorporated into the phospholipids fraction of the yolk lipids, indicating that laying hens were capable of converting LnA into eicosapentaenoic acid (EPA) and DHA (Jiang *et al.*, 1991). The ω6:ω3 ratio was reduced from 19:1 in commercial eggs to 1.4:1 in our ω3 eggs.

Omega-3 PUFA exhibit their protective roles against cardiovascular diseases mainly through lowering plasma triglyceride and cholesterol levels and through modifying the quantity and profile of tissue eicosanoid production (Kinsella *et al.*, 1990). The present study demonstrated that consumption of ω3 PUFA enriched eggs reduced plasma lipids, particularly plasma triglycerides. This observation collaborated previous results that consumption of chicken eggs enriched with EPA and DHA prevented plasma lipids from rising as compared with that of regular eggs (Oh *et al.*, 1991). In animal studies, we have also demonstrated that feeding ω3 yolk powders to rats reduced both plasma and liver cholesterol levels (Jiang and Sim, 1992). In addition, feeding ω3 yolks also increased ω3 PUFA levels in plasma and tissues lipids (Jiang and Sim, 1992), and reduced prostaglandin E_2 production by rat skeletal muscle (Jiang, 1992).

Several lines of evidence suggest that humans evolved on a diet that was low in fat with an ω6:ω3 PUFA ratio of 1:1 (Simopoulos, 1991). The importance of ω3 PUFA in human health, however, was recognised only during the past decade. Modern agribusiness and food industry are often blamed for bringing about the abrupt changes that resulted in the high fat, high ω6:ω3 PUFA ratios (20:1) of today's diet. This drastic deviation from the traditional diet has been implied in the pathogenesis of several degenerative diseases including coronary heart disease. Thus it is up to the agriculture and food industries to rectify the current imbalance of dietary fat.

Conclusion

The present study demonstrates that the chicken egg is an ideal vehicle to deliver the much needed ω3 PUFA in addition to its high quality protein and other nutrients. The concept of modifying fatty acid composition through dietary treatments can also be applied to other food animal products and this might be of great strategic importance in the battle against degenerative diseases.

References

Cruickshank, E.M. (1934) Studies in fat metabolism in the fowl. 1. The composition of the egg fat and depot fat of the fowl as affected by the ingestion of large amounts of different fats. *Biochem. J.* 28:965-977.

Hargis, P.S., van Elswky, M.E., and Hargis, B.M. (1991) Dietary modification of yolk lipid with Menhaden oil. *Poultry Sci.* 70:874-883.

Jiang, Z. (1992) Studies on egg yolk lipids and ovo-cholesterol metabolism. Ph.D. thesis. University of Alberta, Edmonton, Alberta, Canada.

Jiang, Z. and Sim, J.S. (1992) Effects of dietary n-3 PUFA-enriched chicken eggs on plasma and tissue cholesterol and fatty acid composition of rats. *Lipids* 27:279-282.

Jiang, Z., Ahn, D.U., and Sim, J.S. (1991) Effects of feeding flax and two types of sunflower seeds on fatty acid composition of yolk lipid classes. *Poultry Sci.* 70:2467-2475.

Kinsella, J.E., Lokesh, B., and Stone, R.A. (1990) Dietary n-3 polyunsaturated fatty acids and amelioration of cardiovascular disease: possible mechanisms. *Am. J. Clin. Nutr.* 52:1-28.

Noble, R.C. (1987). Egg lipids. In: Wells, R.G. and Belyavin, C.G. (eds), *Egg Quality - Current Problems and Recent Advances.* Butterworth Co. Ltd, London, UK, pp. 159-177.

Oh, S.Y., Ryue, J., Hsieh, C.H., and Bell, D.E. (1991). Eggs enriched in ω3 fatty acids and alterations in lipid concentrations in plasma and lipoproteins and in blood pressure. *Am. J. Clin. Nutr.* 54:689-695.

Sim, J.S. (1990) Flax seed as a high energy/protein/omega-3 fatty acid feed ingredient for poultry. In: Carter, J.F. (ed.) *Proceedings of the 53rd Flax Institute of the United States.* North Dakota State University, Fargo, ND, pp. 65-71.

Simopoulos, A.P. (1991) Omega-3 fatty acids in health and disease and in growth and development. *Am. J. Clin. Nutr.* 54:438-463.

Yu, M.M. and Sim, J.S. (1987) Biological incorporation of n-3 polyunsaturated fatty acids into chicken eggs. *Poultry Sci.* 66 (Suppl. 1):195. (Abstract).

Index

Page numbers in *italics* refer to figures and tables